Comprehending Molecular Targets: Mechanisms and Actions in Drug Development

Comprehending Molecular Targets: Mechanisms and Actions in Drug Development

Guest Editors

Cristina Manuela Dragoi
Ion-Bogdan Dumitrescu

MDPI

Basel • Beijing • Wuhan • Barcelona • Belgrade • Novi Sad • Cluj • Manchester

Guest Editors
Cristina Manuela Dragoi
Department of Biochemistry
Carol Davila University of
Medicine and Pharmacy
Bucharest
Romania

Ion-Bogdan Dumitrescu
Faculty of Pharmacy
University of Medicine and
Pharmacy "Carol Davila"
Bucharest
Romania

Editorial Office
MDPI AG
Grosspeteranlage 5
4052 Basel, Switzerland

This is a reprint of the Special Issue, published open access by the journal *Targets* (ISSN), freely accessible at: https://www.mdpi.com/journal/targets/special_issues/B0MVB63703.

For citation purposes, cite each article independently as indicated on the article page online and as indicated below:

Lastname, A.A.; Lastname, B.B. Article Title. *Journal Name* **Year**, *Volume Number*, Page Range.

ISBN 978-3-7258-7757-7 (Hbk)
ISBN 978-3-7258-7758-4 (PDF)
https://doi.org/10.3390/books978-3-7258-7758-4

Contents

About the Editors vii

Preface ix

Yuki Yoshimatsu, Yomogi Shiota, Julia Osaki, Kazuyoshi Yanagihara and Tadashi Kondo
Tumor-Intrinsic Transcriptional Signatures Linked to Cachexia Induction and Chemotherapy Response in Paired Human Neuroendocrine Carcinoma Cell Lines
Reprinted from: *Targets* **2026**, *4*, 9, https://doi.org/10.3390/targets4010009 . . . . **1**

Catherine Sharo, Jiayu Zhang, Tianhua Zhai, Jingxuan Bao, Andrés Garcia-Epelboim, Elizabeth Mamourian, et al.
Repurposing FDA-Approved Drugs Against Potential Drug Targets Involved in Brain Inflammation Contributing to Alzheimer's Disease
Reprinted from: *Targets* **2024**, *2*, 446–469, https://doi.org/10.3390/targets2040025 . . . . **19**

Shilpa Bisht, Paras Varshney and Abhishek Gupta
Aging in 3D: Organoid Systems as Models to Uncover Cellular Senescence and Therapeutic Targets Across Diseases
Reprinted from: *Targets* **2026**, *4*, 12, https://doi.org/10.3390/targets4020012 . . . . **43**

Susmita Barman, Seetur R. Pradeep and Krishnapura Srinivasan
Comprehensive Roles of ZIP and ZnT Zinc Transporters in Metabolic Inflammation
Reprinted from: *Targets* **2026**, *4*, 5, https://doi.org/10.3390/targets4010005 . . . . **78**

Cristina Manuela Drăgoi, Alina-Crenguța Nicolae and Ion-Bogdan Dumitrescu
The Indole Scaffold in Biochemistry and Therapeutics: A Privileged Structure with Diverse Chemical, Biological, and Clinical Significance
Reprinted from: *Targets* **2026**, *4*, 4, https://doi.org/10.3390/targets4010004 . . . . **110**

Anichavezhi Devendran and Sivasankar Perumal
Navigating the Molecular and Cellular Landscape of Breast Cancer in India: From Unique Pathogenesis to the Promise of Personalized Medicine and Future Technologies
Reprinted from: *Targets* **2025**, *3*, 38, https://doi.org/10.3390/targets3040038 . . . . **144**

Zhenzhen Li, Xiaoli Huang, Xuchi Zhao, Yunxiu Zhang and Ping Li
The Expanding E3 Ligase-Ligand Landscape for PROTAC Technology
Reprinted from: *Targets* **2025**, *3*, 30, https://doi.org/10.3390/targets3040030 . . . . **196**

Alireza Shoari
Potential of MMP-2 and MMP-9 Gelatinase Blockade as a Therapeutic Strategy in Fibrosarcoma Treatment: A Decadal Review
Reprinted from: *Targets* **2024**, *2*, 104–125, https://doi.org/10.3390/targets2020007 . . . . **224**

About the Editors

Cristina Manuela Dragoi

Cristina Manuela Drăgoi is affiliated with the Department of Biochemistry, Faculty of Pharmacy, University of Medicine and Pharmacy "Carol Davila", Bucharest, Romania. Her research interests focus on oxidative stress mechanisms, the evaluation of drug effects on biochemical markers, and the identification of predictive biomarkers in metabolic and chronic diseases. In recent years, her work has increasingly centered on chronobiology and chrononutrition, particularly the role of circadian entrainment in regulating neuroendocrine function, metabolic homeostasis, healthy aging, and the optimization of pharmacological and nutritional interventions. She is actively involved in interdisciplinary research projects that integrate biochemistry, nutrition, and clinical sciences, contributing to the development of innovative approaches in drug research and personalized medicine. Her scientific achievements include the authorship of numerous research articles, books, and book chapters, reflecting a sustained commitment to advancing knowledge in applied biochemistry and translational biomedical research.

Ion-Bogdan Dumitrescu

Ion-Bogdan Dumitrescu is affiliated with the Department of Physics and Informatics, Faculty of Pharmacy, University of Medicine and Pharmacy "Carol Davila", Bucharest, Romania. His research interests focus on pharmaceutical technologies, particularly the application of 3D printing in the design and optimization of pharmaceutical formulations. He is also interested in the study of natural bioactive compounds, with a particular emphasis on those derived from olive oil and their potential health benefits. In recent years, he has also been involved in interdisciplinary research projects exploring chronobiology, contributing to a better understanding of temporal regulation in human physiology and its implications for therapeutic strategies. His scientific activity is reflected in the publication of numerous research articles, books, and book chapters, highlighting his contributions to pharmaceutical sciences and translational biomedical research.

Preface

The present Reprint brings together a diverse and high-quality collection of original research articles and reviews that collectively advance our understanding of molecular targets across a broad spectrum of pathological conditions and therapeutic strategies. The scope of this Reprint encompasses cutting-edge investigations ranging from tumor biology and metabolic inflammation to neurodegeneration, aging, and innovative drug design technologies.

The contributions included in this Reprint highlight the complexity and heterogeneity of molecular mechanisms underlying disease processes, as illustrated by studies on tumor-intrinsic transcriptional programs in cancer cachexia, zinc transporter regulation in metabolic inflammation, and the molecular landscape of breast cancer. At the same time, several articles emphasize translational and therapeutic innovation, including drug-repurposing approaches for Alzheimer's disease, the expanding applications of PROTAC technology, and the therapeutic potential of targeting matrix metalloproteinases in oncology.

Notably, this Reprint also reflects the growing importance of advanced experimental models and molecular scaffolds in drug development, such as organoid systems for studying cellular senescence and the indole nucleus as a privileged structure in medicinal chemistry. Together, these works underscore the paradigm shift toward precision medicine, where detailed molecular characterization informs targeted therapeutic interventions.

The aim of this Reprint is to provide a comprehensive and integrative perspective on current advances in molecular targeting, fostering interdisciplinary dialog between basic research and clinical application. It is addressed to researchers, clinicians, and professionals in biomedical sciences, pharmacy, and related fields, as well as to early-career scientists seeking to deepen their understanding of modern drug development.

Cristina Manuela Dragoi and Ion-Bogdan Dumitrescu
Guest Editors

targets

MDPI

Article

Tumor-Intrinsic Transcriptional Signatures Linked to Cachexia Induction and Chemotherapy Response in Paired Human Neuroendocrine Carcinoma Cell Lines

Yuki Yoshimatsu [1,*], Yomogi Shiota [2], Julia Osaki [2], Kazuyoshi Yanagihara [2] and Tadashi Kondo [2]

1 Medical Seeds Discovery Project, Tochigi Cancer Center Research Institute, 4-9-13 Yohnan, Utsunomiya, Tochigi 320-0834, Japan
2 Division of Rare Cancer Research, National Cancer Center Research Institute, 5-1-1 Tsukiji, Chuo-ku, Tokyo 104-0045, Japan
* Correspondence: yuyoshim@tochigi-cc.jp

Abstract

Cancer cachexia is a multifactorial metabolic syndrome that profoundly impairs treatment tolerance and prognosis; however, how tumor-intrinsic transcriptional programs contribute to cachexia induction and shape responses to chemotherapeutic stress remains poorly understood. In this study, we analyzed a paired human duodenal neuroendocrine carcinoma cell-line model consisting of the non-cachexia-inducing parental line TCC-NECT-2 and its cachexia-inducing derivative AkuNEC. Bulk RNA sequencing was performed under baseline conditions and after doxorubicin treatment (10 μM, 24 h), and transcriptomic differences were assessed using log2 fold-change–based analyses to characterize baseline reprogramming, drug-induced responses, and differential stress adaptation. Despite comparable morphology and similar levels of doxorubicin-induced acute cytotoxicity, AkuNEC cells exhibited extensive baseline transcriptional reprogramming relative to TCC-NECT-2, including coordinated upregulation of inflammatory, secretory, and metabolic regulators previously implicated in cancer cachexia, together with suppression of structural and homeostatic programs. Following doxorubicin exposure, AkuNEC cells showed a distinct transcriptional response characterized by selective reorganization of proliferation-, metabolism-, and stress-related pathways, indicating enhanced transcriptional plasticity rather than uniform stress suppression. Differential response analyses further revealed preferential induction of genes involved in cell cycle control, DNA replication, and metabolic adaptation in AkuNEC under chemotherapeutic stress. These findings indicate that cachexia-inducing capacity is embedded within tumor-intrinsic transcriptional states and is amplified by stress-induced plasticity, supporting a network-level model of cancer cachexia that links systemic host effects with tumor adaptation to therapy. The TCC-NECT-2/AkuNEC model provides a tractable framework for dissecting these tumor-intrinsic mechanisms and their relevance to cachexia-associated cancer biology.

Keywords: cancer cachexia; tumor-intrinsic transcriptional reprogramming; chemotherapy-induced stress response; neuroendocrine carcinoma; transcriptional plasticity; doxorubicin

1. Introduction

Cancer cachexia is a progressive metabolic syndrome characterized by involuntary weight loss, skeletal muscle wasting, and reduced appetite, and these symptoms cannot be fully reversed by nutritional support [1,2]. Cachexia has a substantial impact on physical function, tolerance to anticancer treatment, and overall survival, yet effective therapeutic options remain limited [2]. Because current diagnostic criteria rely mainly on visible and late-stage manifestations, many patients are diagnosed only after systemic homeostasis has already been markedly disrupted [1]. Recent multi-organ studies suggest that tumor–host metabolic and inflammatory interactions begin long before overt wasting becomes apparent [3,4], indicating that cachexia represents a gradual and dynamic pathological process. However, the molecular mechanisms that drive these early changes are still incompletely understood.

There is increasing evidence that cachexia may also influence the responsiveness of tumors to chemotherapy. A recent meta-analysis reported that cachexia is associated with reduced treatment efficacy and poorer survival across several malignancies [5]. Altered body composition, including decreased muscle mass and disrupted lipid metabolism, can affect drug pharmacokinetics, increase toxicity, and lead to dose reductions or early discontinuation [6]. Experimental studies further show that tumor-derived factors such as interleukin-6 (IL-6), PTH-related protein (PTHrP), myostatin/activin, and ATGL-mediated lipolysis can promote adipose tissue browning, muscle catabolism, and systemic metabolic disturbance [7–11]. These findings support the hypothesis that tumors capable of inducing cachexia may also possess distinct drug-response programs. Yet, tumor-intrinsic mechanisms that connect cachexia-inducing ability with altered transcriptional responses to therapeutic stress remain largely unexplored.

To address this gap, suitable experimental models are essential [12]. Human cancer cell lines that reproducibly induce cachexia in vivo are rare, and commonly used xenograft models often show variable or late-onset wasting [12]. The human cell-line pair TCC-NECT-2 [13] and AkuNEC provides a unique opportunity to investigate tumor-intrinsic differences underlying cachexia induction [13,14]. TCC-NECT-2, derived from a duodenal neuroendocrine carcinoma, does not induce cachexia when transplanted into mice, whereas its in vivo–passaged derivative AkuNEC reproducibly causes systemic cachexia. Because these two lines share a common origin, they enable controlled comparison of cachexia-inducing and non-inducing tumor states without confounding genomic background. Such paired models remain uncommon but offer a tractable system for mechanistic studies.

Although interest in tumor–host metabolic crosstalk has grown, relatively few studies have examined tumor-intrinsic transcriptional programs linked to cachexia. Most previous research has focused on host tissues, including skeletal muscle, adipose tissue, and liver [10,11], whereas far less is known about gene-expression programs within cachexia-inducing tumors themselves, either at baseline or under therapeutic stress. This remains an important gap in the field. In this study, we performed RNA sequencing of TCC-NECT-2 and AkuNEC cells before and after exposure to doxorubicin to identify (i) baseline transcriptional differences associated with cachexia-inducing capacity and (ii) differential chemotherapy-induced responses. Through this paired-cell-line approach, we aim to define tumor-intrinsic transcriptional signatures that may contribute to cachexia induction and potentially influence chemotherapeutic sensitivity. Based on prior experimental and clinical studies, a literature-curated reference set of tumor-derived genes previously implicated in cancer cachexia was compiled and used to guide downstream transcriptomic analyses (Supplementary Table S1).

2. Materials and Methods

2.1. Cell Lines

Two human cell lines derived from a duodenal neuroendocrine carcinoma were examined: the parental line TCC-NECT-2 [13] and its cachexia-inducing subline AkuNEC [14]. The procedures for establishing these cell lines and their initial characterization have been

https://doi.org/10.3390/targets4010009

described previously. Cells were routinely maintained in RPMI-1640 medium (Gibco, Thermo Fisher Scientific, Waltham, MA, USA) supplemented with 10% heat-inactivated fetal bovine serum (FBS), 100 IU/mL penicillin, and 100 μg/mL streptomycin. All cultures were kept at 37 °C in a humidified incubator with 5% CO_2. For 2D culture, cell suspensions were seeded onto 10 cm tissue culture dishes (Nunc Easydish; Thermo Fisher Scientific Inc.) and maintained at 37 °C in a humidified incubator with 5% CO_2. For 3D culture, the cells were kept in round-bottom 96-well ultra-low attachment plates (Thermo Scientific™ 96 Well Round (U) Bottom Plate, TC Surface; Thermo Fisher Scientific Inc., Waltham, MA, USA). The cells were seeded at 5000 cells per well in the ultra-low attachment 96-well plates. 3D culture in ultra-low attachment plates was selected for all drug treatments and transcriptomic assays to provide a model of tumor architecture.

2.2. *Treatments with Doxorubicin*

The cells were seeded at 5000 cells per well in the ultra-low attachment 96-well plates. After 24 h of incubation to allow for spheroid stabilization, doxorubicin (doxorubicin hydrochloride, Selleck Chemicals LLC, Houston, TX, USA) was added to each well at a final concentration of 10 μM and the cells were exposed for an additional 24 h. Doxorubicin was selected as a prototypical genotoxic stressor to induce acute DNA damage and capture the resulting transcriptional plasticity, rather than to simulate a specific clinical regimen for neuroendocrine carcinoma. Cell viability following drug treatment was assessed using the Cell Counting Kit-8 (Dojindo Laboratories, Kumamoto, Japan). After adding the reagent to each well, plates were incubated at 37 °C for the recommended period, and absorbance at 450 nm was measured using a Multiskan SkyHigh microplate reader (Thermo Fisher Scientific, Waltham, MA, USA). As a vehicle control, to represent 100% cell viability, the same final concentration of DMSO used to prepare doxorubicin (10 μM) was added to the culture medium. We confirmed in advance that this concentration of DMSO did not affect cell proliferation. The absorbance of the vehicle DMSO control was defined as 100%, and relative cell viability (%) was calculated for the doxorubicin-treated samples. All experiments were performed in duplicate, and mean values were reported. All experiments were performed in duplicate (n = 2 biological replicates, each with technical duplicates).

2.3. *Scanning Electron Microscopy (SEM)*

Scanning electron microscopy was carried out following a previously described procedure [15], with several minor adjustments to suit our experimental conditions. The spheroids were fixed at 4 °C overnight in 2.5% glutaraldehyde prepared in 0.1 M phosphate buffer (pH 7.4). After fixation, the spheroids were gently rinsed with phosphate-buffered saline (PBS) and subsequently post-fixed in 1% osmium tetroxide (OsO_4) for 30 min at room temperature. The samples were then dehydrated through a graded ethanol series, transferred to 100% ethanol, and allowed to dry in air. Once fully dried, the spheroids were coated with a thin platinum film using an MSP-1S sputter coater (Shinku Device, Ibaraki, Japan). SEM imaging was performed using a Phenom Pro desktop scanning electron microscope (Thermo Fisher Scientific, Waltham, MA, USA), and representative images were obtained to document the surface morphology of the spheroids.

2.4. *Transcriptomic Study*

2.4.1. RNA Extraction and Sample Preparation

The cells were washed twice with phosphate-buffered saline (PBS) directly in the plates. Cells were then resuspended in PBS per well, followed by the QIAzol Lysis Reagent (QIAGEN K.K., Tokyo, Japan) for cell lysis. Cell lysates were collected and shipped to Zeon BioSolutions Inc. (Hyogo, Japan), where total RNA extraction and purification were performed using TRIzol reagent according to the manufacturer's protocol.

https://doi.org/10.3390/targets4010009

2.4.2. RNA Quality Assessment

RNA purity was assessed by spectrophotometric measurement of the A260/280 ratio, and RNA integrity was evaluated using an Agilent Bioanalyzer or TapeStation system (Agilent Technologies, Santa Clara, CA, USA). Only RNA samples with an RNA integrity number (RIN) greater than 8.0 were used for downstream analyses. RNA concentration and RNA integrity were applied as the primary quality-control criteria for RNA sequencing. For library preparation, 20 ng of total RNA per sample was used as starting material.

2.4.3. RNA-Seq Library Preparation and Sequencing

RNA-seq libraries were prepared using the Quartz-Seq2 protocol. Neither ribosomal RNA depletion nor poly(A) selection was performed. Libraries were multiplexed and sequenced on an Illumina NextSeq platform. Quartz-Seq2 is designed for low-input transcriptome profiling and has been shown to provide robust relative expression quantification at modest sequencing depth. Quality control of raw sequencing data was performed using FastQC to evaluate read quality, base composition, and potential adapter contamination prior to downstream bioinformatic analyses.

2.5. Statistical and Bioinformatic Analyses

Gene expression analyses were primarily based on effect-size–driven comparisons in this exploratory pilot study. Due to the small sample size ($n = 2$ biological replicates per condition), formal statistical significance testing or multiple-testing correction was not applied to the RNA-seq data. RNA-seq expression values were log2-transformed prior to downstream analyses. For pairwise comparisons between experimental conditions, log2 fold-change (log2FC) values were calculated using the mean expression of two biological replicates per condition. The log2-normalized gene expression matrix for all genes and all experimental conditions is provided in Supplementary Table S2.

Baseline transcriptional differences between TCC-NECT-2 and AkuNEC were evaluated under no-treatment conditions. Doxorubicin-induced transcriptional responses were assessed by comparing doxorubicin-treated samples with DMSO-treated controls within each cell line. Differential drug response (Δresponse) was defined as the difference in log2FC between AkuNEC and TCC-NECT-2, calculated as (AkuNEC_Dox − AkuNEC_DMSO) − (TCC-NECT-2_Dox−TCC-NECT-2_DMSO).

Because of the limited number of biological replicates ($n = 2$ per condition), no formal statistical significance testing or multiple-testing correction was applied. Instead, transcriptomic differences were interpreted based on the magnitude and consistency of log2FC values across comparisons. This approach was chosen to identify robust and reproducible transcriptional patterns while avoiding overinterpretation of statistical significance.

Pathway enrichment analyses were performed using the Enrichr platform implemented via the gseapy Python package. Gene ontology (GO) biological process databases were used to assess functional enrichment among genes upregulated and downregulated in AkuNEC relative to TCC-NECT-2. Enrichment results were ranked by combined score, and representative pathways were selected for visualization. Enrichment results were ranked by "Combined Score", which integrates the p-value from Fisher's exact test with the z-score of the deviation from the expected rank, providing a prioritized list of biological processes even in limited datasets. All enrichment analyses were conducted without statistical filtering to reflect effect-size–driven differences.

All computational analyses and data visualizations were performed using Python 3.14 with the pandas, numpy, matplotlib, seaborn, and gseapy libraries. Figures were generated using log2-transformed expression values and exported at publication-quality resolution.

https://doi.org/10.3390/targets4010009

3. Results

3.1. Morphological Features and Doxorubicin Sensitivity of TCC-NECT-2 and AkuNEC

To establish a phenotypic baseline for subsequent transcriptomic analyses, we first compared the morphological characteristics and doxorubicin sensitivity of the non–cachexia-inducing parental cell line TCC-NECT-2 and its cachexia-inducing derivative AkuNEC. Under standard culture conditions, phase-contrast microscopy revealed no overt morphological differences between the two cell lines (Figure 1a,b). Both TCC-NECT-2 and AkuNEC exhibited similar cell shape and growth patterns at the gross morphological level.

We next examined cell surface architecture using scanning electron microscopy (SEM). Although overall cellular morphology was largely comparable, SEM imaging suggested subtle differences in surface features between TCC-NECT-2 and AkuNEC cells (Figure 1c). These differences were qualitative in nature and were not subjected to quantitative morphometric analysis, but they raised the possibility that cachexia-inducing capacity may be associated with modest alterations in cell surface organization rather than dramatic changes in cell shape. These differences were qualitative in nature and were not subjected to quantitative morphometric analysis, but they raised the possibility that cachexia-inducing capacity may be associated with modest alterations in cell surface organization.

To assess whether cachexia-inducing capacity is accompanied by altered sensitivity to chemotherapeutic stress at the phenotypic level, we evaluated acute cytotoxicity following doxorubicin treatment. Under the tested conditions, doxorubicin treatment resulted in a significantly lower viability in AkuNEC cells compared with TCC-NECT-2 (Figure 1d). Cell viability was reduced to approximately 19% in AkuNEC, whereas TCC-NECT-2 retained ~41% viability relative to DMSO-treated controls. Notably, despite this quantitative difference in drug sensitivity, no overt morphological differences were observed between the two cell lines under standard culture conditions (Figure 1a–c), suggesting that enhanced chemosensitivity in AkuNEC is not accompanied by gross phenotypic alterations.

Together, these observations demonstrate that TCC-NECT-2 and AkuNEC cells are largely indistinguishable at the level of gross morphology and short-term acute cytotoxicity by doxorubicin. These findings suggest that cachexia-inducing capacity is not readily captured by conventional phenotypic assays, thereby motivating a deeper investigation into tumor-intrinsic transcriptional programs and stress-responsive molecular differences addressed in subsequent analyses.

3.2. Literature-Based Identification of Cachexia-Associated Tumor Factors

To identify tumor-derived factors potentially involved in cachexia induction, we first conducted a systematic literature-based survey of genes previously reported to drive cancer cachexia. Using PubMed-indexed studies, we compiled a curated list of tumor-secreted molecules that have been experimentally shown to promote key features of cachexia, including muscle wasting, adipose tissue mobilization, systemic inflammation, and anorexia (Supplementary Table S1 for detailed descriptions and a tabulated summary). This approach ensured that subsequent transcriptomic analyses were anchored in established biological mechanisms rather than relying solely on unbiased differential expression.

https://doi.org/10.3390/targets4010009

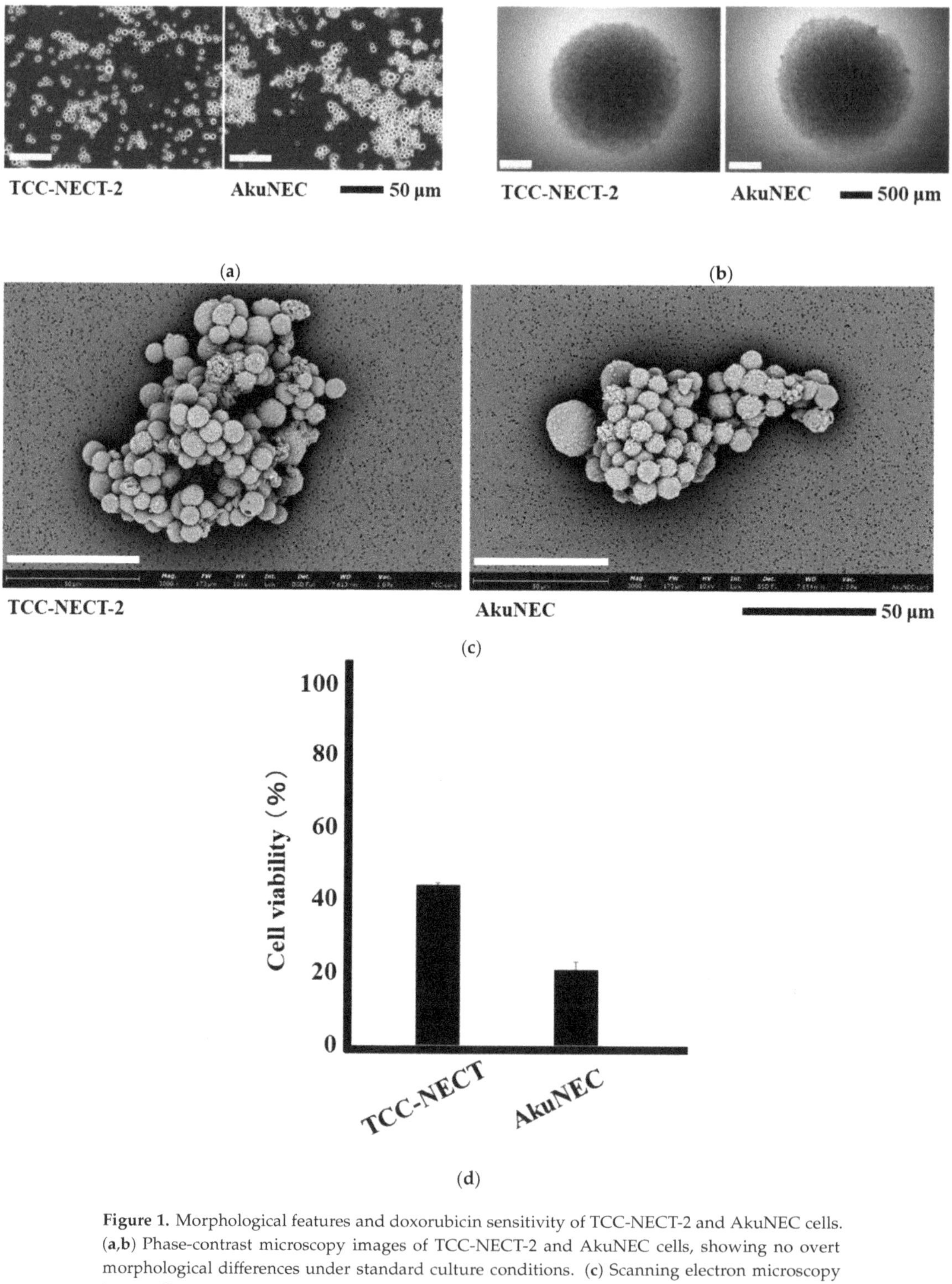

Figure 1. Morphological features and doxorubicin sensitivity of TCC-NECT-2 and AkuNEC cells. (**a**,**b**) Phase-contrast microscopy images of TCC-NECT-2 and AkuNEC cells, showing no overt morphological differences under standard culture conditions. (**c**) Scanning electron microscopy images illustrating subtle differences in cell surface architecture between the two cell lines. Scale bars,

https://doi.org/10.3390/targets4010009

50 μm. (**d**) Cell viability assessed using the Cell Counting Kit-8 (CCK-8) assay after 24 h exposure to doxorubicin (10 μM). Absorbance values were normalized to DMSO-treated controls, which were set to 100%. TCC-NECT-2 cells retained approximately 41% viability, whereas AkuNEC cells exhibited significantly lower viability (~19%), indicating enhanced sensitivity to doxorubicin in the cachexia-inducing AkuNEC line. Data are presented as mean ± SD. Statistical significance was evaluated using a two-tailed Student's *t*-test.

Based on this literature survey, the identified cachexia-associated factors clustered into three major functional categories: (i) pro-inflammatory cytokines, including IL6, TNF, and IL1B, which mediate systemic inflammation and muscle proteolysis; (ii) TGF-β superfamily ligands, such as GDF15, MSTN, and INHBA (encoding activin A), which suppress muscle anabolism and activate catabolic pathways; and (iii) metabolic and lipid-mobilizing factors, notably AZGP1 (ZAG) and LIF, which enhance lipolysis, alter energy expenditure, and promote anorexia in several tumor models. Extensive experimental work has validated the cachexia-inducing activity of these mediators: IL-6 neutralization attenuates wasting in the C26 carcinoma model; TNF-α was identified as the macrophage-derived cachexia factor "cachectin"; GDF15 drives tumor-induced anorexia via its receptor GFRAL; myostatin and activin A potently inhibit muscle growth; AZGP1 functions as a lipid-mobilizing factor; and tumor-derived LIF induces severe cachexia in melanoma-bearing mice. Together, these findings establish a mechanistic framework for interpreting tumor-intrinsic transcriptional changes in cachexia-inducing cancer cells.

Having defined this literature-based reference framework, we next investigated whether acquisition of cachexia-inducing capacity in AkuNEC cells is accompanied by transcriptional upregulation of these known tumor-derived factors. To this end, we analyzed RNA-seq data obtained from the non–cachexia-inducing parental line TCC-NECT-2 and its cachexia-inducing derivative AkuNEC, both at baseline and following exposure to doxorubicin. The list of candidate genes extracted from the literature was used as a predefined reference panel for interpretation of the transcriptomic data (Supplementary Table S1).

3.3. Baseline Transcriptional Differences Between TCC-NECT-2 and AkuNEC

We first compared baseline gene expression profiles between TCC-NECT-2 and AkuNEC under no-treatment conditions to identify tumor-intrinsic transcriptional differences associated with cachexia-inducing capacity. A complete list of baselines, differentially expressed genes and corresponding log2FC values is provided in Supplementary Table S3, which served as the basis for ranked and pathway analyses. Differential expression was evaluated using log2 fold-change (log2FC) values calculated from the mean expression of two biological replicates per condition.

A bidirectional ranked plot demonstrated extensive baseline transcriptional reprogramming in AkuNEC cells relative to TCC-NECT-2, with both upregulated and downregulated gene sets clearly evident (Figure 2a). Genes exhibiting increased expression in AkuNEC included multiple inflammatory, secretory, and metabolic regulators, whereas genes reduced in AkuNEC were enriched for factors associated with cellular structure, differentiation, and homeostatic regulation.

https://doi.org/10.3390/targets4010009

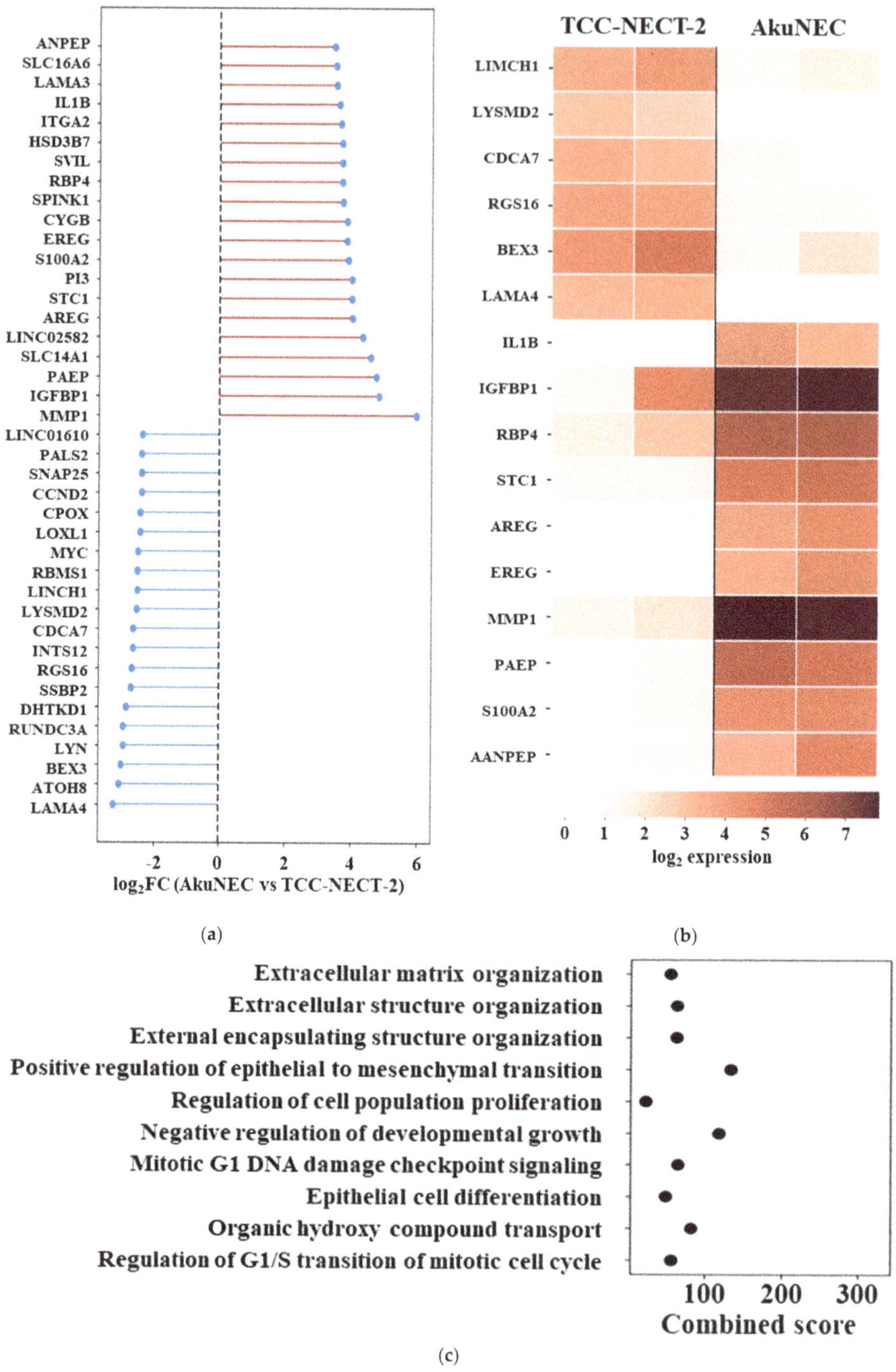

Figure 2. *Cont.*

https://doi.org/10.3390/targets4010009

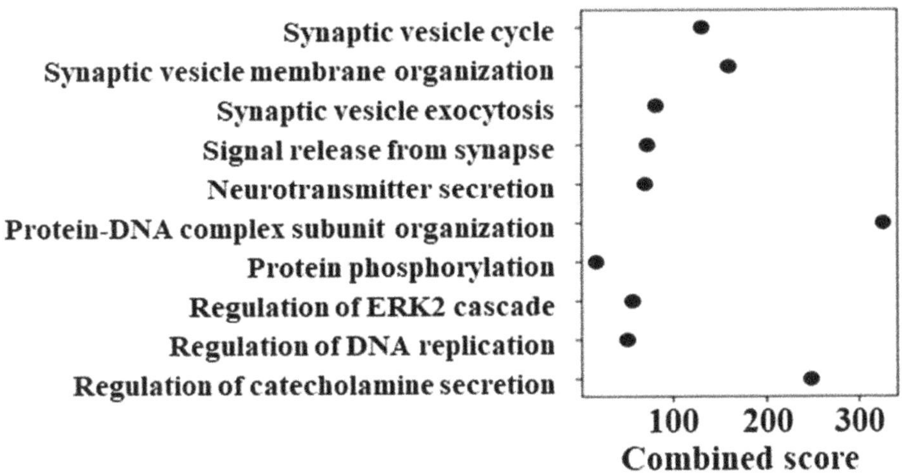

(d)

Figure 2. Baseline transcriptional differences between AkuNEC and TCC-NECT-2. (**a**) Ranked plot showing log2 fold changes (log2FC) of gene expression between AkuNEC and TCC-NECT-2 under no-treatment conditions. Genes are ordered according to log2FC, with positive values indicating higher expression in AkuNEC and negative values indicating lower expression relative to TCC-NECT-2, illustrating bidirectional baseline transcriptional reprogramming associated with cachexia-inducing capacity. Log2FC values were calculated from the mean expression of two biological replicates per condition ($n = 2$). (**b**) Heatmap showing expression patterns of 16 representative genes selected to illustrate bidirectional baseline transcriptional differences. Gene selection was based on a combination of literature-curated cachexia-associated tumor-derived factors (Supplementary Table S1) and data-driven ranking of baseline log2FC. Six genes downregulated in AkuNEC relative to TCC-NECT-2 are shown in the upper panel, while ten upregulated genes are shown in the lower panel. Expression values are log2-transformed, and each column represents an individual biological replicate ($n = 2$ per condition). The vertical line separates TCC-NECT-2 and AkuNEC samples. (**c**,**d**) Gene ontology (GO) biological process enrichment analysis performed separately for genes upregulated (**c**) and downregulated (**d**) in AkuNEC relative to TCC-NECT-2 under no-treatment conditions. Enrichment results are shown as dot plots ranked by combined score. Upregulated genes were primarily associated with inflammatory responses, secretory pathways, and extracellular matrix remodeling, whereas downregulated genes were enriched for processes related to cellular structure, differentiation, and homeostatic regulation. Gene sets were derived from baseline log2FC values and analyzed without statistical filtering to reflect effect-size–driven transcriptional differences ($n = 2$).

To visualize representative features of this bidirectional reprogramming, we constructed a heatmap of 16 selected genes integrating literature-curated cachexia-associated factors with data-driven baseline expression differences (Figure 2b). The log2-normalized expression values of these 16 genes are shown in Supplementary Table S7. Six genes downregulated in AkuNEC (LIMCH1, LYSMD2, CDCA7, RGS16, BEX3, and LAMA4) were displayed in the upper portion of the heatmap, while ten upregulated genes (IL1B, IGFBP1, RBP4, STC1, AREG, EREG, MMP1, PAEP, S100A2, and ANPEP) were shown in the lower portion. This analysis revealed a coordinated loss of baseline cellular and structural programs alongside the acquisition of inflammatory and secretory transcriptional states in AkuNEC cells.

We next examined the functional implications of these baseline transcriptional differences using pathway enrichment analysis. Gene ontology (GO) biological process enrichment was performed separately for genes upregulated and downregulated in AkuNEC based on baseline log2FC rankings. Gene lists used for enrichment analyses are provided in Supplementary Table S5 (upregulated) and Table S6 (downregulated). Pathways enriched among upregulated genes (Figure 2c) were predominantly associated with extracellular matrix organization, epithelial–mesenchymal transition, regulation of cell population proliferation, inflammatory signaling, and stress-related processes. In contrast, pathways enriched among downregulated genes (Figure 2d) were associated with synaptic vesicle organization, neurotransmitter secretion, protein–DNA complex organization, regulation of ERK1/2 signaling, and DNA replication-related processes. The same combined score scale was applied to both panels to allow direct comparison of functional enrichment patterns.

Collectively, these results indicate that acquisition of cachexia-inducing capacity in AkuNEC cells is accompanied by extensive baseline transcriptional reprogramming, characterized by upregulation of inflammatory, extracellular, and secretory gene programs and concomitant suppression of cellular structural and homeostatic pathways. This baseline state provides a molecular context for subsequent analyses of chemotherapeutic stress responses.

3.4. Doxorubicin-Induced Transcriptional Responses in TCC-NECT-2 and AkuNEC

Having established that AkuNEC cells exhibit extensive baseline transcriptional reprogramming associated with cachexia-inducing capacity (Figure 2), we next examined how these intrinsic differences influence transcriptional responses to chemotherapeutic stress. To this end, we analyzed gene expression changes following doxorubicin treatment in TCC-NECT-2 and AkuNEC cells and compared both the individual drug responses and their differential patterns.

In TCC-NECT-2 cells, doxorubicin exposure induced a transcriptional response characterized primarily by activation of general stress- and transcription-associated pathways (Figure 3a). Prominent changes included strong induction of RN7SK, a regulator of transcriptional elongation, along with alterations in genes associated with membrane trafficking, stress signaling, and prostaglandin metabolism, such as PTGS1. Concurrently, several genes related to proliferation and mitochondrial function, including MYC and mitochondrial-encoded transcripts, were markedly downregulated. Overall, this response pattern is consistent with a canonical DNA damage–associated stress response accompanied by suppression of proliferative programs. The full list of doxorubicin-responsive genes in TCC-NECT-2 cells is provided in Supplementary Table S7.

In contrast, AkuNEC cells displayed a distinct transcriptional response to doxorubicin treatment (Figure 3b). Although induction of RN7SK was also observed, the AkuNEC response was notable for the additional upregulation of genes implicated in immune-like stress signaling, apoptosis resistance, and adaptive survival pathways, including SERPINB9, DRAM1, IFIT2, and multiple dual-specificity phosphatases. At the same time, downregulated genes in AkuNEC included regulators of transcriptional control and cellular metabolism, suggesting a coordinated reprogramming rather than uniform suppression of gene expression. These features indicate that doxorubicin elicits a more complex and adaptive transcriptional response in AkuNEC cells than in the non–cachexia-inducing parental line. The full list of doxorubicin-responsive genes in AkuNEC is provided in Supplementary Table S8.

To directly compare drug-induced responses between the two cell lines, we calculated the differential doxorubicin response (Δresponse), defined as the difference in log2 fold-change between AkuNEC and TCC-NECT-2 (Figure 3c). This analysis revealed a set of genes that were preferentially induced in AkuNEC under doxorubicin stress, including CCND2, GINS2, MYC, HMGCS2, FDX1, DHFR, and BRMS1L. These genes are involved in cell cycle regulation, DNA replication, metabolic reprogramming, and mitochondrial

https://doi.org/10.3390/targets4010009

function, suggesting enhanced transcriptional plasticity in AkuNEC cells. Conversely, genes exhibiting relatively attenuated responses in AkuNEC included BAMBI, CDKN2B, FOXF1, SRF, and other regulators associated with differentiation and growth suppression. Genes ranked by Δresponse are provided in Supplementary Table S9.

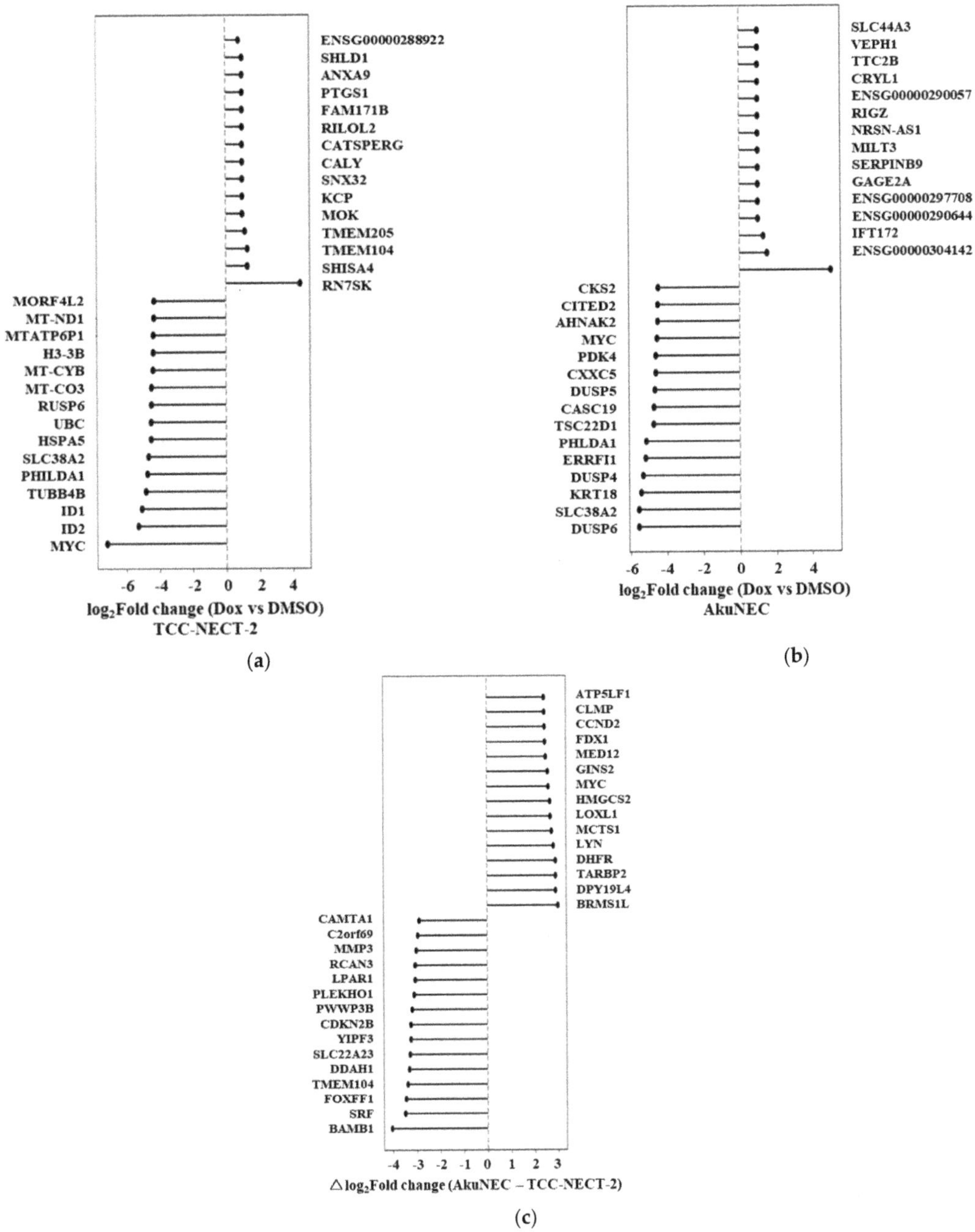

Figure 3. Doxorubicin-induced transcriptional responses in TCC-NECT-2 and AkuNEC. (**a**) Bidirectional ranked plot showing log2 fold changes (log2FC) of gene expression in TCC-NECT-2 cells following dox-

orubicin treatment relative to DMSO-treated controls. Genes are ordered according to log2FC, with positive values indicating induction and negative values indicating suppression upon drug exposure. Log2FC values were calculated from the mean expression of two biological replicates per condition ($n = 2$). Genes displayed represent the top and bottom differentially expressed genes and are listed in Supplementary Table S10. (**b**) Bidirectional ranked plot showing doxorubicin-induced transcriptional changes in AkuNEC cells, calculated in the same manner as in (**a**). Compared with TCC-NECT-2, AkuNEC cells exhibited a distinct pattern of drug-responsive gene expression, reflecting qualitative differences in stress-adaptive transcriptional programs. Genes displayed represent the top- and bottom-ranked genes and are listed in Supplementary Table S11. (**c**) Bidirectional ranked plot of the differential doxorubicin response (Δresponse) between AkuNEC and TCC-NECT-2, defined as: Δlog2FC = (AkuNEC_Dox − AkuNEC_DMSO) − (TCC-NECT-2_Dox − TCC-NECT-2_DMSO). Genes are ranked according to Δlog2FC, with positive values indicating stronger induction in AkuNEC and negative values indicating attenuated or reduced responses relative to TCC-NECT-2. Genes displayed represent the top- and bottom-ranked genes and are listed in Supplementary Table S12.

Taken together, these results demonstrate that the cachexia-inducing cell line AkuNEC not only exhibits a distinct baseline transcriptional state (Figure 2) but also responds to chemotherapeutic stress through a qualitatively different transcriptional program. Rather than displaying a uniformly heightened sensitivity or resistance, AkuNEC cells undergo selective reorganization of proliferation-, metabolism-, and stress-related pathways in response to doxorubicin. This dynamic response contrasts with the more stereotypical stress-induced suppression observed in TCC-NECT-2 cells and suggests that cachexia-associated tumor-intrinsic programs may shape how cancer cells adapt to chemotherapy. Genes displayed in Figure 3a–c are listed in Supplementary Tables S10–S12.

4. Discussion

4.1. Phenotypic Similarity Masks Tumor-Intrinsic Reprogramming Associated with Cachexia-Inducing Capacity

In this study, we found that the cachexia-inducing cell line AkuNEC and its non–cachexia-inducing parental counterpart TCC-NECT-2 were largely indistinguishable at the level of gross cellular morphology (Figure 1). The absence of overt morphological differences is consistent with the concept that cachexia is primarily driven by tumor–host systemic interactions and endocrine-like signaling rather than by easily detectable changes in cancer cell shape under routine culture conditions [1–4]. In addition, the limited ability of conventional phenotypic assays to capture clinically relevant tumor–host pathophysiology has been discussed in broader contexts of translational cancer research and preclinical modeling [16]. From a clinical perspective, the constitutive upregulation of inflammatory and metabolic factors in AkuNEC cells suggests that these signatures could serve as candidate prognostic biomarkers to identify patients at higher risk for therapy intolerance and early-onset wasting.

Although scanning electron microscopy suggested subtle differences in surface features, these changes were qualitative and modest in scale. Such subtle alterations may reflect differences in cell–cell or cell–matrix interactions, which can influence secretion, extracellular matrix remodeling, and host signaling [17–19]. Importantly, the lack of striking phenotypic divergence supports the interpretation that cachexia-inducing capacity may arise from tumor-intrinsic transcriptional reprogramming that is not necessarily mirrored by gross morphology [1–4].

Although AkuNEC cells displayed increased sensitivity to doxorubicin in short-term viability assays, this difference was not accompanied by overt morphological changes, underscoring that quantitative drug sensitivity and qualitative transcriptional reprogramming

https://doi.org/10.3390/targets4010009

represent partially distinct aspects of the cachexia-associated tumor phenotype. In this context, transcriptomic profiling is useful for capturing stress-induced pathway reorganization that may not translate into large differences in short-term viability assays [20,21].

A central finding of this study is that multiple factors previously implicated in cancer cachexia were constitutively upregulated in AkuNEC cells at baseline (Figure 2). The literature-curated cachexia-associated reference gene set used in this analysis is provided in Supplementary Table S1. Inflammatory and secretory mediators such as IL1B, together with metabolic/endocrine-related regulators including IGFBP1, RBP4, STC1, AREG, and EREG, were increased in AkuNEC relative to TCC-NECT-2. These observations are consistent with a large body of literature supporting tumor-derived inflammatory cytokines and secreted factors as drivers or correlates of systemic metabolic dysregulation, anorexia, and tissue wasting [1–4,22–27]. Classical experimental work established causal links between cytokines and wasting phenotypes (e.g., IL-6 in the C26 model; TNF/cachectin; IL-1–associated anorexia pathways), providing a mechanistic basis for interpreting inflammatory gene programs within cachexia-inducing tumor contexts [22–27].

Importantly, cachexia-associated genes were not upregulated in isolation. We observed coordinated induction of additional genes with comparable effect sizes (e.g., MMP1, PAEP, S100A2, ANPEP), many of which are linked to extracellular matrix remodeling, stress responses, and secretory biology. Such coordinated shifts are consistent with the concept that cachexia is rarely attributable to a single "master factor" and instead reflects network-level tumor programs that collectively perturb host metabolism and tissue homeostasis [1–4,22–29]. In support of this view, multiple mechanistic studies have demonstrated that distinct tumor-derived pathways can converge on muscle proteolysis, adipose browning/lipolysis, and systemic inflammation, including ATGL-mediated lipolysis, PTHrP-driven thermogenic programming, and activin/myostatin signaling via ActRIIB [7–10,30,31].

At the pathway level, enrichment among genes upregulated in AkuNEC highlighted inflammatory signaling, extracellular matrix organization, and stress-associated processes (Figure 2c), whereas downregulated programs were enriched for cellular structural and homeostatic regulation (Figure 2d). Gene lists underlying these enrichment analyses are provided in Supplementary Tables S11 and S12. This bidirectional remodeling is compatible with contemporary multi-organ and multi-omics views of cachexia, which emphasize early systemic host abnormalities, broad endocrine and immune changes, and progressive rewiring of metabolism and inflammation across tissues [32–34]. Together, our data support a model in which cachexia-inducing capacity is embedded in coordinated tumor-intrinsic transcriptional states that generate a cachexia-promoting milieu through cumulative or synergistic effects of multiple factors rather than isolated mediators.

4.2. Stress-Induced Transcriptional Reprogramming Builds upon a Cachexia-Associated Baseline Program

Baseline transcriptional reprogramming in AkuNEC provides a framework for interpreting the response to chemotherapeutic stress. Chemotherapy can modulate systemic inflammation and metabolic states, and several studies have suggested that stress-induced endocrine and inflammatory pathways may influence appetite, muscle wasting, and immune competence [35–37]. In our data, doxorubicin elicited general stress-associated changes in both cell lines, including induction of RN7SK, consistent with transcriptional stress and global regulatory responses [38,39]. However, the overall response pattern differed between TCC-NECT-2 and AkuNEC (Figure 3a,b), suggesting that cachexia-associated baseline programs may predispose tumor cells to qualitatively distinct stress-adaptive transcriptional trajectories. Since doxorubicin is not the first-line chemotherapy for neu-

https://doi.org/10.3390/targets4010009

roendocrine carcinoma, the observed responses represent a general model of stress-induced plasticity rather than a direct clinical simulation.

AkuNEC cells showed selective induction of genes linked to immune-like stress signaling, autophagy/apoptosis control, and adaptive survival pathways. Such responses align with broader concepts of transcriptional plasticity under stress, in which tumor cells reorganize regulatory networks to sustain viability and metabolic flexibility [40–44]. The Δresponse analysis further identified preferential induction of genes involved in cell cycle regulation, DNA replication, and metabolic remodeling (Figure 3c). Gene lists used for Δresponse-based enrichment analyses are provided in Supplementary Tables S13 and S14. Notably, metabolic rewiring and mitochondrial programs have been implicated in cachexia-related host metabolism and systemic immune modulation, suggesting possible links between tumor stress responses and host-pathophysiology–relevant signaling [45–48]. Genes commonly responsive to doxorubicin are summarized in Supplementary Table S15, whereas AkuNEC-specific responses are summarized in Supplementary Table S16.

Taken together, our findings support a two-tiered framework: (i) baseline cachexia-associated tumor-intrinsic transcriptional programs that may shape tumor–host interactions and systemic wasting, and (ii) stress-induced transcriptional plasticity that reorganizes proliferation- and metabolism-related pathways under chemotherapeutic challenge. This framework reconciles the limited phenotypic divergence observed in standard assays (Figure 1) with pronounced molecular divergence at the transcriptomic level (Figures 2 and 3) and argues that cachexia-associated tumor evolution may affect not only host physiology but also qualitative aspects of tumor stress adaptation [1–6,20].

4.3. Limitations and Future Perspectives

Several limitations of the present study should be acknowledged. First, the transcriptomic analyses were performed with a limited number of biological replicates (n = 2 per condition). While this design was sufficient to detect robust and reproducible effect-size–driven transcriptional differences between TCC-NECT-2 and AkuNEC cells, it restricts the use of formal statistical testing and precludes definitive conclusions regarding variability across broader biological contexts. The findings should be viewed as hypothesis-generating, and further validation in larger cohorts is required. Future studies using standard-of-care agents, such as etoposide or platinum-based drugs, will be necessary to confirm the clinical relevance of these sensitivity programs.

Second, this study was conducted exclusively using in vitro cell line models. Although the paired nature of the TCC-NECT-2/AkuNEC system provides a controlled framework for dissecting tumor-intrinsic programs associated with cachexia-inducing capacity, it does not fully recapitulate the complex tumor–host interactions that underlie cancer cachexia in vivo. In particular, interactions with immune cells, stromal components, and systemic metabolic organs are not captured in this experimental setting [12].

Third, while our analyses identified coordinated upregulation of multiple inflammatory, secretory, and metabolic regulators in AkuNEC cells, functional validation of individual factors was beyond the scope of this study. We did not directly assess the secretion, bioactivity, or host-modulating effects of candidate cachexia-associated proteins identified at the transcriptomic level. As such, the relative contribution of individual factors versus their combined or synergistic effects remains to be determined. Moreover, this study lacks protein-level validation of the identified factors, and the validation study at the protein-level using the proteomics modalities and the direct assessment of secreted proteins via ELISA or Western blotting, alongside functional assays such as myotube atrophy models, will be critical to establish the causal contribution and should be performed.

https://doi.org/10.3390/targets4010009

Despite these limitations, the present study provides several important perspectives for future research. The observation that cachexia-inducing capacity is associated with coordinated baseline transcriptional reprogramming and enhanced stress-induced plasticity suggests that cancer cachexia may emerge from network-level tumor-intrinsic programs rather than from single dominant mediators. Future studies integrating transcriptomic data with proteomic profiling of tumor-conditioned media, functional assays using muscle or adipose cell models, and in vivo validation in xenograft or organoid-based systems will be critical to establish causal links between tumor-derived programs and systemic cachexia phenotypes.

Moreover, extending this analytical framework to additional cachexia-inducing and non-inducing tumor models, as well as to patient-derived samples, may help clarify whether the transcriptional architecture identified here represents a generalizable feature of cachexia-associated cancers. Such efforts could ultimately inform strategies aimed at disrupting coordinated tumor-derived programs, rather than targeting individual factors in isolation, as a means of mitigating cancer cachexia. Future investigations utilizing confocal microscopy for intracellular localization and organelle-specific staining will be necessary to provide a comprehensive internal morphological assessment under stress.

5. Conclusions

In summary, this study demonstrates that cachexia-inducing capacity in tumor cells is associated with coordinated tumor-intrinsic transcriptional reprogramming rather than with overt phenotypic differences in morphology, suggesting that qualitative transcriptional divergence may be more pronounced than quantitative differences in bulk cell viability. Using a paired duodenal neuroendocrine carcinoma cell line system, we show that the cachexia-inducing subline AkuNEC exhibits constitutive upregulation of inflammatory, secretory, and metabolic gene programs at baseline, alongside suppression of cellular structural and homeostatic pathways.

Importantly, this baseline transcriptional state shapes the response to chemotherapeutic stress. Upon doxorubicin exposure, AkuNEC cells display a qualitatively distinct transcriptional response characterized by selective reorganization of proliferation-, metabolism-, and stress-related pathways, indicative of enhanced transcriptional plasticity. These findings suggest that cachexia-associated tumor programs and stress-induced adaptive responses are functionally linked.

Together, our results support a model in which cancer cachexia emerges from network-level tumor-intrinsic programs that influence both systemic host interactions and tumor adaptability under therapeutic stress. This work provides a framework for exploring cachexia as a consequence of coordinated transcriptional states rather than isolated molecular factors and highlights the importance of integrating baseline and stress-induced tumor programs in future cachexia research.

Supplementary Materials: The following supporting information can be downloaded at: https://www.mdpi.com/article/10.3390/targets4010009/s1. Supplementary Table S1: Literature-curated tumor-derived factors implicated in cancer cachexia. Supplementary Table S2: Log2-normalized gene expression profiles of TCC-NECT-2 and AkuNEC cells. Supplementary Table S3: Baseline differential gene expression between AkuNEC and TCC-NECT-2 cells. Supplementary Table S4: Log2-normalized expression values of the 16 representative genes shown in Figure 2b. Supplementary Table S5: Gene list used for pathway enrichment analysis of baseline upregulated genes. Supplementary Table S6: Gene list used for pathway enrichment analysis of baseline downregulated genes. Supplementary Table S7: Doxorubicin-induced transcriptional changes in TCC-NECT-2 cells. Supplementary Table S8: Doxorubicin-induced transcriptional changes in AkuNEC cells. Supplementary Table S9: Differential doxorubicin response between AkuNEC and TCC-NECT-2

https://doi.org/10.3390/targets4010009

cells. Supplementary Table S10: Top and bottom differentially expressed genes in TCC-NECT-2 cells (Figure 3a). Supplementary Table S11: Top and bottom differentially expressed genes in AkuNEC cells (Figure 3b). Supplementary Table S12: Top and bottom genes ranked by differential doxorubicin response (Figure 3c). Supplementary Table S13: Gene list used for pathway enrichment analysis of genes with increased Δresponse. Supplementary Table S14: Gene list used for pathway enrichment analysis of genes with decreased Δresponse. Supplementary Table S15: Genes commonly responsive to doxorubicin in TCC-NECT-2 and AkuNEC cells. Supplementary Table S16: Genes showing AkuNEC-specific transcriptional responses to doxorubicin.

Author Contributions: Conceptualization, Y.S.; methodology, Y.Y.; formal analysis, T.K.; investigation, J.O.; resources, K.Y.; data curation, Y.Y.; writing—original draft preparation, T.K. and Y.Y.; writing—review and editing, T.K. and Y.Y.; visualization, T.K.; supervision, Y.Y.; project administration, Y.Y. All authors have read and agreed to the published version of the manuscript.

Funding: This research was supported by internal research funding provided to Yuki Yoshimatsu by the Tochigi Cancer Center in fiscal year 2025.

Institutional Review Board Statement: Ethical review and approval were not required for this study because it involved established human cell lines and did not include human participants or identifiable personal data.

Informed Consent Statement: Not applicable. Ethical approval was not required for this study, as it involved the use of established patient-derived cell lines that are publicly available and did not include identifiable human subjects.

Data Availability Statement: The data are available upon reasonable request.

Acknowledgments: The authors thank all members of the participating laboratories for their valuable discussions. RNA sequencing data were generated by Zeon BioSolutions, Inc. (Hyogo, Japan) as a contracted service provider. The authors also acknowledge the use of ChatGPT (OpenAI, San Francisco, CA, USA) for improving the clarity and readability of the English text. The authors remain fully responsible for the scientific content, interpretation, and conclusions presented in this study.

Conflicts of Interest: The authors declare no conflicts of interest associated with this manuscript.

References

1. Fearon, K.; Strasser, F.; Anker, S.D.; Bosaeus, I.; Bruera, E.; Fainsinger, R.L.; Jatoi, A.; Loprinzi, C.; MacDonald, N.; Mantovani, G.; et al. Definition and classification of cancer cachexia: An international consensus. *Lancet Oncol.* **2011**, *12*, 489–495. [CrossRef] [PubMed]
2. Suzuki, H.; Asakawa, A.; Amitani, H.; Nakamura, N.; Inui, A. Cancer cachexia--pathophysiology and management. *J. Gastroenterol.* **2013**, *48*, 574–594. [CrossRef]
3. Schmidt, S.F.; Rohm, M.; Herzig, S.; Berriel Diaz, M. Cancer Cachexia: More Than Skeletal Muscle Wasting. *Trends Cancer* **2018**, *4*, 849–860. [CrossRef]
4. Li, L.; Ling, Z.Q. Mechanisms of cancer cachexia and targeted therapeutic strategies. *Biochim. Biophys. Acta Rev. Cancer* **2024**, *1879*, 189208. [CrossRef]
5. Khalil, M.A.; Jafri, S.A.A.; Zahra Naqvi, M.E.; Jauhar, I.; Kumar, V.; Ali, S.M.S.; Khalil, S.; Raheem, A.; Kumar, R.; Haseeb, A.; et al. Impact of Cachexia on Chemotherapy Efficacy and Survival in Pancreatic Cancer: A Systematic Review and Meta-Analysis. *Cancer Control* **2024**, *31*, 10732748241292784. [CrossRef]
6. Pin, F.; Couch, M.E.; Bonetto, A. Preservation of muscle mass as a strategy to reduce the toxic effects of cancer chemotherapy on body composition. *Curr. Opin. Support. Palliat. Care* **2018**, *12*, 420–426. [CrossRef]
7. Petruzzelli, M.; Schweiger, M.; Schreiber, R.; Campos-Olivas, R.; Tsoli, M.; Allen, J.; Swarbrick, M.; Rose-John, S.; Rincon, M.; Robertson, G.; et al. A switch from white to brown fat increases energy expenditure in cancer-associated cachexia. *Cell Metab.* **2014**, *20*, 433–447. [CrossRef]
8. Kir, S.; White, J.P.; Kleiner, S.; Kazak, L.; Cohen, P.; Baracos, V.E.; Spiegelman, B.M. Tumour-derived PTH-related protein triggers adipose tissue browning and cancer cachexia. *Nature* **2014**, *513*, 100–104. [CrossRef]
9. Kir, S.; Komaba, H.; Garcia, A.P.; Economopoulos, K.P.; Liu, W.; Lanske, B.; Hodin, R.A.; Spiegelman, B.M. PTH/PTHrP Receptor Mediates Cachexia in Models of Kidney Failure and Cancer. *Cell Metab.* **2016**, *23*, 315–323. [CrossRef] [PubMed]

https://doi.org/10.3390/targets4010009

10. Das, S.K.; Eder, S.; Schauer, S.; Diwoky, C.; Temmel, H.; Guertl, B.; Gorkiewicz, G.; Tamilarasan, K.P.; Kumari, P.; Trauner, M.; et al. Adipose triglyceride lipase contributes to cancer-associated cachexia. *Science* **2011**, *333*, 233–238. [CrossRef] [PubMed]
11. Dahlman, I.; Mejhert, N.; Linder, K.; Agustsson, T.; Mutch, D.M.; Kulyte, A.; Isaksson, B.; Permert, J.; Petrovic, N.; Nedergaard, J.; et al. Adipose tissue pathways involved in weight loss of cancer cachexia. *Br. J. Cancer* **2010**, *102*, 1541–1548. [CrossRef] [PubMed]
12. Talbert, E.E.; Cuitiño, M.C.; Ladner, K.J.; Rajasekerea, P.V.; Siebert, M.; Shakya, R.; Leone, G.W.; Ostrowski, M.C.; Paleo, B.; Weisleder, N.; et al. Modeling Human Cancer-induced Cachexia. *Cell Rep.* **2019**, *28*, 1612–1622.e4. [CrossRef]
13. Yanagihara, K.; Kubo, T.; Mihara, K.; Kuwata, T.; Ochiai, A.; Seyama, T.; Yokozaki, H. Establishment of a novel cell line from a rare human duodenal poorly differentiated neuroendocrine carcinoma. *Oncotarget* **2018**, *9*, 36503–36514. [CrossRef] [PubMed]
14. Yanagihara, K.; Kubo, T.; Iino, Y.; Mihara, K.; Morimoto, C.; Seyama, T.; Kuwata, T.; Ochiai, A.; Yokozaki, H. Development and characterization of a cancer cachexia model employing a rare human duodenal neuroendocrine carcinoma-originating cell line. *Oncotarget* **2019**, *10*, 2435–2450. [CrossRef]
15. Minami, F.; Sasaki, N.; Shichi, Y.; Gomi, F.; Michishita, M.; Ohkusu-Tsukada, K.; Toyoda, M.; Takahashi, K.; Ishiwata, T. Morphofunctional analysis of human pancreatic cancer cell lines in 2- and 3-dimensional cultures. *Sci. Rep.* **2021**, *11*, 6775. [CrossRef]
16. Hou, X.; Du, C.; Lu, L.; Yuan, S.; Zhan, M.; You, P.; Du, H. Opportunities and challenges of patient-derived models in cancer research: Patient-derived xenografts, patient-derived organoid and patient-derived cells. *World J. Surg. Oncol.* **2022**, *20*, 37. [CrossRef]
17. Lu, P.; Weaver, V.M.; Werb, Z. The extracellular matrix: A dynamic niche in cancer progression. *J. Cell Biol.* **2012**, *196*, 395–406. [CrossRef]
18. Mina-Osorio, P. The moonlighting enzyme CD13: Old and new functions to target. *Trends Mol. Med.* **2008**, *14*, 361–371. [CrossRef] [PubMed]
19. Pickup, M.W.; Mouw, J.K.; Weaver, V.M. The extracellular matrix modulates the hallmarks of cancer. *EMBO Rep.* **2014**, *15*, 1243–1253. [CrossRef]
20. Kim, J.; DeBerardinis, R.J. Mechanisms and Implications of Metabolic Heterogeneity in Cancer. *Cell Metab.* **2019**, *30*, 434–446. [CrossRef]
21. Hata, A.N.; Engelman, J.A.; Faber, A.C. The BCL2 Family: Key Mediators of the Apoptotic Response to Targeted Anticancer Therapeutics. *Cancer Discov.* **2015**, *5*, 475–487. [CrossRef] [PubMed]
22. Strassmann, G.; Fong, M.; Kenney, J.S.; Jacob, C.O. Evidence for the involvement of interleukin 6 in experimental cancer cachexia. *J. Clin. Investig.* **1992**, *89*, 1681–1684. [CrossRef]
23. Beutler, B.; Greenwald, D.; Hulmes, J.D.; Chang, M.; Pan, Y.C.; Mathison, J.; Ulevitch, R.; Cerami, A. Identity of tumour necrosis factor and the macrophage-secreted factor cachectin. *Nature* **1985**, *316*, 552–554. [CrossRef]
24. Uehara, A.; Sekiya, C.; Takasugi, Y.; Namiki, M.; Arimura, A. Anorexia induced by interleukin 1: Involvement of corticotropin-releasing factor. *Am. J. Physiol.* **1989**, *257*, R613–R617. [CrossRef]
25. Roeland, E.J.; Bohlke, K.; Baracos, V.E.; Bruera, E.; Del Fabbro, E.; Dixon, S.; Fallon, M.; Herrstedt, J.; Lau, H.; Platek, M.; et al. Management of Cancer Cachexia: ASCO Guideline. *J. Clin. Oncol.* **2020**, *38*, 2438–2453. [CrossRef]
26. Argilés, J.M.; Busquets, S.; Stemmler, B.; López-Soriano, F.J. Cancer cachexia: Understanding the molecular basis. *Nat. Rev. Cancer* **2014**, *14*, 754–762. [CrossRef]
27. Baracos, V.E.; Martin, L.; Korc, M.; Guttridge, D.C.; Fearon, K.C.H. Cancer-associated cachexia. *Nat. Rev. Dis. Primers* **2018**, *4*, 17105. [CrossRef] [PubMed]
28. Fearon, K.C.; Glass, D.J.; Guttridge, D.C. Cancer cachexia: Mediators, signaling, and metabolic pathways. *Cell Metab.* **2012**, *16*, 153–166. [CrossRef]
29. Nakamura, Y.; Saldajeno, D.P.; Kawaguchi, K.; Kawaoka, S. Progressive, multi-organ, and multi-layered nature of cancer cachexia. *Cancer Sci.* **2024**, *115*, 715–722. [CrossRef]
30. Zhou, X.; Wang, J.L.; Lu, J.; Song, Y.; Kwak, K.S.; Jiao, Q.; Rosenfeld, R.; Chen, Q.; Boone, T.; Simonet, W.S.; et al. Reversal of cancer cachexia and muscle wasting by ActRIIB antagonism leads to prolonged survival. *Cell* **2010**, *142*, 531–543. [CrossRef] [PubMed]
31. Zimmers, T.A.; Davies, M.V.; Koniaris, L.G.; Haynes, P.; Esquela, A.F.; Tomkinson, K.N.; McPherron, A.C.; Wolfman, N.M.; Lee, S.J. Induction of cachexia in mice by systemically administered myostatin. *Science* **2002**, *296*, 1486–1488. [CrossRef]
32. Flint, T.R.; Janowitz, T.; Connell, C.M.; Roberts, E.W.; Denton, A.E.; Coll, A.P.; Jodrell, D.I.; Fearon, D.T. Tumor-Induced IL-6 Reprograms Host Metabolism to Suppress Anti-tumor Immunity. *Cell Metab.* **2016**, *24*, 672–684. [CrossRef] [PubMed]
33. Hiam-Galvez, K.J.; Allen, B.M.; Spitzer, M.H. Systemic immunity in cancer. *Nat. Rev. Cancer* **2021**, *21*, 345–359. [CrossRef] [PubMed]
34. Skipworth, R.J.; Stewart, G.D.; Dejong, C.H.; Preston, T.; Fearon, K.C. Pathophysiology of cancer cachexia: Much more than host-tumour interaction? *Clin. Nutr.* **2007**, *26*, 667–676. [CrossRef]

https://doi.org/10.3390/targets4010009

35. Lerner, L.; Hayes, T.G.; Tao, N.; Krieger, B.; Feng, B.; Wu, Z.; Nicoletti, R.; Chiu, M.I.; Gyuris, J.; Garcia, J.M. Plasma growth differentiation factor 15 is associated with weight loss and mortality in cancer patients. *J. Cachexia Sarcopenia Muscle* **2015**, *6*, 317–324. [CrossRef]
36. Cheng, L.Y.; Desai, J.; Neeson, P.J. Targeting GDF-15 in cancer cachexia and immunity. *Nat. Med.* **2025**, *31*, 1069–1070. [CrossRef]
37. Patel, H.J.; Patel, B.M. TNF-α and cancer cachexia: Molecular insights and clinical implications. *Life Sci.* **2017**, *170*, 56–63. [CrossRef]
38. Nguyen, V.T.; Kiss, T.; Michels, A.A.; Bensaude, O. 7SK small nuclear RNA binds to and inhibits the activity of CDK9/cyclin T complexes. *Nature* **2001**, *414*, 322–325. [CrossRef]
39. Yang, F.; Kemp, C.J.; Henikoff, S. Doxorubicin enhances nucleosome turnover around promoters. *Curr. Biol.* **2013**, *23*, 782–787. [CrossRef] [PubMed]
40. Amaravadi, R.; Kimmelman, A.C.; White, E. Recent insights into the function of autophagy in cancer. *Genes Dev.* **2016**, *30*, 1913–1930. [CrossRef]
41. Ibáñez-Molero, S.; van Vliet, A.; Pozniak, J.; Hummelink, K.; Terry, A.M.; Monkhorst, K.; Sanders, J.; Hofland, I.; Landeloos, E.; Van Herck, Y.; et al. SERPINB9 is commonly amplified and high expression in cancer cells correlates with poor immune checkpoint blockade response. *Oncoimmunology* **2022**, *11*, 2139074. [CrossRef]
42. Budanov, A.V.; Karin, M. p53 target genes sestrin1 and sestrin2 connect genotoxic stress and mTOR signaling. *Cell* **2008**, *134*, 451–460. [CrossRef]
43. Tisdale, M.J. Cancer cachexia: Metabolic alterations and clinical manifestations. *Nutrition* **1997**, *13*, 1–7. [CrossRef]
44. Goldman, A.; Majumder, B.; Dhawan, A.; Ravi, S.; Goldman, D.; Kohandel, M.; Majumder, P.K.; Sengupta, S. Temporally sequenced anticancer drugs overcome adaptive resistance by targeting a vulnerable chemotherapy-induced phenotypic transition. *Nat. Commun.* **2015**, *6*, 6139. [CrossRef]
45. Puchalska, P.; Crawford, P.A. Multi-dimensional Roles of Ketone Bodies in Fuel Metabolism, Signaling, and Therapeutics. *Cell Metab.* **2017**, *25*, 262–284. [CrossRef] [PubMed]
46. Weinberg, S.E.; Chandel, N.S. Targeting mitochondria metabolism for cancer therapy. *Nat. Chem. Biol.* **2015**, *11*, 9–15. [CrossRef] [PubMed]
47. Tsvetkov, P.; Coy, S.; Petrova, B.; Dreishpoon, M.; Verma, A.; Abdusamad, M.; Rossen, J.; Joesch-Cohen, L.; Humeidi, R.; Spangler, R.D.; et al. Copper induces cell death by targeting lipoylated TCA cycle proteins. *Science* **2022**, *375*, 1254–1261. [CrossRef] [PubMed]
48. Shaffer, S.M.; Dunagin, M.C.; Torborg, S.R.; Torre, E.A.; Emert, B.; Krepler, C.; Beqiri, M.; Sproesser, K.; Brafford, P.A.; Xiao, M.; et al. Rare cell variability and drug-induced reprogramming as a mode of cancer drug resistance. *Nature* **2017**, *546*, 431–435. [CrossRef]

https://doi.org/10.3390/targets4010009

targets

MDPI

Article

Repurposing FDA-Approved Drugs Against Potential Drug Targets Involved in Brain Inflammation Contributing to Alzheimer's Disease

Catherine Sharo [1], Jiayu Zhang [2], Tianhua Zhai [2], Jingxuan Bao [2], Andrés Garcia-Epelboim [2], Elizabeth Mamourian [2], Li Shen [2] and Zuyi Huang [1,*]

[1] Department of Chemical and Biological Engineering, Villanova University, Villanova, PA 19085, USA; csharo@villanova.edu

[2] Department of Biostatistics, Epidemiology and Informatics, University of Pennsylvania, Philadelphia, PA 19104, USA; zjiayu@seas.upenn.edu (J.Z.); tianhua.zhai@pennmedicine.upenn.edu (T.Z.); jingxuan.bao@pennmedicine.upenn.edu (J.B.); garciaan@sas.upenn.edu (A.G.-E.); elizabeth.mamourian@pennmedicine.upenn.edu (E.M.); li.shen@pennmedicine.upenn.edu (L.S.)

* Correspondence: zuyi.huang@villanova.edu; Tel.: +1-610-519-4848

Abstract: Alzheimer's disease is a neurodegenerative disease that continues to have a rising number of cases. While extensive research has been conducted on Alzheimer's disease in the last few decades, only a few drugs have been approved by the FDA for its treatment, and even fewer aim to be curative rather than manage symptoms. There remains an urgent need to understand disease pathogenesis, as well as identify new targets for further drug discovery. Alzheimer's disease (AD) is known to stem from the build-up of amyloid beta (Aβ) plaques, as well as tangles of tau proteins. Furthermore, inflammation in the brain is known to arise from the degeneration of tissue and the build-up of insoluble material. Therefore, there is a potential link between the pathology of AD and inflammation in the brain, especially as the disease progresses to later stages, where neuronal death and degeneration levels are higher. Proteins that are relevant to both brain inflammation and AD, thus, make ideal potential targets for therapeutics; however, the proteins need to be evaluated to determine which targets would be ideal for potential drug therapeutic treatments, or 'druggable' targets. Druggability analysis was conducted using two structure-based methods (i.e., drug-like density analysis and SiteMap), as well as a sequence-based approach, SPIDER. The most druggable targets were then evaluated using single-nucleus sequencing data for their clinical relevance to inflammation in AD. For each of the top five targets, small molecule docking was used to evaluate which FDA approved drugs were able to bind with the chosen proteins. The top targets included DRD2 (inhibits adenylyl cyclase activity), C9 (binds with C5B8 to form the membrane attack complex), C4b (binds with C2a to form C3 convertase), C5AR1 (a GPCR that binds C5a), and GABA-A-R (the GPCR involved in inhibiting neurotransmission). Each target had multiple potential inhibitors from the FDA-approved drug list with decent binding infinities. Among these inhibitors, two drugs were found to be top inhibitors for more than one protein target. They were C15H14N2O2 and v316 (paracetamol), originally used to treat pain/inflammation for cataracts and relieve headaches/fever, respectively. These results provide the groundwork for further experimental investigations or clinical trials.

Keywords: brain inflammation; Alzheimer's disease; druggability analysis; drug-like density; single-nucleus sequencing analysis; small molecule inhibitor; drug repurposing; docking

1. Introduction

The neurodegenerative disease known as Alzheimer's disease leads to progressive memory loss and impaired functions. It is associated with damaged neurons in the brain, which have been linked to the build-up of insoluble amyloid beta (Aβ) plaques and tau protein tangles [1]. The presence of these abnormal accumulations may be directly correlated with the progression of the disease and the increasing severity of symptoms [2]. It is the severity of these symptoms and the increasing number of cases [3] that generate demand for therapeutic treatment options. There currently are very few FDA-approved drugs for the purpose of either diminishing the symptoms of the disease or actually stopping its progression [2,4–7]. Other treatments that were investigated, such as BACE-1 or PSEN-1 inhibitors, proved to not have curative effects and, thus, were not approved [2,8]. Therefore, there is still a need to identify new targets for drug therapeutics that aim to cure Alzheimer's disease.

One potential source of target candidates may come from investigating inflammation in the brain. Inflammation in the brain arises from the combination of abnormal insoluble materials and degenerating tissues. Thus, the increasing levels of neuronal death, Aβ plaques, and tau tangles that occur during the progression of AD provide a clear link between the inflammation of the brain and AD. It has been proven that the cytokines connected to inflammation are chronically upregulated in the areas where AD has impacted the brain [9]. Originally, it was thought that inflammation arose alongside AD due to dead tissue and abnormal proteins; however, it is possible that inflammation that arises due to Alzheimer's disease may increase the rate of neurodegeneration and attack living neural tissues [9]. There have been previous attempts to target Alzheimer's disease through inflammation via the use of non-steroidal anti-inflammatory drugs, but none have been successful so far [2,10,11]. Therefore, there is still a need to identify new Alzheimer's disease inflammation targets.

In order to identify these potential targets, the pathways and proteins relevant to both Alzheimer's disease and inflammation must be explored. Alzheimer's disease-related inflammation is impacted by multiple interdependent pathways in the body. However, it has been shown that the dominant pathway is the classical complement pathway [11–14]. Therefore, the essential genes from this pathway were the basis for this investigation. The search was also expanded to include associate G protein-coupled receptors (GPCRs), as those have been proven to be ideal targets for drug therapeutics, serving as targets for 50–60% of current drugs [15]. However, while all identified potential targets are relevant to both inflammation and Alzheimer's disease, not all may actually be able to bind to a therapeutic. Therefore, the binding capabilities needed to be evaluated for each target. Druggability analysis is the most common approach for this task.

Druggablity analysis approaches are generally categorized into either structure-based or sequence-based approaches. A protein is considered druggable if a drug molecule can bind to one of its binding pockets and impact its typical function [16]. Drug-like density (DLID) is one of the most popular criteria for structure-based draggability analysis [17]. Molsoft ICM-Pro 3.9 software is one of the best platforms for identifying pockets that ligands can dock to and measuring their structure features, which can impact the likelihood of ligand docking [18]. For example, a binding pocket must be accessible to a ligand, and the pocket must be large enough to accommodate the ligand. Protein structures used for analysis in ICM-Pro are obtained from the Protein Data Bank (PDB). Given the protein structures, a pocket searching program (i.e., ICM Pocket Finder) determines the locations of the pockets and extracts their structure features (e.g., aromatic). PockDrug is a similar structure-based method focusing on geometry, hydrophobicity, and aromaticity that is used to evaluate target binding pockets [19,20]. Similar models include Cavity, which assesses pockets based on 3D structures and ranks them by druggability, employing submodels like CavPharmer for pharmacophore features and CorrSite for allosteric sites [21,22]. Other platforms like ProteinPlus/DoGSiteScorer [23–25], fpocket [26], and InDeep [27] analyze pockets and predict druggability through features like pocket geometry and protein–protein interactions. JEDI, which stands for Just Exploring Druggability at protein Interfaces, uses molecular dynamics descriptors [28], while P2Rank leverages machine learning to identify ligand binding sites from known templates [29–33]. SiteMap, with 86–96% accuracy, offers quantitative binding site analysis by enhancing binding affinity assessments [34].

Dynamics-aware methods such as TRAPP (i.e., Transient Pockets with Proteins) [35] and BiteNet [36,37] evaluate transient pockets. Programs, such as DeepSurf [38], CryptoSite [39], DEPTH [40], GNN_pocket [41–43], Kalasanty [44], and PocketMiner [45] were developed to predict the location of cryptic binding sites.

While all the aforementioned tools are valuable for druggability analysis, SiteScore was chosen for this work because it incorporates molecular dynamics information and offers a quantitative approach to provide insights into pocket hydrophobicity and volume. [46] There is some debate regarding the idea that sequence-based methods of druggability analysis are inferior to structure-based methods [47]. However, there are several sequence-based programs that should still be considered. These include DrugMiner [48], GA-Baggin-SVM [49], DrugHybrid_BS [50], XGB-DrugPred [51], Yu's Method [52], SPIDER [53], Quote-Target [54], DrugFinder [55], Sun's Method [56], and Iraji's Method [57]. Among all these valuable programs, SPIDER was chosen for this work because it demonstrated reliable cross-validation and test results in Shoombuatong and colleagues' study [58]. Additionally, SPIDER's accessible webserver interface allows for its practical application in large-scale studies.

This study explored several druggability prediction methods, including two structure-based approaches (DLID and SiteScore) and one sequence-based approach (Stacked PredIc-tor of DruggablE pRoteins—SPIDER). While these methods help to cross-validate the most druggable targets, RNA-sequencing data related to inflammation in Alzheimer's disease, sourced from databases such as ssREAD portal, were analyzed to confirm the value of these targets in AD. The most druggable and valuable targets identified for addressing AD-related inflammation were subsequently used in Molsoft ICM, a leading ligand–protein docking program [59,60], to evaluate the potential binding of pre-FDA-approved drugs to these targets. The identified drug targets, along with their potential inhibitors repurposed from FDA-approved drugs, opened new avenues for combating AD-related inflammation.

2. Materials and Methods

2.1. Structure-Based Druggability Analysis

The initial target list was obtained from [61], which provided a detailed literature review of proteins involved in inflammation that contributed to Alzheimer's disease. The first method evaluated for the structure-based methods was the Molsoft ICM PocketFinder program established by Merck to find the "drug-like density" of each protein [62]. The PocketFinder program first identified the pockets available in the provided protein crystal structure. It then reported the volume, hydrophobicity, buriedness, DLID, and other important metrics for each pocket. Each crystal structure had a unique combination of number of pockets and metric scores that made it more or less viable as a drug therapeutic target. The DLID score for each pocket was shown to be linked to the druggability of the target. Pockets were characterized using three main parameters: volume, hydrophobicity, and buriedness. This allowed the program to predict whether the protein was capable of bonding with a 'drug-like' molecule. The first step in the program was to generate the pockets for the protein and evaluate what protein chains were important to the pockets. Any cofactors were removed so the structure could be properly isolated. The protein residues that were 3.5 angstroms within any surface point of the pocket were then assigned to the pocket's 'shell'. PocketFinder was able to report the volume of each pocket directly, while the buriedness was calculated based on the ratio of the solvent-accessible surface area covered by its shell to the solvent-accessible surface area in isolation. The hydrophobicity was calculated from the fraction of the pocket surface in contact with hydrophobic atoms in the shell of the pocket. The actual DLID score calculation began with the number of drug-like ligand-containing (DLLC) pockets with similar volume, buriedness, and hydrophobicity scores versus the total number of pockets with similar scores to find the density of the DLLC pockets. After applying a log-based correction factor, the DLID was calculated, as shown in Equation (1) [63]:

$$DLID = \log(DLLC\ neighbors/total\ neighors) + 1.71 \quad (1)$$

However, this equation required the pockets to have information available from the pocket database. If this was not the case, then Equation (2) could be used as a substitute [63]:

$$DLID = -8.70 + 1.71 \log(volume) + 3.94\ (buriedness) + 2.27\ (hydrophobicity) \quad (2)$$

Once the scores for volume, hydrophobicity, buriedness, and DLID were calculated for each pocket, they were evaluated to see which targets scored well. It was determined that a DLID score above the 0.5 benchmark was considered "druggable", meaning it would be likely to bond with a drug therapeutic [63].

Similar to DLID, the second structure-based approach, SiteMap, provided several key metrics alongside its druggability score [64]. These included the number of 'site points', which correlated to the size of the site. SiteMap also reported both the hydrophobicity and hydrophilicity characteristics of the site. The 'balance' was the ratio of the hydrophobicity to the hydrophilicity. The average hydrophobicity and hydrophilicity score for a tight-binding site was 1.0, while the average balance score was 1.6. This was due to the fact that binding sites tended to have higher hydrophobicity than hydrophilicity scores. The enclosure score evaluated how open the site was to solvent interaction [65].

The SiteScore of a pocket was based on a weighted sum of several of the properties that are discussed below:

$$SiteScore = 0.0733\sqrt{n} + 0.6688e - 0.20p \quad (3)$$

where n is the number of site points (capped at 100), e is the enclosure score, and p is the hydrophilic score. It is capped at 1.0 to limit the impact of hydrophilicity in charged and highly polar sites. This score was calibrated so that the average SiteScore for 157 investigated submicromolar sites was 1.0. Thus, a score of greater than 1.0 suggested an ideal binding; however, a SiteScore of 0.80 has been found to accurately distinguish between drug binding and non-drug binding sites. SiteScore evaluated the site as it would be available for ligand binding. Therefore, while the SiteScore was important, the Dscore was still essential, as it focused on binding with drugs [64,65].

Dscore uses the same properties as SiteScore but with different coefficients, as seen in Equation (4):

$$Dscore = 0.094\sqrt{n} + 0.60e - 0.324p \quad (4)$$

For Dscore, the hydrophilic score was not capped. This was one of the essential factors in distinguishing undruggable targets from druggable ones [65,66]. The use of different functions for binding site identification, SiteScore, and for classifying druggability, Dscore, could be justified because they had different characteristics and involved different interactions. For example, the PTP1B phosphate pocket could bind with ligands that had nanomolar or even subnanomolar affinity [65,67]. However, these ligands were highly active and had charge structures like that of the natural phosphate substrate, making them not drug-like. SiteMap was capable of recognizing that this site could bind ligands but would not be considered druggable [65].

2.2. Sequence-Based Methods for Druggability Analysis

SPIDER, or Score Protein Interaction Decoys using Exposed Residues, uses machine learning to create a knowledge-based score function for protein–protein interaction decoys. It is based on the geometric similarity of interfacial residues between docking and native poses. The computational geometry approach known as Almost-Delaunay tessellation was used, which transforms protein–protein complexes into a residue contact network [68–70].

The FASTA sequences for each target were uploaded to the SPIDER webpage to obtain the Possibility score of the likelihood that a target could be druggable.

2.3. Small Molecule Docking for Repurposing FDA-Approved Drugs

The structure of each protein was loaded into the Molsoft ICM program, where the top potential binding pocket was identified. Using this pocket, a Monte Carlo simulation was run to assess every possible binding position with the selected drug. This resulted in a binding score that represented the likelihood of the protein binding to that molecule [17,61,71–74]. The standard was set as a score of −20 kcal/mol, and anything below that indicated that binding was likely [62]. The drugs used for this docking simulation were a set of approximately 2500 FDA-approved drugs.

2.4. Upstream RNA-Sequencing Analysis

The raw FASTQ files were obtained from datasets GSE138852 and GSE147528, comprising 16 samples from entorhinal cortex brain tissue (8 healthy and 8 AD individuals) and 20 samples from both entorhinal cortex and prefrontal cortex tissues (6 healthy and 14 AD individuals). The datasets were processed using Cell Ranger (version 5.0.0) with default settings, and further mapping raw reads to GRCh38 reference from 10x Genomics. After quantification, genes without detectable counts in any cells were removed, and cells were filtered to include only those with gene and UMI counts failing between the 5th and 95th percentiles, as well as less than 10% of UMIs attributed to mitochondrial genes. For downstream analysis, data from the same brain tissue with different conditions were integrated using both *FindIntegrationAnchors* with the canonical correlation analysis method to address batch effects and the *IntegrateData* function. The processed seurat objects were normalized with a scaling factor of 10,000 and further log-transformed. Highly variable genes were defined using *FindVariableGenes*, and *ScaleData* was applied to center gene expression. After quality control, the resulting data consisted of 82,007 cells with 26,754 genes for data in entorhinal cortex brain tissue and 94,558 cells with 27,590 genes for data in prefrontal cortex brain tissue. We performed principal component analysis to identify clusters based on significant variability, and *FindClusters* at a resolution of 0.5 revealed 21 clusters in the entorhinal cortex and 17 in the prefrontal cortex, respectively.

To categorize these clusters identified for major brain cell types, the following marker genes were used: SLC17A7 and SNAP25 for excitatory neurons, GAD1 and GAD2 for inhibitory neurons, SLC1A3 or GFAP and AQP4 for astrocytes, PECAM1 and VWF for endothelial cells, PTPRC and P2RY12 for microglia cells, PDGFRA and CSPG4 for oligodendrocyte precursor cells, and CLDN11 and MBP for oligodendrocytes. Non-linear dimensional reduction (uniform manifold approximation and projection UMAP) calculation was used for visualizing these datasets.

To identify inflammation-related targets differentially expressed in a cell-type subpopulation by comparing control and AD conditions. 'pseudo-bulk' samples were generated by summing all cells from the same condition and further analyzed using *FindMarkers*. Heatmaps were constructed to present the relative gene expression across cell types, where the value of log-transformed normalized fold change for each gene was transformed to z-score within each cluster to perform differential expression. The significance of differentially expressed genes was determined using p-values adjusted by the Benjamini–Hochberg method, with values less than 0.05 being the threshold for statistical significance.

3. Results

3.1. Druggability Analysis of Targets via Drug-like Density

For the drug targets recommended by the literature [61] for their involvement in inflammation in Alzheimer's disease, the results for the DLID scores are summarized below in Table 1. All of the targets with viable structures had at least one pocket that returned a score, except for CD59. While there was a protein structure available, the ICM *PocketFinder* program did not identify any pockets in CD59 that could bind to potential drugs. The majority of pockets that were identified did not have DLID scores that met the druggability threshold of 0.5. The target with the highest number of viable pockets was C5 with 11 pockets, followed by GABA-A-R with 9. C3 had six, while C4B, RAB7, and C9 each had five. DRD2 had two pockets. The following targets each had one pocket above 0.5: C1S, VTN, and C1QB. The remaining targets each had identifiable pockets but did not score high enough due to a combination of poor volume, hydrophobicity, and buriedness scores. However, it is important to note that this did not mean that there would be no drugs that could bind to these molecules. Having a low druggability score meant that the chances were lower than those that did reach the 0.5 threshold. This was ultimately reflected when the small molecule docking program was run, as those targets were able to bind some of the FDA-approved drugs but much less than those with higher DLID scores. The results from both the *PocketFinder* run and the DLID assessment are summarized in Table 1.

Table 1. Summary of ICM PocketFinder and DLID results.

Target	Number of Pockets	Largest Volume	Highest Hydrophobicity	Buriedness Score	Number of Druggable Pockets
C1R	17	386.7	0.682	0.995	0
C3	37	1024	0.88	0.993	6
C1S	11	1761	0.586	0.824	1
C4B	27	879.1	0.7595	0.9128	5
C4A	4	295.1	0.631	0.859	0
CFH	8	467.5	0.566	0.778	0
Serping1	3	259.5	0.575	0.803	0
C2A	9	334.9	0.63	0.82	0
C2B	2	393.5	0.568	0.87	0
C5AR1	5	348.8	0.771	0.865	0
RAB7	23	1368	0.593	0.884	5
BACE1	2	428.8	0.623	0.833	0
RAB11	4	387.9	0.529	0.805	0
ARF6	3	278.4	0.544	0.834	0
GABA-A R	62	1923	0.893	0.998	9
OPRD1	13	452.4	0.685	0.917	0
VTN	4	779.5	0.728	0.954	1
C1QB	10	542.9	0.595	0.983	1
C6	16	715.4	0.626	0.87	0
Clusterin	24	1035	0.885	0.949	0
C5	108	1686	0.838	0.98	11
C4BPA	11	250.3	0.693	0.816	0
CD59	0	-	-	-	0
C9	23	614.1	0.895	0.996	5
DRD2	8	620.8	0.832	0.882	2

The highest DLID score for each target is shown in Figure 1. The trend indicated that having the highest number of pockets did not correlate to having the top-scoring pockets. C5, which had the greatest number of pockets, only had the fifth-highest DLID score. C9 ultimately had the highest singular pocket score, followed by C4B, C1S, and GABA-A R. ARF6 had the lowest DLID score. Seven of the targets had top scores in the negatives, with another seven having positive scores that were just below the 0.5 threshold.

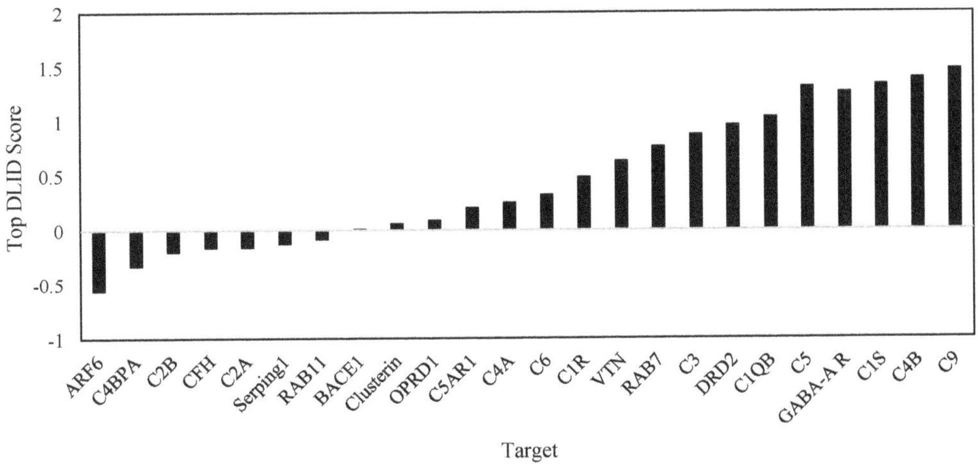

Figure 1. The DLID score returned by the ICM PocketFinder program to evaluate the druggability of each target.

3.2. *Druggability Analysis of Targets via SITEMAP*

Although SiteMap reported that both the SiteScore and DScore reflected the binding ability of the top pockets, the DScore was specifically calibrated to reflect the druggability of the pocket. The results from the SiteMap analysis were thus analyzed around the DScore in order to compare it to the other programs. The top results of the SiteScore, DScore, volume, and balance are reported below in Table 2. C5 did not have any pockets with SiteScores that met the threshold to be reported by the program. However, unlike the DLID scores, CD59 did have one pocket that met the threshold. Additionally, SiteMap reported the top five pockets for each target if enough pockets met the criteria. C4A, C2B, BACE1, RAB11, ARF6, VTN, and CD59 each had less than five viable pockets. This generally matched the trends from the DLID results, as the majority of those targets did not have high-scoring pockets for the DLID scores either. The highest DScore result for each target is depicted in Figure 2.

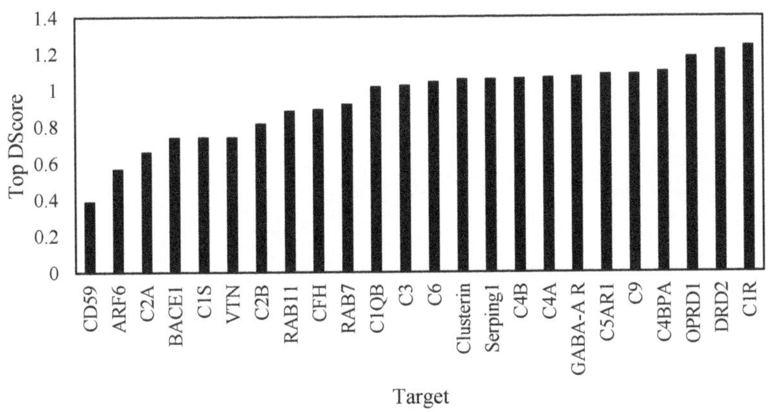

Figure 2. Top DScore returned by SiteMap to evaluate the druggability of each target.

Table 2. A summary of the SiteScore, DScore, volume, and balance results for the pocket with the highest DScore by each target.

Target	DScore	SiteScore	Volume	Balance
C1R	1.235	1.134	332.367	6.057
DRD2	1.213	1.103	167.384	12.82
OPRD1	1.176	1.096	1353.135	3.447
C4BPA	1.093	1.016	343	1.479
C9	1.079	1.048	270.284	0.91
C5AR1	1.078	1.032	228.095	5.406
GABAAR	1.065	1.035	2247.679	0.759
C4A	1.061	1.015	370.44	0.906
C4B	1.056	1.015	673.309	0.708
SERPING1	1.053	1.005	286.748	0.7
CLU	1.052	1.007	294.98	0.732
C6	1.038	0.994	822.857	0.473
C3	1.02	1.034	439.726	0.294
C1QB	1.011	1.126	479.171	0.549
RAB7	0.917	0.987	531.65	0.0005
CFH	0.888	0.903	248.332	0.119
RAB11	0.88	0.881	164.983	0.435
C2B	0.812	0.907	230.839	0.08
C1S	0.738	0.947	98.441	0.223
VTN	0.738	0.821	146.118	0.165
BACE1	0.735	0.806	244.559	0.034
C2A	0.656	0.778	200.655	0.081
ARF6	0.564	0.756	86.093	0.087
CD59	0.384	0.51	45.276	0.021
C5	-	-	-	-

3.3. Druggability Analysis of Targets via SPIDER

Once each FASTA file was uploaded to SPIDER, the Possibility score was reported for each target. All of the scores fell between 0.085 and 0.089, as shown in Figure 3. CFH had the lowest score at 0.085, while C1R and VTN both had the highest score of 0.089. The SPIDER results did not match the trends of the DLID and SiteMap scores. This was seen in targets such as ARF6, which was one of the lowest-scoring molecules for both DLID and SiteMap but returned the second highest score in SPIDER. Conversely, C9 was a top-scoring target for the two structure-based programs but received the second-lowest score from SPIDER. However, this likely stemmed from the fact that the results did not have enough variation to reflect major trends. All but six of the targets scored either 0.087 or 0.088. Additionally, CD59, which scored low for SiteMap and was undetectable for DLID, scored towards the high end at 0.088, indicating this program may not be one of the ideal means of evaluating this set of targets.

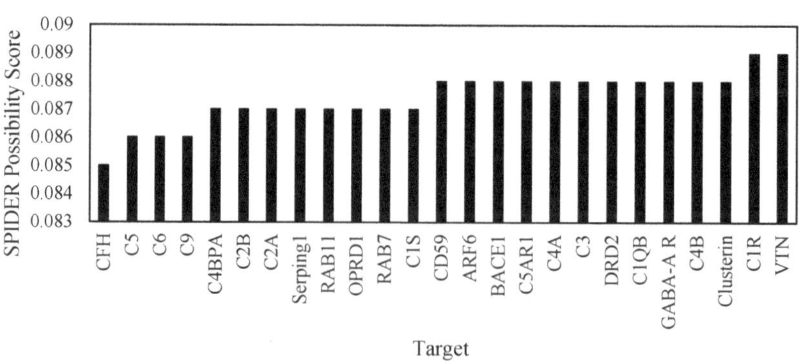

Figure 3. The possibility scores returned by SPIDER to evaluate the druggability of each target.

3.4. Comparison of Druggability Analysis Results

In order to effectively compare the druggability results across the three platforms, the data had to be normalized. Each of the methods resulted in a different range of scores that meant trends could not be determined by comparing the original numerical results. The following formula was used to normalize each set of results:

$$x_{new} = \frac{x_{original} - minimum}{maximum - minimum} \quad (5)$$

This resulted in three new scores for each target. The two targets that were missing a score from one of the methods (C5 and CD59) were removed from the set since they could not effectively be included. Each target's three scores were then added together to obtain a composite score that could be used for comparison. The three normalized scores and the composite score for each target are reported in Table 3. The targets were then ranked using the composite score, as shown in Figure 4. The top five molecules in ascending order were C1QB, GABA-A R, C4B, DRD2, and C1R. However, these top five were not all ideal for the small molecule docking that would be performed next. Both C1QB and C1R were, from very early on, in the classical complement pathway. This made them non-ideal targets for inhibition, as that could have an overly negative impact on the cascade started by those molecules that would affect essential functions in the brain. In moving down the ranking of the targets, C3 was then next but had similar concerns. Therefore, the next two targets, C9 and C5AR1, were selected for small molecule docking.

Table 3. Comparison of druggability analysis using normalized results.

Target	DLID Score	Spider Score	SiteMap Score	Composite
ARF6	0	0.75	0	0.750
C4BPA	0.113	0.5	0.788	1.401
C2B	0.178	0.5	0.370	1.047
CFH	0.196	0	0.483	0.679
C2A	0.198	0.5	0.137	0.835
Serping1	0.213	0.5	0.729	1.442
RAB11	0.234	0.5	0.471	1.205
BACE1	0.277	0.75	0.255	1.282
Clusterin	0.302	0.75	0.727	1.779
OPRD1	0.317	0.5	0.912	1.729
C5AR1	0.373	0.75	0.766	1.889
C4A	0.396	0.75	0.741	1.887
C6	0.432	0.25	0.706	1.388
C1R	0.512	1	1	2.512
VTN	0.585	1	0.259	1.844
RAB7	0.651	0.5	0.526	1.677
C3	0.705	0.75	0.680	2.135
DRD2	0.746	0.75	0.967	2.464
C1QB	0.783	0.75	0.666	2.199
GABA-A R	0.896	0.75	0.747	2.393
C1S	0.931	0.5	0.259	1.690
C4B	0.959	0.75	0.733	2.442
C9	1	0.25	0.768	2.018

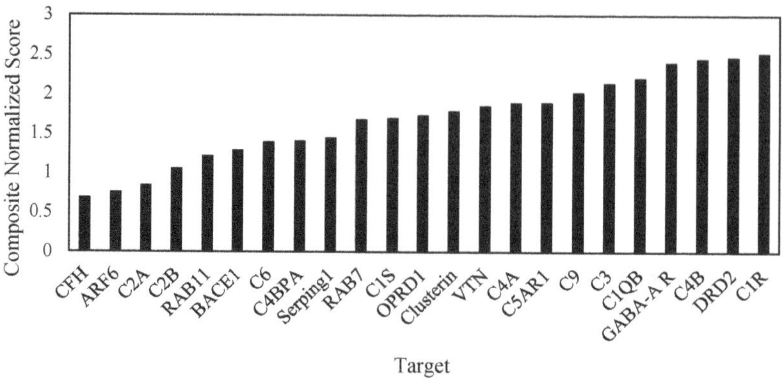

Figure 4. The ranking of targets by combining the normalized results from all three approaches, i.e., the DLID approach, the Spider approach, and the SiteMap approach.

3.5. Validation of Certain Targets Identified from Druggability Analysis by Single-Nucleus Sequencing Data for Inflammation Contributing to Alzheimer's Disease

To investigate the expression of protein targets involved in inflammation at the mRNA level in AD, single-nucleus RNA-sequencing data from the 10× Genomic platform available on ssREAD portal, i.e., GSE138852 and GSE147528, was used. This dataset comprised 26 samples from the entorhinal cortex and 10 from the prefrontal cortex, collected from eight AD and four healthy individuals aged 50–91. After quality control filtering, 81,994 cells from the entorhinal cortex (26,754 genes) and 94,558 cells from the prefrontal cortex (27,590 genes) were retained (Figures 5A–C and 6A–C). The visualization of single-nucleus transcriptomes in uniform manifold approximation and projection (UMAP) space revealed a clear separation into seven cell types from the entorhinal cortex (astrocytes, microglia, oligodendrocytes, endothelial cells, oligodendrocytes precursor cells (OPC), excitatory and inhibitory neurons) (Figure 5D) and six from the prefrontal cortex (astrocytes, microglia, oligodendrocytes, OPC, excitatory and inhibitory neurons) (Figure 6D).

Differential analysis identified notable differences in cell-type-specific gene expression between the control and AD samples, revealing critical patterns that align with disease pathology (Figures 5E and 6E). Endothelial cells in AD exhibited downregulation of the GABRA1 gene, suggesting altered GABA signaling, a pathway known to regulate neuronal activity and implicated in AD pathology [75]. Additionally, neuron cells including inhibitory neurons, and excitatory neurons in Alzheimer's disease displayed downregulated GABRA1 expression and upregulated CD59, a complement regulatory protein, indicating increased complement activation [76].

Astrocytes in AD showed upregulated expression of subcomponent of Complement Component 1 (C1R, C1S, and C1QB), Complement Component 3, Complement Component 6, and the complement cytolysis inhibitor (CLU) in both entorhinal and prefrontal cortices, highlighting the involvement of the complement system in neuroinflammation. This finding aligns with the role of the complement system in neuroinflammation, synaptic pruning, and cognitive decline in AD. C1QB is able to complement cascade activation, contributing to neuronal damage, and C1r and C1s form a complex with C1QB, triggering complement activation. C3 plays a role in mediating neuroinflammation and attracts astrocytes to amyloid plaques, while C6 participates in membrane attack complex (MAC) formation [77]. The role of CLU in promoting Aβ aggregation reinforces its significance in AD pathology. Similarly, both oligodendrocytes and OPCs demonstrated increased expression of complement proteins, which are CR1 and C6. These findings are consistent with the known role of complement activation in synaptic pruning and neuronal damage in AD. Similarly, the upregulation of inflammation-related genes (CR1, C2, CLU, C1S, SERPING1, C6, ARF6, C3, C5AR1, C1QB, CD59) was also observed across neurons in

AD. Moreover, the RAB11A gene, involved in recycling amyloid species, was upregulated across multiple cell types in AD except for endothelial cells, emphasizing its potential role in amyloid pathology [78]. These results collectively indicate that the complement pathway and GABA signaling are disrupted in AD, contributing to inflammation and neuronal injury. The identification of these differentially expressed genes supports the hypothesis that targeting inflammation could be a promising therapeutic strategy for AD.

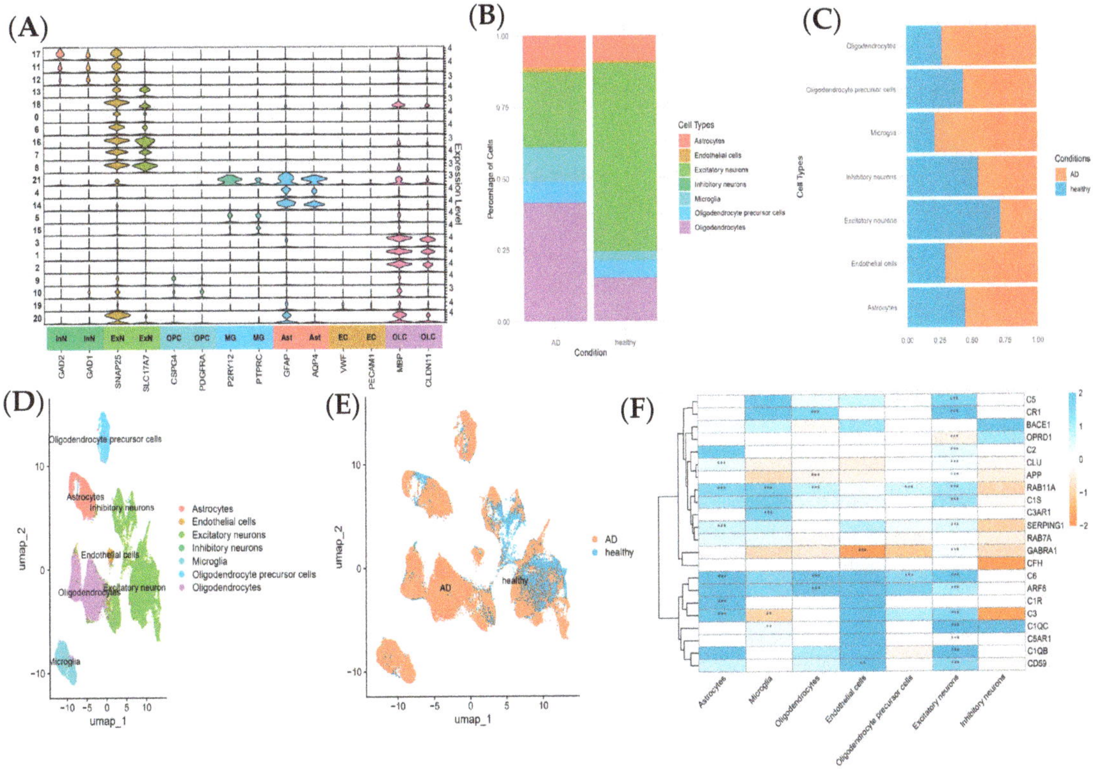

Figure 5. Single-nucleus sequencing data GSE138852 of human entorhinal cortex characterize identified cell types and differential inflammation target expression in AD and healthy brain samples. (**A**) Stacked violin plot for the marker gene expression of seven cell types, including inhibitory neurons, excitatory neurons, oligodendrocyte precursor cells, microglia, astrocytes, endothelial cells, and oligodendrocytes. (**B**,**C**) Stacked bar plot representing the cell-type distribution in AD and healthy conditions, as well as cell-type proportions between two conditions. (**D**,**E**) UMAP visualization for cell distribution of both cell types and different conditions (AD versus healthy). (**F**) Heatmap of inflammation targets expression across cell types and color gradient indicating expression level (blue for upregulation, orange for downregulation) with statistical significance (* p-value < 0.05; ** p-value < 0.01; *** p-value < 0.001).

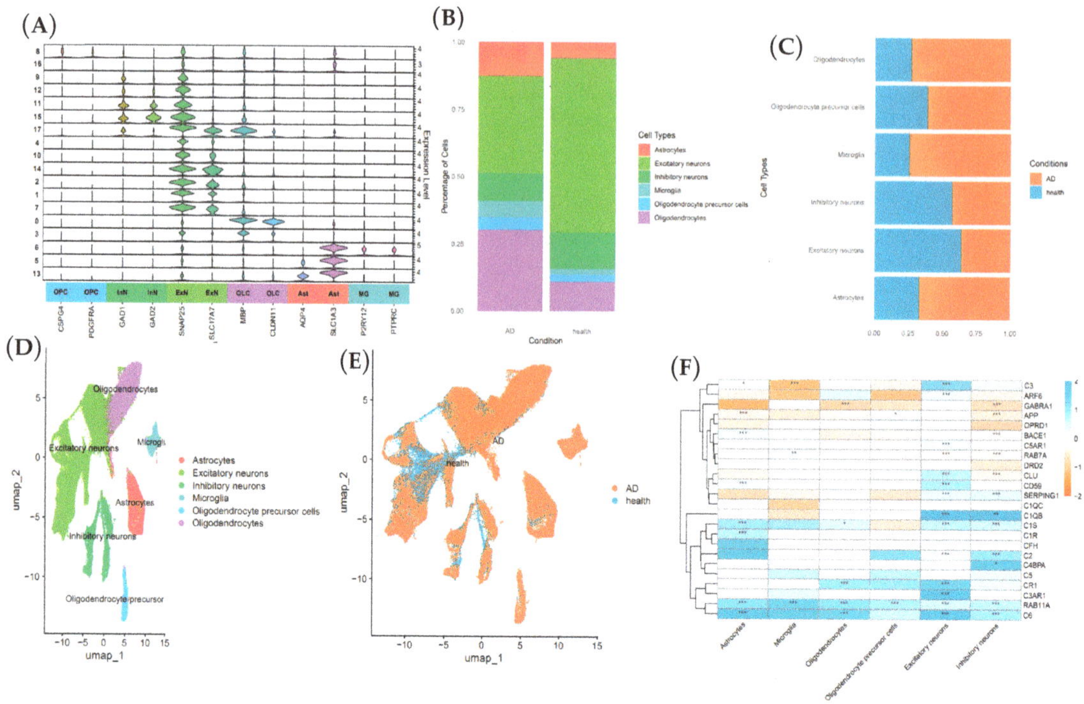

Figure 6. Single-nucleus sequencing data GSE147528 of human prefrontal cortex characterize identified cell types and differential inflammation target expression in AD and healthy brain samples. (**A**) Stacked violin plot for marker gene expression of seven cell types, including inhibitory neurons, excitatory neurons, oligodendrocyte precursor cells, microglia, astrocytes, and oligodendrocytes. (**B**,**C**) Stacked bar plot representing cell-type distribution in AD and healthy conditions, as well as cell-type proportions between two conditions. (**D**,**E**) UMAP visualization for cell distribution of both cell types and different conditions (AD versus healthy). (**F**) Heatmap of inflammation targets expression across cell types and color gradient indicating expression level (blue for upregulation, orange for downregulation) with statistical significance (* p-value < 0.05; ** p-value < 0.01; *** p-value < 0.001).

3.6. Small Molecule Docking for Drug Repurposing for the Targets with the Highest Druggability

Among the top targets identified from the druggability analysis, four of them (i.e., DRD2, C4B, GABA-A-R, and C5AR1) were shown in the single-nucleus sequencing data because of their clinical relevance to inflammation involved in Alzheimer's disease. This validated the clinical values of these targets. The computational docking program was then further implemented below to identify potential FDA-approved drugs as the inhibitors for these targets.

3.6.1. Repurposing FDA-Approved Drugs to Inhibit DRD2

Despite having the highest druggability score of the examined targets, DRD2 had the least amount of small molecule inhibitors that reached the −20 kcal/mol threshold at eight molecules. The top-scoring molecule was $C_{20}H_{15}F_3N_4O_3$, also known as Trovafloxacin or Trovan. Trovafloxacin is a broad-spectrum antibiotic that works by inhibiting the uncoiling of bacterial DNA that is supercoiled through the blocking of DNA gyrase and topoisomerase IV [79,80]. However, Trovafloxacin was recently taken off of the market because it had been shown to increase the risk of liver failure [79,81]. Therefore, the next option would be $C_{23}H_{22}ClN_5O_3$, also known as Betrixaban, which is an anti-coagulant that works by inhibiting factor Xa. It is used for patients at risk of venous thromboembolism [82,83].

DRD2 shared two common inhibitors overall, though neither of them were in the top 10 molecules for the other targets. v1099 met the threshold for C9 and is also known as Favipiravir, an antiviral agent used to treat influenza [84]. The other molecule was $C_{23}H_{22}ClN_5O_3$ (Betrixaban), which was below −20 kcal/mol for C5AR1. Table 4 summarizes the eight inhibitors resulting from binding simulation with the protein DRD2.

Table 4. Small molecule docking results with DRD2.

Molecule	Name	Binding Score
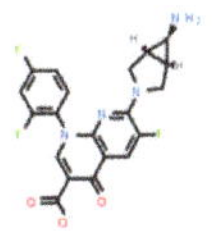	$C_{20}H_{15}F_3N_4O_3$	−24.0
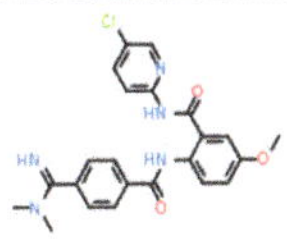	$C_{23}H_{22}ClN_5O_3$	−23.94
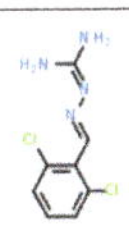	v629	−20.87
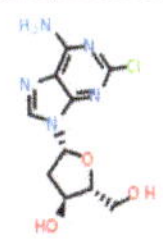	v242	−20.3
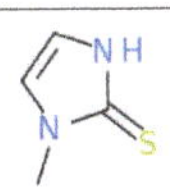	v763	−20.28
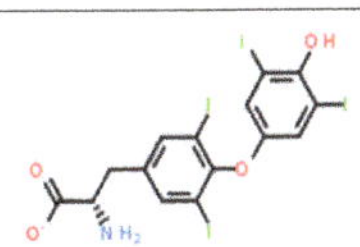	v451	−20.15
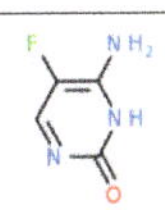	v1099	−20.1
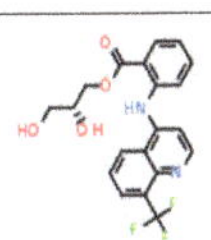	$C_{20}H_{17}F_3N_2O_4$	−20.01

3.6.2. Repurposing FDA-Approved Drugs to Inhibit C4B

C4b had 10 results that scored below −20, the second least amount of all of the targets. The top-scoring molecule for C4b was $C_{21}H_{18}F_3N_3O_5$, also known as Bictegravir. It scored −26.13 kcal/mol. It is a second-generation integrase strand transfer inhibitor that has been approved for the treatment of HIV-1 [85]. C4b had three molecules in common with the other targets. The first was m, or Sulfasalazine ($C_{14}H_{14}N_4O_5S$), which was shared with C9 and C5AR1 but was not in the top 10 for either of those targets. It is an NSAID used to treat several diseases with chronic inflammation [86]. The second was v316, which was a top molecule for C9. v316 is also known as paracetamol, which is an alternate form of acetaminophen, commonly known as Tylenol [87]. It is a cyclooxygenase inhibitor that can reduce pain and fever [87], whose exact mechanisms are controversial but likely targets the central nervous system [88]. The last molecule was v487, which met the threshold for C9 and GABA-A-R. However, information on this molecule was lacking. Table 5 summarizes the results from the docking simulation for protein C4b.

Table 5. The small molecule docking results for C4b.

Molecule	Name	Binding Score
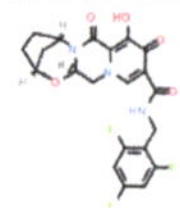	$C_{21}H_{18}F_3N_3O_5$	−26.13
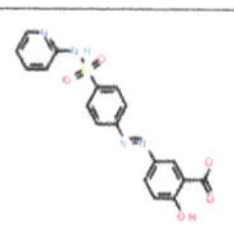	m	−24.36
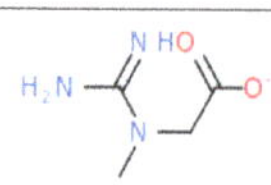	$C_4H_9N_3O_2$	−21.65
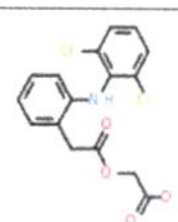	$C_{16}H_{13}Cl_2$	−21.51
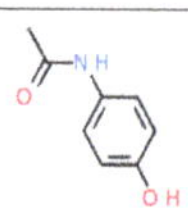	v316	−21.45
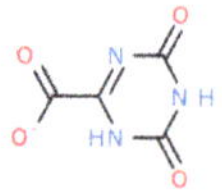	$C_4H_3N_3O_4$	−21.21
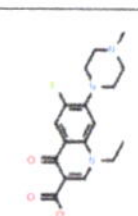	v487	−20.36
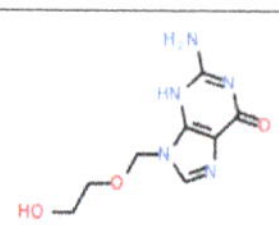	v787	−20.26
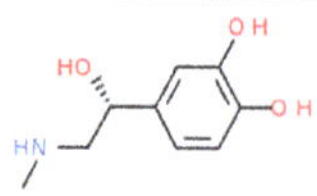	v668	−20.18
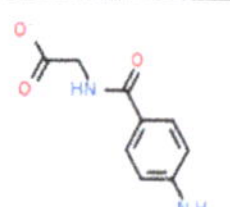	v2148	−20.13

3.6.3. Repurposing FDA-Approved Drugs to Inhibit GABA-A-R

GABA-A-R had 28 molecules that scored below −20 kcal/mol, with the top 10 represented below (Table 6). The top-scoring molecule was v555, which was not identifiable beyond the fact that it was a covalent organic polymer [89]. GABA-A-R shared eight total inhibitors, all of which were shared with C5AR1 and C9, except for v487, which was shared with C4b as well. These included $C_{21}H_{19}ClN_4O_4$, v963, v855, v951, $C_{17}H_{15}N_3O_6$, v461, and $C_{11}H_6ClN_3O_6$.

Table 6. Small molecule docking for GABA-A-R.

Molecule	Name	Binding Score
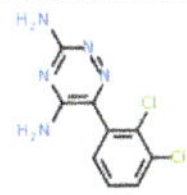	v555	−29.3
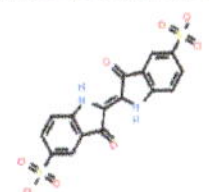	$C_{16}H_{10}N_2O_8S_2$	−26.8
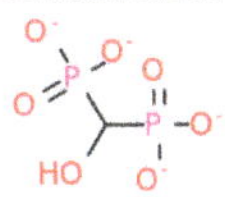	$CH_6O_7P_2$	−26.4
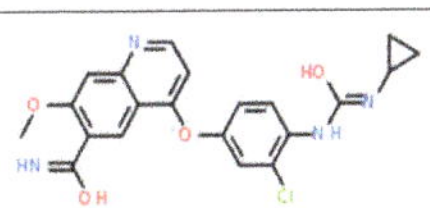	$C_{21}H_{19}ClN_4O_4$	−25.3
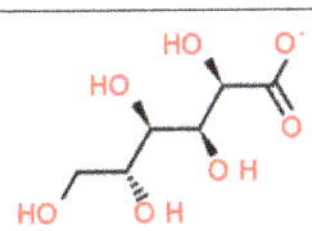	$C_6H_{11}KO_7$	−23.8
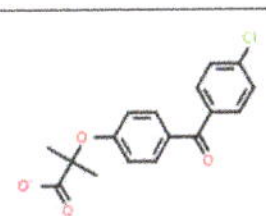	$C_{17}H_{15}ClO_4$	−23.7
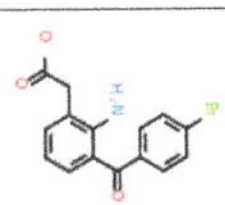	v963	−23.1
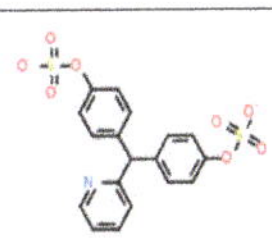	$C_{18}H_{15}NO_8S_2$	−22.8
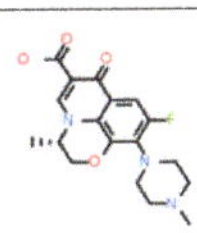	v1165	−22.8
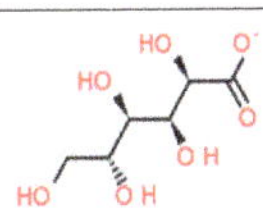	$C_{12}H_{22}MnO_{14}$	−22.3

3.6.4. Repurposing FDA-Approved Drugs to Inhibit C9

C9 had a large number of molecules that scored below −20 kcal/mol, with a total of 89. Due to the space constraint, only the top 10 potential inhibitors are shown in Table 7. The top-scoring molecule was v174 at −33.2. v174 is also known as asparaginate, a conjugate base of asparagine [90]. Asparagine is essential for the metabolic control of cell functions in nerve and brain tissue [91]. C9 had 28 inhibitors in common with the other targets.

Table 7. Small molecule docking for C9.

Molecule	Name	Binding Score
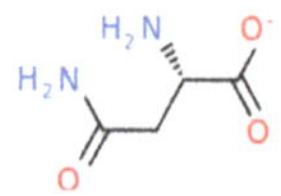	v174	−33.2
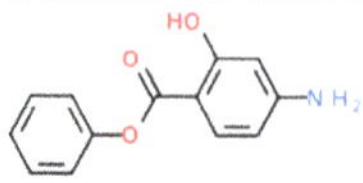	$C_{13}H_{11}NO_3$	−32.7
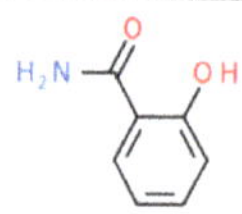	v5147	−31.2
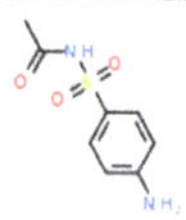	v634	−31.2
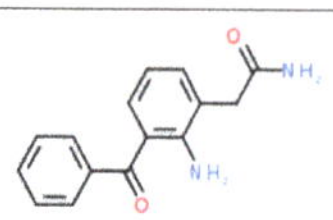	$C_{15}H_{14}N_2O_2$	−29.2
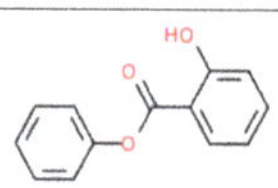	$C_{13}H_{10}O_3$	−29.1
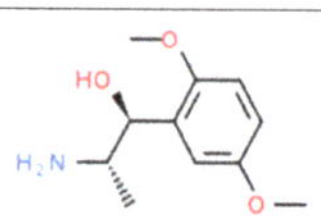	v723	−28.7
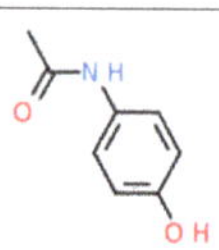	v316	−28.6
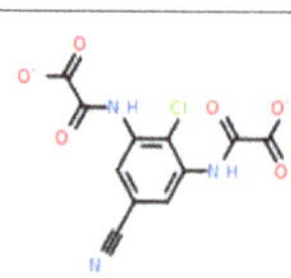	$C_{11}H_6ClN_3O_6$	−28.3
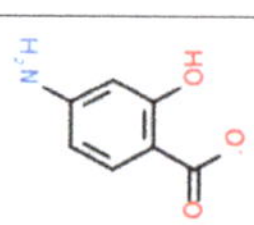	v233	−28.1

3.6.5. Repurposing FDA-Approved Drugs to Inhibit C5AR1

C5AR1 had not only the largest number of molecules that scored below −20 kcal/mol but also the best-scoring molecule. This target had 131 molecules with acceptable scores, with the top one being $C_{10}H_9N_5O$ at −35.3 kcal/mol (Table 8). This molecule is also known kinetin, which is a proven anti-aging agent [92]. C5AR1 had 29 inhibitors in common with the other top targets.

Table 8. Small molecule docking for C5AR1.

Molecule	Name	Binding Score
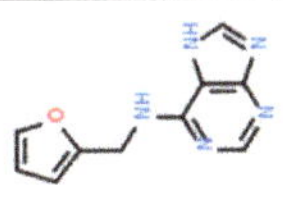	$C_{10}H_9N_5O$	−35.3
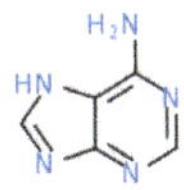	v173	−30.1
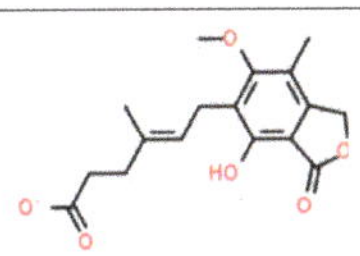	v1024	−29
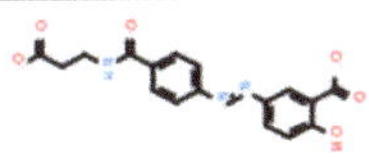	$C_{17}H_{15}N_3O_6$	−26.5
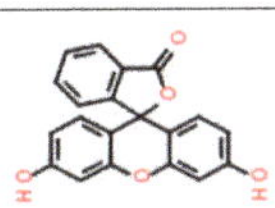	$C_{20}H_{12}O_5$	−26.5
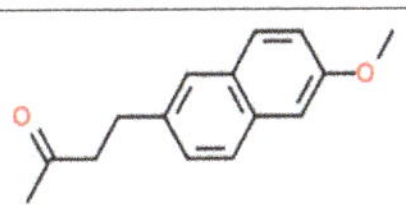	v461	−26.5
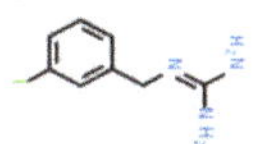	$C_8H_{10}IN_3$	−25.9
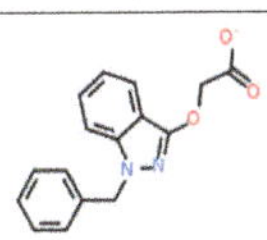	$C_{16}H_{14}N_2O_3$	−25.8
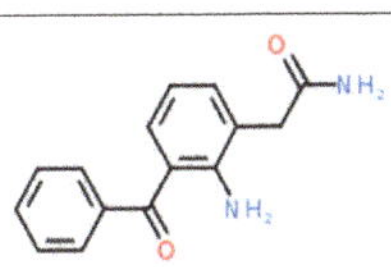	$C_{15}H_{14}N_2O_2$	−25.7
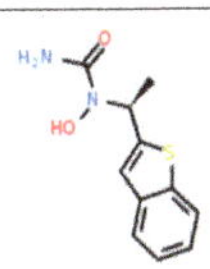	v744	−25.5

4. Discussion

4.1. Comprehensive Druggability Assessment of Inflammation-Related Alzheimer's Disease Targets

The druggability analysis of inflammation-related AD targets using three programs (i.e., DLID, SiteMap, and SPIDER) revealed varying degrees of druggability across the targets. The DLID analysis identified druggable pockets in most targets, but certain targets did not meet the druggability threshold, with CD59 being particularly undruggable as no pockets were identified. Targets like C5, GABA-A R, and C9 had higher numbers of viable pockets, while some had lower scores due to unfavorable pocket characteristics. The SiteMap analysis confirmed the druggability of certain targets, with pockets in C5 and C9 scoring highly, although CD59 had a pocket that met the threshold, unlike DLID. The SPIDER scores, however, showed minimal variation across targets. This suggested that it might not be sensitive enough to reflect major druggability trends. Normalizing the results from all three platforms provided a clearer ranking, with targets like DRD2, C4B, GABA-A-R, and C5AR1 emerging as top candidates for small molecule docking. Although some targets such as C1R and C1QB ranked highly, they were deemed unsuitable for inhibition due to their roles in essential biological pathways. The consistency between DLID and SiteMap in identifying druggable pockets in targets like C9 and GABA-A R supported the validity of our results, as different methods with distinct algorithms pointed to similar conclusions. While SPIDER showed less variability, its findings still aligned with other platforms for most high-ranking targets, offering an additional layer of cross-validation. These variations across approaches make sense because DLID focuses on pocket properties like hydrophobicity, while SiteMap evaluates structural features more holistically, and SPIDER emphasizes protein sequences

4.2. Drug Repurposing for the Top Drug Targets Related to Inflammation for Alzheimer's Disease

Comparing the top 10 highest binding potential inhibitors across targets shown in Tables 4–8 revealed two contenders: v316 and $C_{15}H_{14}N_2O_2$. v316 is used to reduce headaches and fever, while $C_{15}H_{14}N_2O_2$ is used to reduce pain and inflammation for cataract surgery. The results showed that only two small molecule inhibitors appeared in the top 10 docking results for more than one target. The first molecule was v316, which was a top result for C4B and C9. This molecule treats pain and fever [87]. The second molecule was $C_{15}H_{14}N_2O_2$, which was a top result for C5AR1 and C9. This molecule treats pain and inflammation associated with cataracts [93]. It inhibits COX1 and COX2 activity [94]. DRD2 and GABA-A-R did not have any top molecules overlapping with the other targets. However, outside of the top 10, there was significantly more overlap. C4B had 3 of its 10 results overlap with the others. C5AR1 had 29 of its 131 results overlap, while 28 of the 89 C9 results overlapped. Moreover, 8 of the 28 GABA-A-R were present in the results for the other targets. DRD2 had 2 out of 8 in common with the other targets.

4.3. The Role of the Complement Cascade in Alzheimer's Disease Progression

While this work focused on protein targets suggested by the literature [61], our approach can be applied to explore additional targets in the future. The involvement of the studied targets in AD-related inflammation was validated through our sequencing data analysis. Further enrichment analysis confirmed that targets such as C1QC, C1QB, CFH, C4BPA, CR1, C9, C6, C4A, CLU, C5, CD59, SERPING1, C1S, C1R, C3AR1, C3, and C5AR1 play a critical role in the complement cascade, which is essential for neuronal development and plasticity [95]. Specifically, complement proteins C1q, C3, and C4 showed increased levels in AD patients brain tissues compared with those of healthy people [96]. The mRNA levels of C3 and C4 were observed to be elevated in the temporal cortex of AD, and increased levels of the protein of C3b and the products of the terminal MAC (C5b-C9) in AD brain tissues have also been reported, indicating that that MAC can potentially cause neuronal loss and neurodegeneration in AD [97]. This study suggested that significant differences in the mRNA levels of C1QC, C1QB, C4BPA, CR1, C9, C6, C4A, CLU, CD59, SERPING1, C1S, C1R, C3AR1, C3, and C5AR1 exist among AD patients when compared with healthy people through single-cell RNA-seq analysis. Complement proteins C1q, C3, and C4 were detected in amyloid plaques and co-localized with neurofibrillary tangles in the brains of AD individuals [98]. Complement proteins C1q, C3b, C3c, and C3d are linked to amyloid plaques and the associated dystrophic neurites during AD's progress. Complement C3/C3b has become a crucial factor in synapse elimination by tagging weak synapses and serving as a "eat me" signal. Microglia activated by amyloid-β and tau protein exhibit morphological alterations and the expression of complement receptors CR1, CR3, C3aR, and C5aR1, and they secrete a range of immune components, including complement proteins like C1q. The activated microglia with a CR3 receptor could identify the C3-tagged synapses and remove them through phagocytosis [99]. Furthermore, the binding of these complement proteins to their corresponding receptors triggered inflammation and cytokine secretion. C1q released by activated microglia stimulated A1-reactive astrocytes. A1 astrocytes exhibited elevated expression and the release of C3. C3 produced from A1-reactive astrocytes accumulated on amyloid plaques, diminished synapses, and contributed to neurofibrillary tangles [100]. Additionally, complement C3 has been suggested to be an astroglia initiator of nuclear factor-κB (NFκB) signaling via neuronal C3aR, a G protein-coupled receptor present in various cell types [101]. The NFκB/C3/C3aR signaling pathway induces dendritic and synaptic degeneration in neurons; hence, it is logical that APP/PS1 animals treated with a C3aR antagonist demonstrated enhanced cognitive function [101]. Comprehending the interaction between the complement system and AD is essential for elucidating the molecular mechanisms underlying AD. This work repurposed FDA-approved drugs to regulate druggable components from the complement system for the advancement of targeted therapy strategies.

4.4. Limitation and Future Work

Since the X-ray crystal structures from PDB were available for all protein targets, those structures were used for structure-based druggability analysis in this work to ensure consistency across our selected targets. Although cryo-electron microscopy (cryo-EM) is valuable for visualizing large and flexible proteins, corresponding cryo-EM structures were not available for all target proteins in the PDB. This limitation restricted the use of cryo-EM structures in this work. In addition, certain available cryo-EM structures for the proteins of interest had resolutions worse than the X-ray structures. In future work, we may consider incorporating cryo-EM structures as they become available and achieve higher resolutions that allow us to evaluate any structural differences that may influence druggability results across methods.

Early-stage proteins C1QB, C1R, and C3 were not included for further small molecule docking analysis, because targeting these initial components in the complement pathway could lead to broad and unintended immunosuppressive effects. Although early-stage proteins, like G protein-coupled receptors (GPCRs), are often effective drug targets due to their regulatory roles in signaling cascades, there are instances—particularly within the immune system—where early pathway proteins are avoided to prevent widespread effects. Specifically, in the complement system, targeting proteins such as C1QB and C1R may suppress the entire pathway, which may compromise immune function in a non-specific manner. Future studies may explore the potential of early-stage complement proteins as drug targets, but such strategies require careful consideration of potential immune-related side effects.

In this study, certain FDA-approved drugs were identified with similar binding scores for potential repurposing against Alzheimer's disease (AD) inflammation (as shown in Table 4). The similarity in binding scores arises because these scores are based on seven binding energy terms, including hydrogen bonds, van der Waals interactions, and electrostatic forces, as calculated in Molsoft ICM. As a result, compounds with comparable interactions across these energy terms can yield similar binding scores. This static binding screening serves as a preliminary step to enable the rapid assessment of compounds and reduce resource expenditure. Future works will involve detailed molecular dynamics

(MD) simulations to assess dynamic binding stability, which allows us to differentiate compounds further based on their interaction profiles over time. This will be followed by experimental validation of the top-performing candidates to confirm efficacy. Overall, our work establishes a foundation for using MD simulations and experimental studies to refine drug repurposing candidates for AD.

5. Conclusions

Alzheimer's disease has been proven to have key connections to brain inflammation, meaning that novel targets may be found through the study of essential proteins and pathways relevant to both diseases. However, in order for these proteins to be viable targets, the druggability of the targets had to be analyzed. This first involved conducting a comprehensive review of both structure-based and sequence-based methods in order to find the best programs for analysis. Once these programs were identified, each of the targets was run through the DLID, SiteMap, and SPIDER programs. The top-scoring potential targets were DRD2, C4b, GABA-A-R, C9, and C5AR1. Single-nucleus sequencing data were used to validate their clinical relevance to Alzheimer' disease inflammation. For these five most druggable targets, the computational ligand–protein docking program was implemented to evaluate FDA-approved inhibitors against each selected target to confirm the binding capabilities of the targets. DRD2 had 8 inhibitors that had scores indicating high potential for binding, while C4b had 10, GABA-A-R had 28, C9 had 89, and C5AR1 had 131. Comparing the top 10 highest binding potential inhibitors across targets revealed two contenders: v316 and $C_{15}H_{14}N_2O_2$. v316 is used to reduce headaches and fever, while $C_{15}H_{14}N_2O_2$ is used to reduce pain and inflammation for cataract surgery. For future consideration, in vitro experiments for druggability should be conducted to determine if these inhibitors actually affect the protein targets.

Author Contributions: Conceptualization, T.Z., L.S. and Z.H.; methodology, C.S., J.Z., T.Z., J.B., A.G.-E., E.M., L.S. and Z.H.; formal analysis, C.S., J.Z., T.Z., J.B., A.G.-E. and E.M.; writing—original draft preparation, C.S., J.Z., T.Z., J.B., A.G.-E. and E.M.; writing—review and editing, L.S. and Z.H.; funding acquisition, L.S. All authors have read and agreed to the published version of the manuscript.

Funding: This work was supported in part by NIH Grants U19 AG074879, R01 AG071470, U01 AG066833, and U01 AG068057.

Institutional Review Board Statement: Not applicable.

Informed Consent Statement: Not applicable.

Data Availability Statement: Data will be provided upon request.

Conflicts of Interest: The authors declare no conflicts of interest.

Abbreviations

Alzheimer's disease (AD); G protein-coupled receptors (GPCRs); drug-like density (DLID); Protein Data Bank (PDB); SPIDER (Stacked PredIctor of DruggablE pRoteins); drug-like ligand-containing (DLLC); oligodendrocytes precursor cells (OPCs); uniform manifold approximation and projection (UMAP); oligodendrocytes precursor cells (OPCs); membrane attack complex (MAC).

References

1. Kinney, J.W.; Bemiller, S.M.; Murtishaw, A.S.; Leisgang, A.M.; Salazar, A.M.; Lamb, B.T. Inflammation as a central mechanism in Alzheimer's disease. *Alzheimer's Dement. Transl. Res. Clin. Interv.* **2018**, *4*, 575–590. [CrossRef] [PubMed]
2. Tatulian, S.A. Challenges and hopes for Alzheimer's disease. *Drug Discov. Today* **2022**, *27*, 1027–1043. [CrossRef] [PubMed]
3. Finder, V.H. Alzheimer's Disease: A General Introduction and Pathomechanism. *J. Alzheimer's Dis.* **2010**, *22*, S5–S19. [CrossRef] [PubMed]
4. Chohan, M.O.; Khatoon, S.; Iqbal, I.-G.; Iqbal, K. Involvement of in the abnormal hyperphosphorylation of tau and its reversal by Memantine. *FEBS Lett.* **2006**, *580*, 3973–3979. [CrossRef]

5. Wang, L.; Bharti; Kumar, R.; Pavlov, P.F.; Winblad, B. Small molecule therapeutics for tauopathy in Alzheimer's disease: Walking on the path of most resistance. *Eur. J. Med. Chem.* **2021**, *209*, 112915. [CrossRef]
6. Mahase, E. Alzheimer's disease: FDA approves lecanemab amid cost and safety concerns. *BMJ* **2023**, *380*, 73. [CrossRef]
7. Canady, V.A. FDA approves new treatment for Alzheimer's disease. *Ment. Health Wkly.* **2023**, *33*, 6–7. [CrossRef]
8. Liu, J.; Xiao, Q.; Xiao, J.; Niu, C.; Li, Y.; Zhang, X.; Zhou, Z.; Shu, G.; Yin, G. Wnt/β-catenin signalling: Function, biological mechanisms, and therapeutic opportunities. *Signal Transduct. Target. Ther.* **2022**, *7*, 3. [CrossRef]
9. Akiyama, H.; Barger, S.; Barnum, S.; Bradt, B.; Bauer, J.; Cole, G.M.; Cooper, N.R.; Eikelenboom, P.; Emmerling, M.; Fiebich, B.L.; et al. Inflammation and Alzheimer's disease. *Neurobiol. Aging* **2000**, *21*, 383–421. [CrossRef]
10. Perry, V. The influence of systemic inflammation on inflammation in the brain: Implications for chronic neurodegenerative disease. *Brain Behav. Immun.* **2004**, *18*, 407–413. [CrossRef]
11. Zhu, M.; Wang, X.; Sun, L.; Schultzberg, M.; Hjorth, E. Can inflammation be resolved in Alzheimer's disease? *Ther. Adv. Neurol. Disord.* **2018**, *11*, 175628641879110. [CrossRef] [PubMed]
12. Garcez, M.L.; Jacobs, K.R.; Guillemin, G.J. Microbiota Alterations in Alzheimer's Disease: Involvement of the Kynurenine Pathway and Inflammation. *Neurotox. Res.* **2019**, *36*, 424–436. [CrossRef] [PubMed]
13. Perry, V.H.; Teeling, J. Microglia and macrophages of the central nervous system: The contribution of microglia priming and systemic inflammation to chronic neurodegeneration. *Semin. Immunopathol.* **2013**, *35*, 601–612. [CrossRef]
14. Ryu, J.K.; Cho, T.; Choi, H.B.; Jantaratnotai, N.; McLarnon, J.G. Pharmacological antagonism of interleukin-8 receptor CXCR2 inhibits inflammatory reactivity and is neuroprotective in an animal model of Alzheimer's disease. *J. Neuroinflammation* **2015**, *12*, 144. [CrossRef]
15. Lundstrom, K. An Overview on GPCRs and Drug Discovery: Structure-Based Drug Design and Structural Biology on GPCRs. In *G Protein-Coupled Receptors in Drug Discovery*; Humana Press: Totowa, NJ, USA, 2009; pp. 51–66. [CrossRef]
16. Sheridan, R.P.; Maiorov, V.N.; Holloway, M.K.; Cornell, W.D.; Gao, Y.-D. Drug-like Density: A Method of Quantifying the "Bindability" of a Protein Target Based on a Very Large Set of Pockets and Drug-like Ligands from the Protein Data Bank. *J. Chem. Inf. Model.* **2010**, *50*, 2029–2040. [CrossRef]
17. Hanes, R.; Liu, Y.; Huang, Z. Druggability Analysis of Protein Targets for Drug Discovery to Combat *Listeria monocytogenes*. *Microorganisms* **2024**, *12*, 1073. [CrossRef]
18. An, J.; Totrov, M.; Abagyan, R. Pocketome via Comprehensive Identification and Classification of Ligand Binding Envelopes. *Mol. Cell. Proteom.* **2005**, *4*, 752–761. [CrossRef]
19. Borrel, A.; Regad, L.; Xhaard, H.; Petitjean, M.; Camproux, A.-C. PockDrug: A Model for Predicting Pocket Druggability That Overcomes Pocket Estimation Uncertainties. *J. Chem. Inf. Model.* **2015**, *55*, 882–895. [CrossRef]
20. Rodriguez-Salarichs, J. PockDrug Descriptors (Druggability). Available online: https://github.com/rodriguez-salarichs/PockDrug_descriptors (accessed on 5 May 2024).
21. Wang, S.; Xie, J.; Pei, J.; Lai, L. CavityPlus 2022 Update: An Integrated Platform for Comprehensive Protein Cavity Detection and Property Analyses with User-friendly Tools and Cavity Databases. *J. Mol. Biol.* **2023**, *435*, 168141. [CrossRef]
22. Xu, Y.; Wang, S.; Hu, Q.; Gao, S.; Ma, X.; Zhang, W.; Shen, Y.; Chen, F.; Lai, L.; Pei, J. CavityPlus: A web server for protein cavity detection with pharmacophore modelling, allosteric site identification and covalent ligand binding ability prediction. *Nucleic Acids Res.* **2018**, *46*, W374–W379. [CrossRef]
23. Schöning-Stierand, K.; Diedrich, K.; Ehrt, C.; Flachsenberg, F.; Graef, J.; Sieg, J.; Penner, P.; Poppinga, M.; Ungethüm, A.; Rarey, M. Proteins*Plus*: A comprehensive collection of web-based molecular modeling tools. *Nucleic Acids Res.* **2022**, *50*, W611–W615. [CrossRef] [PubMed]
24. Schöning-Stierand, K.; Schöning-Stierand, K.; Diedrich, K.; Diedrich, K.; Fährrolfes, R.; Fährrolfes, R.; Flachsenberg, F.; Flachsenberg, F.; Meyder, A.; Meyder, A.; et al. ProteinsPlus: Interactive analysis of protein–ligand binding interfaces. *Nucleic Acids Res.* **2020**, *48*, W48–W53. [CrossRef] [PubMed]
25. Fährrolfes, R.; Bietz, S.; Flachsenberg, F.; Meyder, A.; Nittinger, E.; Otto, T.; Volkamer, A.; Rarey, M. ProteinsPlus: A web portal for structure analysis of macromolecules. *Nucleic Acids Res.* **2017**, *45*, W337–W343. [CrossRef]
26. Schmidtke, P.; Le Guilloux, V.; Shorkar, M. fpocket. Available online: https://github.com/Discngine/fpocket (accessed on 5 May 2024).
27. InDeep. Available online: https://gitlab.pasteur.fr/InDeep/InDeep (accessed on 5 May 2024).
28. Cuchillo, R.; Pinto-Gil, K.; Michel, J. A Collective Variable for the Rapid Exploration of Protein Druggability. *J. Chem. Theory Comput.* **2015**, *11*, 1292–1307. [CrossRef] [PubMed]
29. Krivák, R.; Hoksza, D. P2Rank: Machine learning based tool for rapid and accurate prediction of ligand binding sites from protein structure. *J. Chemin.* **2018**, *10*, 39. [CrossRef] [PubMed]
30. Jakubec, D.; Skoda, P.; Krivak, R.; Novotny, M.; Hoksza, D. PrankWeb 3: Accelerated ligand-binding site predictions for experimental and modelled protein structures. *Nucleic Acids Res.* **2022**, *50*, W593–W597. [CrossRef]
31. Jendele, L.; Krivak, R.; Skoda, P.; Novotny, M.; Hoksza, D. PrankWeb: A web server for ligand binding site prediction and visualization. *Nucleic Acids Res.* **2019**, *47*, W345–W349. [CrossRef]
32. Krivák, R.; Hoksza, D. P2RANK: Knowledge-Based Ligand Binding Site Prediction Using Aggregated Local Features. In *Algorithms for Computational Biology*; Springer: Berlin/Heidelberg, Germany, 2015; pp. 41–52. [CrossRef]

33. Krivák, R.; Hoksza, D. Improving protein-ligand binding site prediction accuracy by classification of inner pocket points using local features. *J. Chemin.* **2015**, *7*, 12. [CrossRef]
34. Halgren, T.A. Identifying and Characterizing Binding Sites and Assessing Druggability. *J. Chem. Inf. Model.* **2009**, *49*, 377–389. [CrossRef]
35. Yuan, J.-H.; Han, S.B.; Richter, S.; Wade, R.C.; Kokh, D.B. Druggability Assessment in TRAPP Using Machine Learning Approaches. *J. Chem. Inf. Model.* **2020**, *60*, 1685–1699. [CrossRef]
36. Kozlovskii, I.; Popov, P. Spatiotemporal identification of druggable binding sites using deep learning. *Commun. Biol.* **2020**, *3*, 618. [CrossRef] [PubMed]
37. Kozlovskii, I.; Popov, P. Protein–Peptide Binding Site Detection Using 3D Convolutional Neural Networks. *J. Chem. Inf. Model.* **2021**, *61*, 3814–3823. [CrossRef] [PubMed]
38. Cimermancic, P.; Weinkam, P.; Rettenmaier, T.J.; Bichmann, L.; Keedy, D.A.; Woldeyes, R.A.; Schneidman-Duhovny, D.; Demerdash, O.N.; Mitchell, J.C.; Wells, J.A.; et al. CryptoSite: Expanding the Druggable Proteome by Characterization and Prediction of Cryptic Binding Sites. *J. Mol. Biol.* **2016**, *428*, 709–719. [CrossRef] [PubMed]
39. Aggarwal, R.; Gupta, A.; Chelur, V.; Jawahar, C.V.; Priyakumar, U.D. DeepPocket: Ligand Binding Site Detection and Segmentation using 3D Convolutional Neural Networks. *J. Chem. Inf. Model.* **2022**, *62*, 5069–5079. [CrossRef] [PubMed]
40. Jiménez, J.; Doerr, S.; Martínez-Rosell, G.; Rose, A.S.; De Fabritiis, G. DeepSite: Protein-binding site predictor using 3D-convolutional neural networks. *Bioinformatics* **2017**, *33*, 3036–3042. [CrossRef]
41. Mylonas, S.K.; Axenopoulos, A.; Daras, P. DeepSurf: A surface-based deep learning approach for the prediction of ligand binding sites on proteins. *Bioinformatics* **2021**, *37*, 1681–1690. [CrossRef]
42. Tan, K.P.; Nguyen, T.B.; Patel, S.; Varadarajan, R.; Madhusudhan, M.S. Depth: A web server to compute depth, cavity sizes, detect potential small-molecule ligand-binding cavities and predict the pKa of ionizable residues in proteins. *Nucleic Acids Res.* **2013**, *41*, W314–W321. [CrossRef]
43. Wang, X.; Flannery, S.T.; Kihara, D. Protein Docking Model Evaluation by Graph Neural Networks. *Front. Mol. Biosci.* **2021**, *8*, 647915. [CrossRef]
44. Kawabata, T. Detection of multiscale pockets on protein surfaces using mathematical morphology. *Proteins: Struct. Funct. Bioinform.* **2010**, *78*, 1195–1211. [CrossRef]
45. Li, B.; Turuvekere, S.; Agrawal, M.; La, D.; Ramani, K.; Kihara, D. Characterization of local geometry of protein surfaces with the visibility criterion. *Proteins Struct. Funct. Bioinform.* **2008**, *71*, 670–683. [CrossRef]
46. Stepniewska-Dziubinska, M.M.; Zielenkiewicz, P.; Siedlecki, P. Improving detection of protein-ligand binding sites with 3D segmentation. *Sci. Rep.* **2020**, *10*, 5035. [CrossRef] [PubMed]
47. Jing, B.; Eismann, S.; Suriana, P.; Townshend, R.J.L.; Dror, R. Learning from Protein Structure with Geometric Vector Perceptrons. *arXiv* **2021**, arXiv:2009.01411.
48. Di Palma, F.; Abate, C.; Decherchi, S.; Cavalli, A. Ligandability and druggability assessment via machine learning. *WIREs Comput. Mol. Sci.* **2023**, *13*, e1676. [CrossRef]
49. Liao, J.; Wang, Q.; Wu, F.; Huang, Z. In Silico Methods for Identification of Potential Active Sites of Therapeutic Targets. *Molecules* **2022**, *27*, 7103. [CrossRef]
50. Jamali, A.A.; Ferdousi, R.; Razzaghi, S.; Li, J.; Safdari, R.; Ebrahimie, E. DrugMiner: Comparative analysis of machine learning algorithms for prediction of potential druggable proteins. *Drug Discov. Today* **2016**, *21*, 718–724. [CrossRef]
51. Lin, J.; Chen, H.; Li, S.; Liu, Y.; Li, X.; Yu, B. Accurate prediction of potential druggable proteins based on genetic algorithm and Bagging-SVM ensemble classifier. *Artif. Intell. Med.* **2019**, *98*, 35–47. [CrossRef]
52. Gong, Y.; Liao, B.; Wang, P.; Zou, Q. DrugHybrid_BS: Using Hybrid Feature Combined With Bagging-SVM to Predict Potentially Druggable Proteins. *Front. Pharmacol.* **2021**, *12*, 771808. [CrossRef]
53. Sikander, R.; Ghulam, A.; Ali, F. XGB-DrugPred: Computational prediction of druggable proteins using eXtreme gradient boosting and optimized features set. *Sci. Rep.* **2022**, *12*, 5505. [CrossRef]
54. Yu, L.; Xue, L.; Liu, F.; Li, Y.; Jing, R.; Luo, J. The applications of deep learning algorithms on in silico druggable proteins identification. *J. Adv. Res.* **2022**, *41*, 219–231. [CrossRef]
55. Charoenkwan, P.; Schaduangrat, N.; Lio', P.; Moni, M.A.; Shoombuatong, W.; Manavalan, B. Computational prediction and interpretation of druggable proteins using a stacked ensemble-learning framework. *iScience* **2022**, *25*, 104883. [CrossRef]
56. Chen, J.; Gu, Z.; Xu, Y.; Deng, M.; Lai, L.; Pei, J. QuoteTarget: A sequence-based transformer protein language model to identify potentially druggable protein targets. *Protein Sci.* **2023**, *32*, e4555. [CrossRef] [PubMed]
57. Zhang, M.; Wan, F.; Liu, T. DrugFinder: Druggable Protein Identification Model Based on Pre-Trained Models and Evolutionary Information. *Algorithms* **2023**, *16*, 263. [CrossRef]
58. Sun, T.; Lai, L.; Pei, J. Analysis of protein features and machine learning algorithms for prediction of druggable proteins. *Quant. Biol.* **2018**, *6*, 334–343. [CrossRef]
59. Iraji, M.S.; Tanha, J.; Habibinejad, M. Druggable protein prediction using a multi-canal deep convolutional neural network based on autocovariance method. *Comput. Biol. Med.* **2022**, *151*, 106276. [CrossRef]
60. Shoombuatong, W.; Schaduangrat, N.; Nikom, J. Empirical comparison and analysis of machine learning-based approaches for druggable protein identification. *EXCLI J.* **2023**, *22*, 915–927. [CrossRef]

61. Chilingaryan, G.; Abelyan, N.; Sargsyan, A.; Nazaryan, K.; Serobian, A.; Zakaryan, H. Combination of consensus and ensemble docking strategies for the discovery of human dihydroorotate dehydrogenase inhibitors. *Sci. Rep.* **2021**, *11*, 11417. [CrossRef]
62. Lam, P.C.-H.; Abagyan, R.; Totrov, M. Hybrid receptor structure/ligand-based docking and activity prediction in ICM: Development and evaluation in D3R Grand Challenge 3. *J. Comput. Mol. Des.* **2019**, *33*, 35–46. [CrossRef]
63. Sharo, C.; Zhai, T.; Huang, Z. Investigation of Potential Drug Targets Involved in Inflammation Contributing to Alzheimer's Disease Progression. *Pharmaceuticals* **2024**, *17*, 137. [CrossRef]
64. ICM User's Guide 5.6 3D Predict. Molsoft, LLC. Available online: https://molsoft.com/gui/3d-predict.html (accessed on 10 December 2023).
65. Agoni, C.; Olotu, F.A.; Ramharack, P.; Soliman, M.E. Druggability and drug-likeness concepts in drug design: Are biomodelling and predictive tools having their say? *J. Mol. Model.* **2020**, *26*, 120. [CrossRef]
66. Target Analysis with SiteMap and WaterMap. Schrodinger. Available online: https://learn.schrodinger.com/private/edu/release/current/Documentation/html/tutorials/target_analysis/target_analysis.htm (accessed on 5 May 2024).
67. Schrodinger. How SiteMap Evaluates Sites. Available online: https://learn.schrodinger.com/private/edu/release/current/Documentation/html/sitemap/sitemap_user_manual/sitemap_methodology_evaluate_sites.htm (accessed on 5 May 2024).
68. Zhou, W.; Yan, H. Alpha shape and Delaunay triangulation in studies of protein-related interactions. *Brief. Bioinform.* **2014**, *15*, 54–64. [CrossRef]
69. Thérien, M.; Skorey, K.; Zamboni, R.; Li, C.S.; Lau, C.K.; LeRiche, T.; Truong, V.L.; Waddleton, D.; Ramachandran, C. Synthesis of a novel peptidic photoaffinity probe for the PTP-1B enzyme. *Bioorganic Med. Chem. Lett.* **2004**, *14*, 2319–2322. [CrossRef] [PubMed]
70. Fleishman, S.J.; Whitehead, T.A.; Strauch, E.-M.; Corn, J.E.; Qin, S.; Zhou, H.-X.; Mitchell, J.C.; Demerdash, O.N.; Takeda-Shitaka, M.; Terashi, G.; et al. Community-Wide Assessment of Protein-Interface Modeling Suggests Improvements to Design Methodology. *J. Mol. Biol.* **2011**, *414*, 289–302. [CrossRef] [PubMed]
71. Khashan, R.; Zheng, W.; Tropsha, A. Scoring protein interaction decoys using exposed residues (SPIDER): A novel multibody interaction scoring function based on frequent geometric patterns of interfacial residues. *Proteins Struct. Funct. Bioinform.* **2012**, *80*, 2207–2217. [CrossRef] [PubMed]
72. Khashan, R. SPIDER—About. Available online: https://www.bioinformatics.org/spider/ (accessed on 5 May 2024).
73. Krass, E.; Zhai, T.; Huang, Z. Identification of Apolipoprotein E4 inhibitors for Alzheimer's Disease Therapy through a Large-Scale Virtual Screening. *IFAC-PapersOnLine* **2022**, *55*, 27–32. [CrossRef]
74. Zhang, F.; Graham, J.; Zhai, T.; Liu, Y.; Huang, Z. Discovery of MurA Inhibitors as Novel Antimicrobials through an Integrated Computational and Experimental Approach. *Antibiotics* **2022**, *11*, 528. [CrossRef]
75. Zhai, T.; Zhang, F.; Haider, S.; Kraut, D.; Huang, Z. An Integrated Computational and Experimental Approach to Identifying Inhibitors for SARS-CoV-2 3CL Protease. *Front. Mol. Biosci.* **2021**, *8*, 661424. [CrossRef]
76. Zhang, F.; Zhai, T.; Haider, S.; Liu, Y.; Huang, Z.J. Synergistic Effect of Chlorogenic Acid and Caffeic Acid with Fosfomycin on Growth Inhibition of a Resistant *Listeria monocytogenes* Strain. *ACS Omega* **2020**, *5*, 7537–7544. [CrossRef]
77. Mori, H.; Yoshino, Y.; Iga, J.-I.; Ochi, S.; Funahashi, Y.; Yamazaki, K.; Kumon, H.; Ozaki, Y.; Ueno, S.-I. Aberrant Expression of GABA-Related Genes in the Hippocampus of 3xTg-AD Model Mice from the Early to End Stages of Alzheimer's Disease. *J. Alzheimer's Dis.* **2023**, *94*, 177–188. [CrossRef]
78. Yang, L.-B.; Li, R.; Meri, S.; Rogers, J.; Shen, Y. Deficiency of Complement Defense Protein CD59 May Contribute to Neurodegeneration in Alzheimer's Disease. *J. Neurosci.* **2000**, *20*, 7505–7509. [CrossRef]
79. Shah, A.; Kishore, U.; Shastri, A. Complement System in Alzheimer's Disease. *Int. J. Mol. Sci.* **2021**, *22*, 13647. [CrossRef]
80. Jordan, K.L.; Koss, D.J.; Outeiro, T.F.; Giorgini, F. Therapeutic Targeting of Rab GTPases: Relevance for Alzheimer's Disease. *Biomedicines* **2022**, *10*, 1141. [CrossRef]
81. National Center for Biotechnology Information. PubChem Compound Summary for CID 62959, Trovafloxcin. 2024. Available online: https://pubchem.ncbi.nlm.nih.gov/compound/Trovafloxacin (accessed on 8 May 2024).
82. DRUGBANK Online. Trovafloxacin. Available online: https://go.drugbank.com/drugs/DB00685 (accessed on 8 May 2024).
83. ChEBI. CHEBI:9763—Trovafloxacin. Available online: https://www.ebi.ac.uk/chebi/searchId.do?chebiId=CHEBI:9763 (accessed on 8 May 2024).
84. DRUGBANK Online.Betrixaban. Available online: https://go.drugbank.com/drugs/DB12364 (accessed on 8 May 2024).
85. National Center for Biotechnology. PubChem Compound Summary for CID 10275777, Betrixaban. 2024. Available online: https://pubchem.ncbi.nlm.nih.gov/compound/Betrixaban (accessed on 8 May 2024).
86. DRUGBANK Online. Favipiravir. Available online: https://go.drugbank.com/drugs/DB12466 (accessed on 8 May 2024).
87. PubChem Compound Summary for CID 90311989. Bictegravir. National Center for Biotechnology Information. Available online: https://pubchem.ncbi.nlm.nih.gov/compound/90311989 (accessed on 15 October 2024).
88. National Center for Biotechnology. PubChem Compound Summary for CID 5339, Sulfasalazine. Available online: https://pubchem.ncbi.nlm.nih.gov/compound/Sulfasalazine (accessed on 8 May 2024).
89. National Center for Biotechnology Information. PubChem Compound Summary for CID 1983, Acetaminophen. Available online: https://pubchem.ncbi.nlm.nih.gov/compound/Acetaminophen (accessed on 8 May 2024).
90. Shaughnessy, A. *How Does Acetaminophen Work?* Tufts University School of Medicine: Boston, MA, USA, 2022.

91. Mines, P.D.; Byun, J.; Hwang, Y.; Patel, H.A.; Andersen, H.R.; Yavuz, C.T. Nanoporous networks as effective stabilisation matrices for nanoscale zero-valent iron and groundwater pollutant removal. *J. Mater. Chem. A* **2016**, *4*, 632–639. [CrossRef]
92. National Center for Biotechnology Information. PubChem Compound Summary for CID 5460875, Asparaginate. Available online: https://pubchem.ncbi.nlm.nih.gov/compound/5460875 (accessed on 15 October 2024).
93. National Center for Biotechnology Information. PubChem Compound Summary for CID 6267, Asparagine. Available online: https://pubchem.ncbi.nlm.nih.gov/compound/6267 (accessed on 15 October 2024).
94. National Center for Biotechnology Information. PubChem Compound Summary for CID 3830, Kinetin. Available online: https://pubchem.ncbi.nlm.nih.gov/compound/3830 (accessed on 15 October 2024).
95. National Center for Biotechnology Inofrmation. PubChem Compound Summary for CID 151075, Nepafenac. Available online: https://pubchem.ncbi.nlm.nih.gov/compound/151075 (accessed on 15 October 2024).
96. Lira, R.P.C.; Zanetti, F.R.; Fulco, E.A.M.; Chaves, F.R.P.; Pinto, A.P.d.C.; Arieta, C.E.L. Effect of preoperative use of topical prednisolone acetate, ketorolac tromethamine, nepafenac and placebo, on the maintenance of intraoperative mydriasis during cataract surgery: A randomized trial. *Indian J. Ophthalmol.* **2012**, *60*, 277. [CrossRef] [PubMed]
97. Stevens, B.; Allen, N.J.; Vazquez, L.E.; Howell, G.R.; Christopherson, K.S.; Nouri, N.; Micheva, K.D.; Mehalow, A.K.; Huberman, A.D.; Stafford, B.; et al. The Classical Complement Cascade Mediates CNS Synapse Elimination. *Cell* **2007**, *131*, 1164–1178. [CrossRef] [PubMed]
98. Rogers, J.; Cooper, N.R.; Webster, S.; Schultz, J.; McGeer, P.L.; Styren, S.D.; Civin, W.H.; Brachova, L.; Bradt, B.; Ward, P.; et al. Complement activation by beta-amyloid in Alzheimer disease. *Proc. Natl. Acad. Sci. USA* **1992**, *89*, 10016–10020. [CrossRef]
99. Walker, D.G.; McGeer, P.L. Complement gene expression in human brain: Comparison between normal and Alzheimer disease cases. *Mol. Brain Res.* **1992**, *14*, 109–116. [CrossRef]
100. Batista, A.F.; Khan, K.A.; Papavergi, M.-T.; Lemere, C.A. The Importance of Complement-Mediated Immune Signaling in Alzheimer's Disease Pathogenesis. *Int. J. Mol. Sci.* **2024**, *25*, 817. [CrossRef]
101. Hong, S.; Beja-Glasser, V.F.; Nfonoyim, B.M.; Frouin, A.; Li, S.; Ramakrishnan, S.; Merry, K.M.; Shi, Q.; Rosenthal, A.; Barres, B.A.; et al. Complement and microglia mediate early synapse loss in Alzheimer mouse models. *Science* **2016**, *352*, 712–716. [CrossRef]

targets

MDPI

Review

Aging in 3D: Organoid Systems as Models to Uncover Cellular Senescence and Therapeutic Targets Across Diseases

Shilpa Bisht [1,*], Paras Varshney [2] and Abhishek Gupta [3,*]

1 Cancer Genetics and Epigenetics, Oncology, Sidney Kimmel Comprehensive Cancer Center, Johns Hopkins University School of Medicine, CRB1, Baltimore, MD 21205, USA
2 Cupertino High School, Cupertino, CA 95014, USA; pvarshney200910@gmail.com
3 Department of Biophysics, Johns Hopkins University, 3400 North Charles Street, Baltimore, MD 21218, USA
* Correspondence: shilpa.mhg.bhu11@gmail.com (S.B.); agupt117@jh.edu (A.G.)

Abstract

Aging is a complex biological process characterized by progressive loss of cellular homeostasis, impaired regenerative capacity, and accumulation of senescent cells that collectively predispose tissues to disease. Traditional two-dimensional culture systems and animal models have provided valuable insights but fail to fully recapitulate the spatial organization, cellular heterogeneity, and microenvironmental cues of aging human tissues. Organoid technology—three-dimensional self-organizing structures derived from adult stem cells or pluripotent stem cells has emerged as a transformative platform to model aging in vitro. These mini-tissues retain the architecture, signaling dynamics, and lineage hierarchy of native organs, making them powerful systems to interrogate age-associated cellular phenotypes, DNA damage responses, and senescence programs. This review discusses how organoid models are advancing our understanding of aging biology across multiple organ systems, from the intestines and liver to the brain and lung. We highlighted key molecular pathways driving cellular senescence within organoids—including p16INK4a/p21CIP1 signaling, SASP activation, mitochondrial dysfunction, and epigenetic drift—and how these can be targeted to restore tissue homeostasis. We further discussed how organoids derived from aged tissues, induced pluripotent stem cells, and engineered oncogene systems reveal new therapeutic opportunities to modulate senescence in age-related disorders, cancer, and regenerative medicine. Finally, we discussed emerging integrative tools such as organoid co-cultures, single-cell omics, and senolytics drug screening that are expanding the potential of organoids as translational platforms for anti-aging and disease intervention.

Keywords: organoids; cellular senescence; aging; regenerative medicine; disease modeling; senolytics; molecular targets; 3D culture

1. Introduction

Aging is a universal biological process characterized by the progressive decline of tissue integrity, stem cell function, and regenerative capacity across organ systems [1,2]. As organisms age, cells accumulate DNA damage, undergo metabolic and epigenetic alterations, and experience chronic stress that together contribute to impaired homeostasis and heightened risk for disease [3,4]. A central hallmark linking aging to disease susceptibility is cellular senescence, a stable cell-cycle arrest program accompanied by widespread chromatin remodeling, mitochondrial dysfunction, and secretion of pro-inflammatory factors collectively known as the senescence-associated secretory phenotype (SASP) [5,6].

Although senescence initially acts as a tumor-suppressive mechanism, its chronic accumulation during aging contributes to tissue dysfunction, inflammation, fibrosis, and cancer initiation [7,8].

Traditional in vitro models such as two-dimensional (2D) monolayer cultures have provided foundational insights into senescence biology but fail to capture the three-dimensional (3D) architecture, multi-lineage complexity, and native signaling gradients of aging tissues [9]. Animal models, while physiologically relevant, are limited by species-specific differences and difficulty in parsing human-specific aging phenomena [10]. These limitations have driven the emergence of organoid technology as a transformative platform to model human aging and senescence in vitro. Organoids are self-organizing, stem cell-derived 3D structures that recapitulate key features of native tissues, including epithelial polarity, cellular heterogeneity, spatial organization, and lineage hierarchy [11]. Organoids derived from adult stem cells, pluripotent stem cells, or patient tissues have enabled researchers to study aging signatures such as stem cell exhaustion, impaired niche signaling, metabolic decline, altered stress responses, and senescence accumulation in controlled experimental settings [12]. Importantly, organoids retain donor-specific epigenetic and transcriptional landscapes, enabling direct comparison of young and aged tissues, modeling organismal aging, and identification of cell-intrinsic drivers of senescence [13]. Furthermore, organoids can be genetically engineered—using tools such as CRISPR/Cas9 to induce senescence through DNA damage, oncogene activation, telomere dysfunction, or mitochondrial stress, allowing mechanistic interrogation of senescence pathways including p53/p21, p16INK4a/Rb, and SASP regulation [14] (Table 1).

Table 1. Key cellular senescence pathways and markers studied using organoid models, their functional outcomes, therapeutic implications, and relevant references.

Pathway/Marker	Organoid Model	Functional Outcome	Therapeutic Implication	References
p16INK4a/Rb	Intestinal, hepatic	Cell cycle arrest, SASP induction	Senolytic targeting, tissue rejuvenation	[15–17]
p53/p21CIP1	Intestinal, cerebral	DNA damage response, apoptosis regulation	Chemotherapy response prediction, OIS modeling	[18]
mTOR signaling	Hepatic, pancreatic	Metabolic regulation, stem cell function	Rapamycin/mTOR modulators, stem cell rejuvenation	[19]
SASP factors (IL-6, IL-8, MMPs)	Multiple tissues	Paracrine tissue remodeling, inflammation	Senomorphic interventions	[20]
DNA damage/γ-H2AX	Intestinal, cerebral	Genomic instability, aging phenotype	Senescence biomarkers, therapeutic monitoring	[21]

Because senescence plays roles in diverse pathologies—spanning cancer, neurodegeneration, metabolic disorders, pulmonary fibrosis, and chronic inflammation—organoid systems have become powerful tools for targeted discovery and drug screening aimed at modulating aging and senescence [4,21]. Recent advances integrating organoids with single-cell omics, spatial transcriptomics, immune co-cultures, and organoid-on-chip platforms further enhance their ability to model the complex microenvironmental cues that govern aging in vivo [22–24]. By leveraging organoids as physiologically relevant, experimentally tractable models, researchers are gaining novel insights into human aging biology and accelerating the development of interventions aimed at improving health span and combating age-associated diseases.

https://doi.org/10.3390/targets4020012

In this review, we discuss how organoid systems provide unprecedented opportunities to dissect the mechanisms of aging and cellular senescence across diverse tissues, highlighting key molecular and cellular pathways uncovered through organoid-based studies. These platforms enable direct interrogation of epithelial-intrinsic aging programs, stem cell dysfunction, and microenvironmental regulation within physiologically relevant 3D contexts. We further examine how integration of organoid models with emerging technologies, including single-cell and spatial omics, functional genomics, and organoid-on-chip systems, has advanced our understanding of senescence heterogeneity and dynamics. Finally, we discuss emerging therapeutic strategies, including senolytic and senomorphic interventions, that are enabled and refined through organoid-based platforms, underscoring their potential to accelerate the translation of aging biology into targeted and personalized therapies.

2. Overview of Organoid Systems

Organoid technology has revolutionized in vitro modeling by enabling the generation of self-organizing, 3D structures that recapitulate key features of native tissues, including multicellular composition, spatial architecture, and functional signaling networks [25]. Organoids are typically derived from two major sources: adult stem cells (ASCs) residing in tissue-specific niches, and pluripotent stem cells (PSCs) such as embryonic stem cells (ESCs) or induced pluripotent stem cells (iPSCs) (Table 2). Each system offers unique advantages for modeling development, homeostasis, disease mechanisms, and aging (Figure 1).

Table 2. Organoid platforms for modeling aging-associated mechanisms and cellular senescence across tissues. This table summarizes major organoid systems, their stem cell sources, and the specific aging-related mechanisms they enable investigators to interrogate. Rather than broadly listing disease applications, the table highlights key hallmarks of aging—including stem cell exhaustion, genomic instability, mitochondrial dysfunction, altered niche signaling, and inflammaging—and outlines mechanistic questions that can be addressed using each organoid platform. These models facilitate age-stratified analysis and provide controlled systems to dissect epithelial-intrinsic aging processes, microenvironmental remodeling, and age-modified responses to oncogenic or metabolic stress.

Organoid Type	Stem Cell Source	Aging-Associated Mechanisms Modeled	Key Aging Questions Addressed	Representative Applications	References
Intestinal	Adult stem cells	Stem cell exhaustion; Wnt signaling decline; increased TGF-β signaling; SASP accumulation; genomic instability	How does aging alter niche responsiveness and regenerative capacity? How does epithelial age modify oncogene-induced senescence (OIS)?	Colorectal tumor initiation; inflammaging; regenerative decline	[26]
Hepatic	iPSC/Adult stem cells	DNA damage accumulation; metabolic inflexibility; mitochondrial dysfunction; impaired detoxification	Does mitochondrial decline drive age-associated metabolic disease? How does genomic instability influence fibrosis and cancer risk?	Liver fibrosis; metabolic syndrome; hepatocellular carcinoma	[27]
Cerebral	iPSC	Proteostasis decline; mitochondrial dysfunction; oxidative stress; neuroinflammatory signaling	How does aging alter neuronal resilience and protein aggregation? How does inflammaging contribute to neurodegeneration?	Alzheimer's disease; Parkinson's disease	[28]
Pancreatic	iPSC/Adult stem cells	β-cell senescence; oxidative stress; impaired insulin secretion; ER stress	How does aging reduce endocrine regenerative capacity? Can niche modulation reverse β-cell dysfunction?	Type 2 diabetes; pancreatic tumorigenesis	[29]

https://doi.org/10.3390/targets4020012

Table 2. *Cont.*

Organoid Type	Stem Cell Source	Aging-Associated Mechanisms Modeled	Key Aging Questions Addressed	Representative Applications	References
Renal	Adult stem cells/iPSC	Tubular senescence; oxidative stress; reduced repair capacity; fibrotic signaling	How does epithelial aging impair nephron regeneration? What is the role of senescence in fibrosis progression?	Chronic kidney disease; polycystic kidney disease	[30]
Multi-lineage/Co-culture systems	ASC/iPSC + immune/stromal cells	Inflammaging; immune–epithelial crosstalk; niche deterioration; SASP-mediated remodeling	How do age-associated immune changes influence epithelial stem cell fitness and tumor initiation?	Early tumorigenesis; fibrosis; chronic inflammatory disease	[31–33]

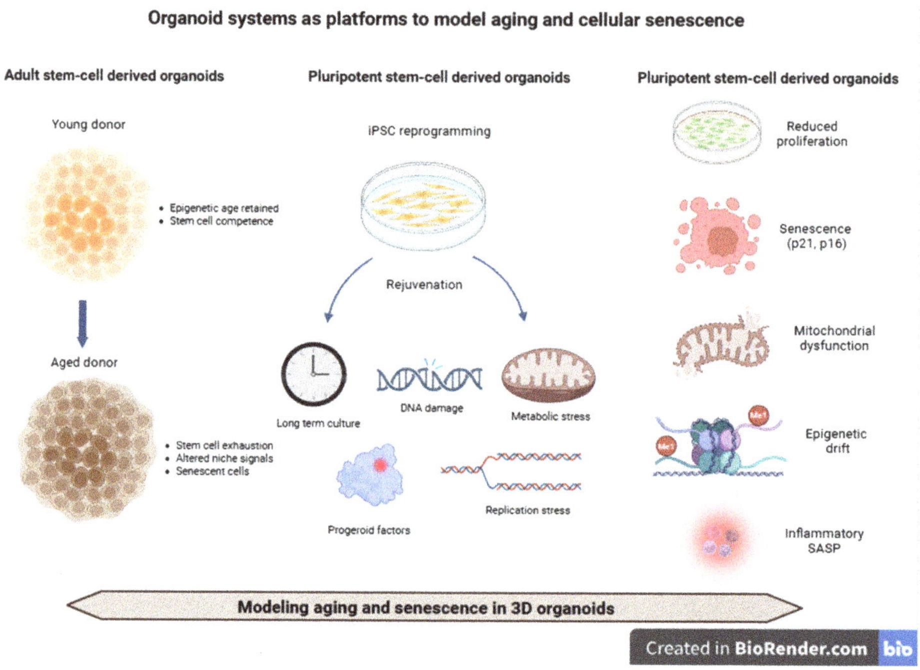

Figure 1. Organoid systems as platforms to model aging and cellular senescence. Organoids derived from adult stem cells (ASCs) and pluripotent stem cells (PSCs) provide complementary approaches to study aging-related processes. ASC-derived organoids retain donor-specific genetic, epigenetic, and aging-associated phenotypes, enabling direct interrogation of stem cell exhaustion, senescence accumulation, and altered niche signaling in aged tissues. In contrast, PSC-derived organoids recapitulate developmental programs and typically resemble fetal tissues but can be engineered to exhibit aging-like features through prolonged culture, DNA damage, metabolic stress, or progeroid manipulation. Together, these systems enable controlled modeling of key hallmarks of aging, including reduced regenerative capacity, mitochondrial dysfunction, genomic instability, epigenetic drift, and senescence-associated secretory phenotypes (SASP), within physiologically relevant 3D tissue contexts. https://app.biorender.com/illustrations/canvas-beta/69753460cde6a4242668b183 (accessed on 17 March 2026).

2.1. Adult Stem Cell-Derived Organoids

ASC-derived organoids are established from tissue-resident stem or progenitor cells that retain intrinsic instructive cues for lineage specification and tissue regeneration. These organoids closely mimic the physiology of the original tissue, including epithelial po-

https://doi.org/10.3390/targets4020012

larity, adult transcriptional identity, and organ-specific functional properties [34]. The reliance on defined niche signals such as Wnt, R-spondin, EGF, BMP inhibition, and Notch modulation enables the ex vivo reconstruction of stem cell behavior and allows precise manipulation of signaling pathways relevant to aging and disease. Importantly, ASC-derived organoids maintain donor-specific epigenetic signatures and aging phenotypes, making them particularly valuable for studying tissue-specific aging, stem cell exhaustion, and senescence accumulation.

2.2. Pluripotent Stem Cell-Derived Organoids

PSC-derived organoids recapitulate stages of embryonic development and are widely used to model organogenesis, developmental defects, and early-onset diseases [34]. These systems rely on stepwise differentiation protocols that mimic embryonic signaling pathways such as Wnt, FGF, TGF-β, and Hedgehog, leading to the generation of highly organized structures such as brain, retinal, intestinal, hepatic, and kidney organoids. Although PSC-derived organoids offer unparalleled developmental modeling capabilities, they often resemble fetal rather than adult tissues and lack aging-associated epigenetic marks, which can limit their application in aging research [35]. Recent strategies, including prolonged culture, DNA damage induction, and metabolic manipulation, have been used to induce aging-like features in PSC-derived organoids, expanding their utility for studying senescence [36].

2.3. Modeling Multicellular Complexity and the Tissue Niche

A critical advantage of organoid systems is their ability to capture aspects of the tissue niche and multicellular interactions. Many organoids retain intrinsic paracrine circuits such as Wnt-producing Paneth cells in intestinal organoids or hepatocyte–cholangiocyte interactions in liver organoids that are essential for maintaining tissue homeostasis and are known to deteriorate with age [37,38]. Advances in co-culture systems have enabled the inclusion of immune cells, fibroblasts, endothelial cells, and microbiota, providing a more physiologically relevant microenvironment to study aging-associated inflammation, stromal remodeling, and immune–epithelial crosstalk [38].

2.4. Comparison with Traditional 2D and Animal Models

Organoid systems overcome several inherent limitations of conventional 2D cell culture and animal models, making them powerful platforms for studying aging, senescence, and complex human diseases. Traditional 2D cultures lack the spatial organization, mechanical cues, and cellular heterogeneity necessary to faithfully model tissue architecture and age-associated functional decline. In contrast, organoids preserve 3D structure, stem cell hierarchies, and lineage relationships that more closely resemble native tissues, enabling investigation of niche-dependent regulation of stem cell behavior and tissue homeostasis [39,40]. Compared with animal models, organoids better capture human-specific aging signatures, including species-specific transcriptional programs, senescence-associated gene expression, and human-relevant drug responses. Their tractability, scalability, and compatibility with genome editing and high-throughput screening make organoids an indispensable platform for aging and senescence research.

Another important advantage of organoid systems is their compatibility with advanced experimental approaches. Organoids can be readily integrated with genome-editing technologies, high-throughput drug screening platforms, and functional genomics tools, enabling systematic interrogation of pathways regulating senescence, stem cell exhaustion, and tissue regeneration [41]. In addition, organoids are highly amenable to multi-omics and imaging-based analyses, including single-cell transcriptomics, epigenomics, spatial

https://doi.org/10.3390/targets4020012

profiling, and live-cell imaging, allowing detailed mechanistic dissection of aging processes at cellular and subcellular resolution [41,42].

Organoids also offer practical and ethical advantages compared with animal models. Their scalability and experimental tractability enable rapid hypothesis testing, precise genetic manipulation, and repeated sampling over time without the confounding effects of whole-organism variability or interspecies differences. While animal models remain essential for studying systemic physiology and organism-level aging, organoids provide a complementary and often more human-relevant system for dissecting cellular and molecular mechanisms underlying age-associated diseases.

Together, these features position organoid systems as an important bridge between simplified in vitro models and in vivo studies, enabling more accurate modeling of human tissue biology and improving the translational relevance of experimental findings.

2.5. Challenges and Limitations

Despite their transformative potential, organoid systems face several conceptual and technical challenges that are particularly relevant for modeling aging and cellular senescence. A major limitation lies in standardization and reproducibility, as organoid derivation efficiency, growth dynamics, and cellular composition can vary substantially depending on donor source, tissue type, passage number, and culture conditions [43].

Long-term culture, which is often required to model chronic aging processes, introduces additional concerns. Extended passaging can lead to culture-induced stress, clonal selection, epigenetic drift, and accumulation of somatic mutations, potentially mimicking or obscuring bona fide aging-associated phenotypes [44–46]. Distinguishing true chronological aging from in vitro–induced senescence, therefore, remains a major conceptual and technical challenge. To address this challenge, aging-associated stem cell exhaustion in organoid systems is defined using comparative, molecular, and functional criteria rather than reliance on single phenotypic markers [47]. In practice, importantly, these transcriptional and epigenetic features align with signatures observed in aged tissues in vivo and are retained across passages, supporting their intrinsic origin. In contrast, diffusion-limited hypoxia or nutrient stress in static cultures typically produces spatially restricted necrosis, acute metabolic stress responses, and reversible growth impairment. Assessment of hypoxia-responsive pathways, combined with the use of perfusion-based microfluidic systems, further aids in distinguishing culture-induced artifacts from stable, age-associated stem cell dysfunction. Moreover, cellular diversity and niche complexity are still incompletely captured in most organoid systems. Although recent co-culture approaches have enabled partial incorporation of immune cells, fibroblasts, and endothelial cells, maintaining stable, long-term multicellular interactions that resemble aged tissue environments remain technically demanding [48–50]. Faithfully modeling age-associated inflammatory niches and SASP requires careful control to avoid non-physiological activation states.

Although animal models remain the preferred systems for studying systemic immune aging, endocrine crosstalk, and pharmacokinetics, the organoid field has made substantial advances in addressing these complex biological processes. Organoid platforms provide a physiologically relevant and dynamically controlled microenvironment that more closely recapitulates human tissue architecture and function compared to traditional in vivo models. Recent developments integrating organoids with immune components, hormonal signaling cues, and microfluidics-based systems have expanded their utility in modeling systemic interactions [51]. Organoids have been successfully applied to study diseases such as cancer and inflammatory bowel disease, while enabling precise experimental manipulation and high-resolution functional readouts. Importantly, their patient-derived

https://doi.org/10.3390/targets4020012

nature allows for individualized disease modeling and therapeutic testing, thereby offering significant potential for personalized medicine approaches.

Emerging technologies provide promising avenues to overcome these limitations. Synthetic and tunable biomaterials allow precise control of matrix stiffness and biochemical signaling, both of which change with age and influence senescence programs. Microfluidic organoid-on-chip platforms enable dynamic perfusion, controlled gradients, and inter-organ communication, improving physiological relevance and long-term stability [52]. Together, these advances are expected to enhance the fidelity and translational relevance of organoid-based aging models. Microfluidic organoid-on-chip platforms enable dynamic perfusion, mechanical cues, and controlled nutrient gradients that improve physiological relevance compared to static cultures [52,53]. However, challenges remain, including device complexity, limited scalability for high-throughput screening, variability in fabrication, and incomplete integration of systemic immune and endocrine components. Addressing these bottlenecks will be essential for translating fluidic organoid systems into robust platforms for aging biology and pharmacokinetic modeling.

2.6. Organoid Systems as Models of Age-Associated Stem Cell Decline, Niche Deterioration, and Senescence Accumulation

A central hallmark of tissue aging is the progressive decline in adult stem cell function, often termed stem cell exhaustion, characterized by reduced self-renewal capacity, altered lineage commitment, and diminished regenerative potential [54,55]. Organoid systems derived from adult stem cells provide a unique platform to model this decline in a controlled ex vivo setting, as they retain donor-specific transcriptional and epigenetic signatures reflective of chronological aging [56]. Comparative studies using organoids derived from young and aged donors have demonstrated persistent reductions in clonogenic efficiency, impaired responsiveness to Wnt and Notch signaling, and increased expression of cell-cycle inhibitors such as CDKN2A/p16INK4a in aged stem cell populations. These intrinsic defects are maintained across passages under identical culture conditions, supporting the interpretation that they reflect bona fide age-associated stem cell dysfunction rather than transient culture stress.

Beyond cell-intrinsic decline, aging is accompanied by deterioration of the stem cell niche, including altered paracrine signaling, extracellular matrix remodeling, and chronic low-grade inflammation ("inflammaging") [57]. Organoid platforms allow experimental manipulation of niche components—such as Wnt ligands, R-spondin, BMP inhibitors, and stromal-derived factors—to causally test how aged microenvironmental cues contribute to stem cell exhaustion [57]. For example, restoration of Wnt signaling in aged intestinal organoids partially rescues proliferative defects, demonstrating that niche deterioration actively drives age-associated functional decline rather than representing an irreversible intrinsic program. Synthetic and tunable extracellular matrices further enable modeling of age-related changes in matrix stiffness and mechanotransduction, both of which influence YAP/TAZ activity, metabolic fitness, and senescence induction.

Organoid systems also permit direct interrogation of senescence accumulation within epithelial stem and progenitor compartments. Aging-associated senescent cells are characterized by stable cell-cycle arrest, activation of p16INK4a/p21 pathways, persistent DNA damage signaling, mitochondrial dysfunction, and acquisition of a SASP. In aged tissue-derived organoids, increased expression of SASP cytokines, accumulation of γH2AX foci, and reduced regenerative capacity have been observed even in the absence of systemic immune or endocrine inputs. This isolation of epithelial-intrinsic aging programs allows discrimination between organism-level influences and cell-autonomous senescence mechanisms.

https://doi.org/10.3390/targets4020012

Importantly, organoids enable distinction between chronological aging signatures and culture-induced stress artifacts. Diffusion-limited hypoxia and nutrient gradients in static 3D cultures can induce stress responses that superficially resemble aging phenotypes [58]. However, comparison of young and aged donor-derived organoids under identical environmental conditions, combined with transcriptomic alignment to in vivo aged tissues, allows identification of conserved aging-specific gene expression programs [59]. Integration of microfluidic perfusion systems further reduces diffusion-related confounders and strengthens modeling fidelity. Together, these approaches position organoids as mechanistically powerful systems to dissect the interplay between intrinsic stem cell decline, niche deterioration, and senescence accumulation that underlie tissue aging.

3. Modeling Cellular Aging in Organoids

Aging is accompanied by progressive and interconnected changes in cellular physiology, stem cell competence, and the integrity of tissue-regenerating niches, ultimately impairing organ function and increasing susceptibility to chronic diseases, degenerative disorders, and cancer. Across multiple tissues, including the intestine, liver, pancreas, and brain, adult stem cells exhibit age-associated declines in proliferative capacity, altered lineage commitment, impaired regenerative responses, and heightened sensitivity to metabolic, inflammatory, and oncogenic stress [60]. Historically, mechanistic insights into tissue aging have largely relied on bulk tissue profiling or in vivo animal models, which, while invaluable, obscure epithelial-intrinsic aging programs due to confounding influences from systemic factors such as immune aging, circulating cytokines, endocrine signaling, and changes in the stromal compartment [61]. These approaches also limit the ability to disentangle cause-and-effect relationships between intrinsic stem cell aging and extrinsic niche alterations. Organoid systems overcome many of these constraints by providing a controlled, genetically tractable, and physiologically relevant ex vivo platform to interrogate how aging reshapes the intrinsic properties of adult stem cells and their immediate microenvironment.

ASC-derived organoids, in particular, have emerged as powerful tools for modeling cellular aging because they retain donor-specific genetic, epigenetic, and transcriptional signatures, including age-associated chromatin states and transcriptional programs [62]. Intestinal organoids derived from aged donors exhibit reduced clonogenicity, delayed growth kinetics, altered differentiation trajectories, and impaired responsiveness to niche-derived signals such as Wnt and Notch, faithfully recapitulating in vivo aging phenotypes [63,64]. Similarly, hepatic and pancreatic organoids have revealed age-related declines in regenerative capacity and increased stress responses that are otherwise masked in whole-organ analyses [65].

Organoid systems further enable high-resolution dissection of hallmarks of cellular aging, including genomic instability, mitochondrial dysfunction, altered nutrient sensing, epigenetic drift, and activation of senescence-associated transcriptional programs. Importantly, age-associated cellular senescence can be studied in a cell-type–specific manner within organoids, allowing direct assessment of senescence markers, cell-cycle arrest, and SASP without confounding contributions from systemic inflammation [66].

In addition to modeling intrinsic aging, organoids offer unique opportunities to investigate how aging alters stem cell–niche interactions. Changes in paracrine signaling, extracellular matrix composition, and mechanical properties of the niche are increasingly recognized as drivers of stem cell dysfunction during aging. By precisely manipulating niche factors and matrix properties ex vivo, organoids enable causal testing of how aged microenvironments contribute to stem cell exhaustion, lineage bias, and senescence accumulation. Collectively, organoid-based aging models provide an unprecedented platform to

https://doi.org/10.3390/targets4020012

dissect epithelial-intrinsic aging mechanisms, separate them from systemic influences, and establish direct links between cellular aging, senescence, and disease susceptibility. These systems, therefore, serve as a critical foundation for identifying age-specific vulnerabilities and therapeutic targets across diverse age-associated diseases.

A major strength of organoid systems in aging research is their ability to preserve cell-autonomous hallmarks of aging. Organoids derived from aged tissues display increased DNA damage accumulation, reduced clonogenic potential, impaired Wnt responsiveness, elevated reactive oxygen species, and molecular signatures of senescence, including upregulated Cdkn2a/p16INK4a and SASP-associated cytokines [67,68]. These changes persist across passages, indicating that stem cell–intrinsic aging programs are retained ex vivo and can be dissected in isolation from immune or endocrine cues. Conversely, organoids derived from young donors provide a rejuvenated baseline against which age-dependent phenotypes can be directly compared, enabling systematic evaluation of pathways such as Wnt, Notch, EGFR, BMP, and YAP that are altered during aging. Organoid-based aging models also allow the controlled introduction of genetic, epigenetic, environmental, and pharmacological perturbations to identify pathways that drive epithelial decline. Studies using intestinal and hepatic organoids have shown that age-related loss of niche-derived Wnt and R-spondin signaling restricts stem cell renewal, while aged stem cells exhibit amplified TGF-β signaling and chromatin remodeling that limit regenerative fitness [69,70]. Additionally, long-term culture systems using induced iPSC-derived organoids enable modeling of premature or accelerated aging syndromes by manipulating telomere maintenance, lamin processing, mitochondrial function, or DNA repair mechanisms [71,72].

Importantly, organoid platforms also facilitate the study of interactions between aging and disease susceptibility, including infection, inflammation, metabolic dysfunction, and cancer initiation. For example, aged intestinal organoids exhibit increased sensitivity to inflammatory cytokines and heightened transcriptional activation of NF-κB and JAK–STAT signaling upon stimulation, which may contribute to age-associated inflammatory bowel disease risk [73]. Similarly, aged colon or pancreatic organoids show differential responses to oncogenic mutations (e.g., *BRAFV600E*, *KRASG12D*), reflecting age-dependent differences in MAPK activation, senescence induction, and differentiation trajectories [74,75]. These discoveries underscore how organoid models can reveal context-specific vulnerabilities and therapeutic targets that arise with epithelial aging. Together, these advances position organoids as a transformative platform for aging research, enabling precise mechanistic dissection of epithelial decline, niche dysregulation, senescence programs, and age-related disease initiation. By integrating organoid-based assays with single-cell transcriptomics, CRISPR editing, and high-Throughput compound screening, researchers can now identify reshapes tissue homeostasis and identify strategies to restore regenerative capacity or prevent age-associated disorders.

4. Senescence Pathways Revealed Through Organoid Models

Cellular senescence is an irreversible growth-arrest program triggered by diverse stressors including DNA damage, telomere attrition, oxidative stress, chromatin disruption, and oncogene activation. While senescence acts as an essential tumor-suppressive barrier, its long-term persistence contributes to chronic inflammation, impaired tissue regeneration, and age-related pathologies through the secretion of pro-inflammatory cytokines, chemokines, and matrix-remodeling enzymes collectively termed the SASP. Despite its importance, studying senescence in vivo has historically been challenging due to tissue heterogeneity, systemic influences, and limited access to early-stage senescent cells. Organoid systems now provide a unique opportunity to interrogate the cell-intrinsic and microenvironment-modulated mechanisms of senescence with unprecedented resolution.

https://doi.org/10.3390/targets4020012

Importantly, the term "senescence marker" refers not to a single definitive molecule but to a constellation of phenotypic and molecular features that collectively define a senescent state. Canonical hallmarks include stable cell-cycle arrest, activation of the p53/p21CIP1 and/or p16INK4a/Rb pathways, persistent DNA damage signaling (e.g., γH2AX foci), senescence-associated β-galactosidase (SA-β-gal) activity, chromatin remodeling, and acquisition of a SASP [76,77]. However, no single marker is sufficient to establish senescence, as individual features such as p16INK4a expression or transient DNA damage responses may also occur in non-senescent stress contexts [78]. In organoid systems, rigorous identification of senescence therefore relies on combinatorial and functional criteria—including durable growth arrest and sustained secretory remodeling—rather than isolated marker expression. Moreover, the relative prominence of specific senescence features may vary across tissue types and organoid platforms, necessitating context-dependent interpretation rather than universal extrapolation.

Organoids faithfully recapitulate epithelial lineage hierarchies and retain the capacity to undergo senescence in response to intrinsic aging, environmental insults, or targeted genetic perturbations. Aged intestinal and hepatic organoids exhibit canonical senescence features including elevated p16INK4a and p21CIP1 expression, increased DNA damage markers (γH2AX), mitochondrial dysfunction, and diminished regenerative capacity [79]. Importantly, these phenotypes persist ex vivo, demonstrating that stem cell-intrinsic aging programs are maintained independent of systemic inflammatory or hormonal inputs. This ability to isolate epithelial-intrinsic drivers makes organoids valuable for dissecting transcriptional, epigenetic, and metabolic mechanisms underlying age-associated senescence (Figure 2).

Organoid models have also been instrumental in uncovering how oncogene-induced senescence (OIS) manifests within epithelial tissues. Organoid-based co-culture systems incorporating fibroblasts, immune cells, or engineered extracellular matrices are illuminating the microenvironmental regulation of senescence. Senescent epithelial organoids secrete SASP factors such as IL-6, IL-8, MMPs, and GDF15, which can remodel the niche, alter stromal behavior, and reinforce senescence in neighboring cells [80,81]. Conversely, niche-derived signals such as Wnt, R-spondin, BMP, and TGF-β profoundly influence the induction, maintenance, or reversal of senescence. For instance, loss of Wnt/R-spondin responsiveness in aged intestinal organoids accelerates senescence, whereas restoring niche cues improves proliferation and reduces stress signaling. These systems highlight how epithelial senescence is a dynamic interplay between intrinsic damage responses and extrinsic niche determinants.

Organoid models also support therapeutic discovery efforts, particularly for senolytic (senescence-clearing) and senomorphic (SASP-modulating) interventions. High-throughput drug screening using organoids derived from aged tissues or OIS-induced systems has identified candidate compounds targeting PI3K/AKT, BCL-2 family proteins, and metabolic vulnerabilities unique to senescent cells [82]. Additionally, CRISPR-based genetic screens in organoids are uncovering regulators of senescence escape, including chromatin modifiers, MAPK modulators, and DNA repair factors, which represent potential targets for preventing early neoplastic transformation [83,84].

Collectively, organoids have emerged as a powerful platform for investigating the full spectrum of senescence biology—from intrinsic aging-related decline and oncogene-induced senescence to microenvironment-driven remodeling and therapeutic targeting. By enabling mechanistic dissection at cellular and tissue levels, organoid models are accelerating the discovery of senescence-associated vulnerabilities relevant to aging, cancer initiation, and regenerative failure.

https://doi.org/10.3390/targets4020012

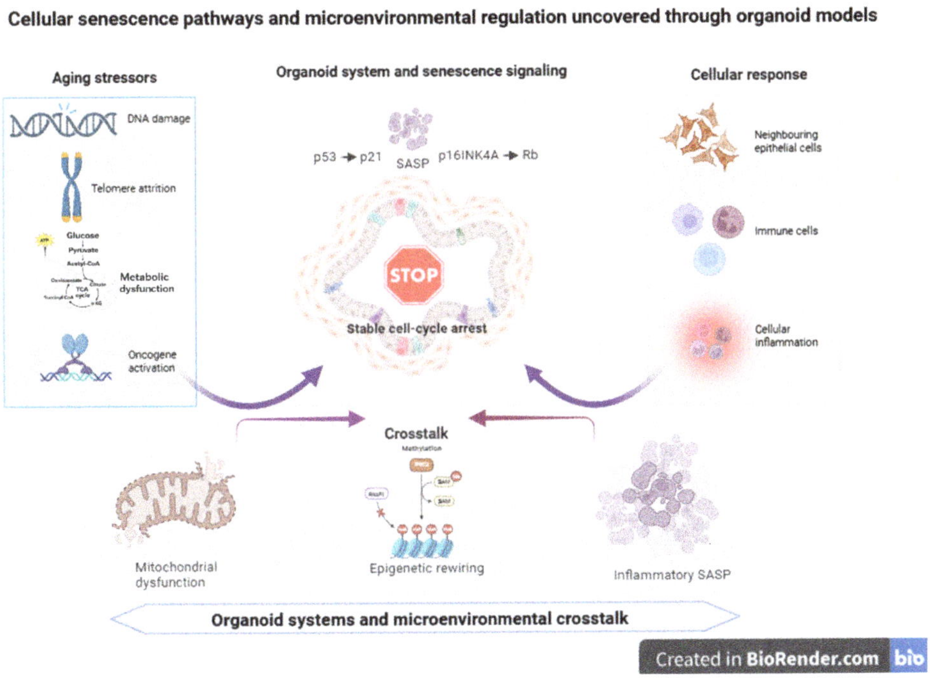

Figure 2. Cellular senescence pathways and microenvironmental regulation uncovered through organoid models. Organoid systems enable detailed dissection of senescence-inducing stimuli, core signaling pathways, and downstream functional consequences within epithelial tissues. Aging-associated stressors, including DNA damage, telomere attrition, metabolic dysfunction, and oncogene activation, converge on canonical p53/p21 and p16INK4a/Rb tumor suppressor pathways to induce stable cell-cycle arrest. Senescent cells exhibit mitochondrial dysfunction, chromatin remodeling, and secretion of senescence-associated secretory phenotype (SASP) factors that remodel the local microenvironment. Organoid co-culture systems further reveal dynamic crosstalk between senescent epithelial cells, stromal fibroblasts, and immune components, highlighting context-dependent roles of senescence in tissue dysfunction, inflammation, and early tumorigenesis. https://app.biorender.com/illustrations/canvas-beta/697543791e425e67f693e5c2 (accessed on 17 March 2026).

5. Organoids as Platforms for Target Discovery

Organoid systems have emerged as powerful and versatile platforms for therapeutic target discovery, effectively bridging the gap between reductionist mechanistic studies and clinically relevant translational research. Their defining features—3D architecture, preservation of tissue-specific stem cell hierarchies, and retention of donor-specific genetic and epigenetic information—enable the interrogation of disease-relevant pathways within a physiologically meaningful context that is difficult to achieve in conventional 2D cultures. These attributes are particularly advantageous for studying aging-related processes, where subtle, cell-intrinsic alterations interact with niche-dependent signals to shape disease susceptibility and therapeutic response. A key strength of organoid-based systems lies in their ability to capture inter-individual heterogeneity, which is especially pronounced in aging tissues. Organoids derived from young versus aged donors, or from patients with age-associated diseases, retain differences in transcriptional programs, epigenetic landscapes, and stress responses, allowing systematic identification of age-specific vulnerabilities and candidate targets [85]. This donor-specific fidelity makes organoids well suited for

https://doi.org/10.3390/targets4020012

uncovering pathways that drive stem cell exhaustion, senescence accumulation, and altered differentiation trajectories during aging.

Organoids also provide a tractable platform to study OIS and early tumorigenic events, processes that are increasingly recognized as critical windows for therapeutic intervention. By introducing defined genetic alterations using genome-editing approaches, such as CRISPR/Cas9, organoids enable direct comparison of oncogenic signaling outcomes in young versus aged epithelial contexts. These studies have revealed that aging can fundamentally alter cellular responses to oncogenic insults, shifting the balance between growth arrest, senescence, and sustained proliferation, thereby uncovering age-dependent therapeutic targets and resistance mechanisms. Importantly, organoids are highly compatible with functional genomics and high-throughput-screening approaches. Pooled or arrayed genetic perturbation screens can be combined with phenotypic readouts such as growth dynamics, lineage composition, senescence markers, and secretory profiles to identify regulators of aging-associated decline and disease progression [86]. Similarly, drug screening in organoid cultures has enabled the discovery of compounds that selectively target senescent cells, modulate senescence-associated secretory phenotypes, or restore aged stem cell function, thereby accelerating the translation of aging biology into therapeutic strategies [87].

Beyond epithelial-intrinsic targets, emerging organoid co-culture systems incorporating immune cells, fibroblasts, and endothelial components allow interrogation of microenvironment- and niche-dependent vulnerabilities. These platforms are particularly relevant for identifying targets linked to inflammaging, stromal remodeling, and immune evasion—hallmarks of age-associated diseases that are poorly modeled in traditional systems. By enabling controlled manipulation of both epithelial and niche compartments, organoids facilitate the discovery of combinatorial and context-dependent therapeutic targets. Collectively, organoid-based platforms provide a scalable and human-relevant framework for identifying therapeutic targets across aging-associated diseases. Their ability to integrate genetic perturbation, functional screening, and patient-specific biology positions organoids as a cornerstone technology for precision target discovery and drug development in the context of aging and cellular senescence.

5.1. Genetic and Molecular Target Identification

Organoids allow precise manipulation of genes, signaling pathways, and microenvironmental factors to identify targets that regulate aging, senescence, and disease progression. CRISPR/Cas9 and RNA interference (RNAi) screens in organoids have been applied to systematically interrogate the roles of tumor suppressors, oncogenes, DNA repair genes, and chromatin regulators in organ-specific contexts. For example, genome-wide CRISPR screens in intestinal organoids have identified modulators of Wnt, Notch, and MAPK signaling that govern stem cell renewal and resistance to senescence, revealing potential targets for rejuvenation strategies and early cancer prevention. Moreover, organoid models derived from aged tissues or engineered to harbor oncogenic mutations (e.g., *BRAFV600E*, *KRASG12D*) enable direct identification of context-dependent vulnerabilities. Studies in aged colon organoids demonstrated that Wnt pathway augmentation can restore proliferative capacity in senescent stem cells, whereas MAPK inhibition selectively suppresses the growth of oncogene-primed cells. These insights are invaluable for designing targeted therapies that selectively modulate senescent or pre-malignant cells while sparing healthy tissue.

https://doi.org/10.3390/targets4020012

5.2. High-Throughput Drug Screening and Senotherapeutics

Organoid cultures are highly compatible with high-throughput drug screening, allowing systematic evaluation of small molecules, biologics, and senolytic or senomorphic compounds in physiologically relevant human tissue models [88]. For instance, screening libraries of BCL-2 inhibitors, PI3K/AKT modulators, or metabolic regulators in aged or senescence-induced organoids have identified candidate compounds that selectively eliminate senescent cells or suppress SASP-associated inflammatory signaling [89]. Importantly, these findings are directly translating to preclinical or clinical studies, as organoids preserve human-specific responses that may not be captured in murine models. In addition to senolytic compounds, organoids have been used to discover regenerative targets that rejuvenate aged tissue stem cells. Manipulating niche factors (e.g., Wnt agonists, BMP inhibitors) in aged intestinal organoids restores stem cell proliferation and mitigates senescence phenotypes, demonstrating the utility of organoids in identifying actionable pathways for tissue rejuvenation.

5.3. Multi-Omics Integration for Target Discovery

The combination of organoid technology with multi-omics profiling—including single-cell transcriptomics, spatial transcriptomics, proteomics, and epigenomics—enables high-resolution mapping of senescence and aging signatures [90]. Such analyses can reveal gene regulatory networks, post-translational modifications, and metabolic dependencies that are unique to senescent or oncogene-primed cells. For example, single-cell RNA sequencing of aged intestinal organoids has uncovered subpopulations of stem cells with elevated inflammatory signaling, chromatin remodeling factors, and SASP-associated genes, which can serve as candidate therapeutic targets [91]. Spatial transcriptomics further allows the identification of cell–cell interaction networks and niche-derived signals that modulate senescence or stem cell function, enhancing the precision of target selection.

5.4. Personalized and Precision Medicine Applications

Patient-derived organoids (PDOs) extend the potential of organoid-based target discovery into personalized medicine. By deriving organoids from healthy, aged, or diseased tissues of individual patients, researchers can identify patient-specific vulnerabilities and screen for drugs tailored to the individual's molecular and epigenetic context [92]. For aging-related interventions, PDOs allow evaluation of senolytic or rejuvenating compounds across genetically diverse human backgrounds, thereby improving the translational relevance of preclinical discoveries.

5.5. Aging-Dependent Vulnerabilities, Senolytic Strategies, and Age-Modulated Oncogenic Responses

A major strength of organoid systems in aging research is their ability to reveal aging-dependent cellular vulnerabilities that are not evident in young tissues or conventional culture systems. Aging is accompanied by alterations in nutrient sensing, mitochondrial metabolism, chromatin organization, and stress response pathways, generating distinct molecular dependencies in aged epithelial stem cells. Organoids derived from aged donors retain these transcriptional and epigenetic features, enabling identification of pathways selectively required for survival or regeneration in aged tissues. For example, aged intestinal organoids display impaired Wnt responsiveness and increased inflammatory signaling, highlighting context-specific therapeutic targets [93].

Organoid models allow tissue-specific testing of senolytic compounds targeting anti-apoptotic regulators such as BCL-2 family proteins or PI3K/AKT signaling, while distinguishing effects on senescent versus functional stem cell compartments. In parallel, modulation of niche factors—including Wnt and BMP signaling—has demonstrated partial

https://doi.org/10.3390/targets4020012

restoration of regenerative capacity in aged organoids. Organoids also uniquely enable interrogation of age-modulated oncogenic responses. The same oncogenic mutation can elicit distinct outcomes depending on epithelial age, with aged tissues exhibiting altered thresholds for senescence induction, DNA damage responses, and proliferative escape [94,95]. These findings position aging as an active modifier of oncogenic fitness landscapes and therapeutic vulnerability. By directly comparing young and aged epithelial backgrounds under controlled genetic perturbations, organoids provide a powerful platform to define age-specific targets and optimize intervention strategies.

5.6. Limitations and Future Directions

Despite their growing utility in therapeutic target discovery, organoid-based systems face several conceptual and technical limitations that must be carefully considered when modeling aging, senescence, and disease. One major challenge is variability and reproducibility, as differences in donor material, tissue source, culture medium composition, extracellular matrix properties, and passage number can substantially influence organoid growth behavior, differentiation state, and drug responses [96]. Such variability complicates cross-study comparisons and poses challenges for standardization in large-scale screening and translational pipelines.

A related limitation concerns the dependence on animal-derived matrices, such as Matrigel, which exhibit batch-to-batch variability and contain ill-defined biochemical and mechanical cues. These matrices can obscure subtle aging-associated phenotypes and confound interpretation of therapeutic effects, particularly when studying mechanosensitive pathways or senescence-related changes in extracellular matrix signaling [97]. The development and adoption of synthetic, tunable biomaterials represent an important future direction for improving reproducibility and physiological relevance.

Another fundamental constraint is the lack of systemic and circulatory inputs. Conventional organoid cultures are largely isolated from endocrine signals, immune surveillance, neuronal inputs, and vascular perfusion, all of which profoundly shape aging trajectories and therapeutic responses in vivo [98]. In particular, the absence of functional vasculature limits long-term culture and drug delivery kinetics, while the lack of immune components restricts the study of immunosenescence, inflammaging, and immune-mediated clearance of senescent cells.

Recent technological advances are beginning to address these challenges. Microfluidic organoid-on-chip platforms enable dynamic perfusion, controlled nutrients and oxygen gradients, and real-time monitoring of tissue responses, thereby improving physiological relevance and long-term stability [99]. Similarly, the generation of vascularized organoids through endothelial co-culture or self-assembly approaches has enhanced tissue maturation, metabolic function, and drug response fidelity [100]. These innovations are particularly promising for modeling chronic aging processes and pharmacokinetics in a human-relevant context. In parallel, advanced co-culture and assembled systems incorporating immune cells, fibroblasts, and other stromal components are expanding the scope of organoid-based target discovery. Such systems enable interrogation of aging-associated inflammatory niches, senescence-associated secretory phenotypes, and stromal–epithelial crosstalk that drive disease progression and therapeutic resistance [101]. Integrating these multicellular platforms with single-cell and spatial multi-omics approaches will further enhance mechanistic resolution and target validation.

Looking ahead, the convergence of organoid technology with functional genomics, artificial intelligence–driven analysis, and patient-derived biobanks is expected to transform target discovery in aging and age-related diseases. Large-scale, standardized organoid repositories representing diverse ages and disease states will enable systematic

https://doi.org/10.3390/targets4020012

identification of age-specific vulnerabilities and facilitate precision therapeutic development [102,103]. In summary, while organoid-based platforms for target discovery face important limitations, ongoing advances in bioengineering, co-culture strategies, and systems integration are rapidly expanding their capabilities. By enabling genetic perturbation, high-throughput drug screening, multi-omics profiling, and patient-specific modeling within a physiologically relevant framework, organoids are poised to play a central role in identifying interventions aimed at restoring tissue homeostasis, mitigating cellular senescence, and preventing age-associated disease and early tumorigenesis.

6. Applications for Disease Modeling

Organoid technology has transformed our ability to model human disease by recapitulating tissue architecture, cellular heterogeneity, and functional responses in vitro. Unlike traditional 2D cultures or animal models, organoids retain patient-specific genetic, epigenetic, and phenotypic characteristics, providing unprecedented opportunities to study aging, cancer, metabolic disorders, infectious diseases, and neurodegeneration [104]. This section highlights key applications of organoids in modeling human disease.

6.1. Modeling Age-Associated Diseases

Aging is a major risk factor for multiple pathologies, including epithelial decline, fibrosis, and cancer. Organoids derived from aged tissues preserve stem cell-intrinsic aging features such as reduced proliferative potential, altered lineage allocation, DNA damage accumulation, and SASP expression. Intestinal organoids from aged donors, for example, have revealed diminished Wnt responsiveness and increased TGF-β signaling, highlighting potential targets for rejuvenation or regenerative therapies [105]. Similarly, hepatic and pancreatic organoids capture age-related declines in metabolic function, stress responses, and regenerative capacity, providing models to study organ-specific aging mechanisms (Table 3).

Table 3. Organoid-based models for investigating aging-associated diseases table summarizes key organoid platforms used to study age-related diseases and highlights the aging-associated mechanisms that can be investigated using these systems.

Disease Area	Organoid Model	Aging-Associated Mechanism Studied	Key Insights from Organoid Studies
Colorectal cancer	Intestinal organoids	Stem cell exhaustion, altered Wnt signaling, oncogene-induced senescence	Reveals age-dependent susceptibility to tumor initiation and altered responses to oncogenic mutations
Liver diseases	Hepatic organoids	Mitochondrial dysfunction, metabolic decline, genomic instability	Enables study of metabolic aging and fibrosis progression
Neurodegenerative diseases	Cerebral organoids	Proteostasis decline, oxidative stress, neuroinflammation	Model age-related protein aggregation and neuronal vulnerability
Diabetes and metabolic disorders	Pancreatic organoids	β-cell senescence, oxidative stress, metabolic dysregulation	Allows investigation of endocrine aging and impaired insulin secretion
Chronic kidney disease	Renal organoids	Tubular senescence, inflammatory signaling	Provides insight into impaired regenerative capacity in aging kidneys
Infectious diseases	Intestinal/airway organoids	Age-related immune decline and epithelial barrier dysfunction	Enables study of increased infection susceptibility in aged tissues

https://doi.org/10.3390/targets4020012

6.2. Cancer Modeling

Patient-derived and genetically engineered organoids are widely used to study tumor initiation, progression, and therapy resistance. In colorectal, pancreatic, gastric, and liver cancers, organoids accurately recapitulate histological features, driver mutations, and signaling pathway dependencies. Importantly, organoids derived from aged or oncogene-primed tissues demonstrate age-dependent differences in responses to mutations, revealing how aging reshapes tumor suppressive mechanisms, senescence induction, and microenvironmental interactions. These models enable dissection of OIS, identification of escape pathways, and preclinical testing of targeted therapies. Importantly, aging alters not only mutation frequency but also the cellular response to oncogenic activation. In aged epithelial organoids, oncogene induction (e.g., *BRAFV600E* or *KRASG12D*) produces distinct senescence thresholds, stress signaling dynamics, and proliferative outcomes compared with young counterparts, reflecting age-dependent remodeling of tumor suppressive barriers. These findings indicate that aging reshapes oncogenic fitness landscapes, influencing whether cells undergo durable senescence, apoptosis, or malignant progression. Organoid systems uniquely enable age-stratified genetic perturbation experiments within controlled human epithelial contexts, allowing direct interrogation of how chronological aging modifies tumor initiation trajectories.

6.3. Infectious Disease Modeling

Organoids have become powerful tools for studying host–pathogen interactions. Human intestinal, airway, hepatic, and brain organoids have been used to model viral, bacterial, and parasitic infections, capturing tissue-specific infection patterns, immune responses, and pathogen-induced cellular remodeling [106]. For example, intestinal organoids have been used to study enteric viruses like rotavirus and SARS-CoV-2, providing insights into epithelial tropism, replication dynamics, and antiviral responses while preserving donor-specific variability [107]. Aging organoids further allow examination of age-dependent susceptibility to infection, reflecting increased vulnerability observed in older populations.

6.4. Metabolic and Genetic Disorders

Organoids derived from patients with monogenic or metabolic diseases faithfully recapitulate disease-specific phenotypes and facilitate functional assays for drug screening. Hepatic organoids from patients with alpha-1 antitrypsin deficiency, familial hypercholesterolemia, or Wilson's disease exhibit defective protein processing, lipid accumulation, or copper metabolism, providing platforms to test gene-corrective therapies or pharmacological interventions. Similarly, pancreatic and kidney organoids have been used to model cystic fibrosis, diabetes, and polycystic kidney disease, allowing evaluation of therapeutic responses in a patient-specific context.

6.5. Neurodegenerative Diseases

Cerebral organoids derived from human iPSCs allow modeling of neurodevelopmental and neurodegenerative disorders, including Alzheimer's disease, Parkinson's disease, and microcephaly. These organoids capture critical features such as neuronal subtype specification, cortical layering, synapse formation, and disease-relevant protein aggregation. Integration with aging or stress-inducing protocols enables modeling of late-onset neurodegenerative processes, which are difficult to replicate in conventional cell cultures or animal models.

https://doi.org/10.3390/targets4020012

6.6. Immune and Microbiome Interactions

Co-culture of organoids with immune cells, fibroblasts, or microbiota components enables the study of tissue–immune interactions and host–microbe dynamics. Immune-competent intestinal and lung organoids have been used to examine inflammatory responses, immune-mediated tissue damage, and the impact of aging on immune surveillance [108,109]. Similarly, microbial colonization of intestinal organoids allows investigation of microbiome-driven metabolic and inflammatory alterations, revealing mechanisms by which aging and dysbiosis contribute to disease susceptibility.

6.7. Mechanistic Modeling of Age-Associated Disease Pathways

Beyond modeling individual diseases, organoid systems provide a platform to dissect shared mechanisms that drive age-associated pathologies across tissues. Aging is characterized by genomic instability, mitochondrial dysfunction, altered nutrient sensing, epigenetic drift, and chronic low-grade inflammation, which collectively increase disease susceptibility. Organoids derived from aged tissues retain many of these intrinsic hallmarks, enabling mechanistic analysis of how accumulated cellular damage translates into organ-specific dysfunction. For example, aged intestinal and hepatic organoids exhibit increased DNA damage signaling and reduced repair capacity, reflecting age-related genomic instability. Similarly, pancreatic and liver organoids demonstrate mitochondrial and metabolic alterations that correlate with diminished regenerative potential and stress resilience. Because organoids isolate epithelial-intrinsic aging from systemic influences, they allow direct testing of whether such metabolic and genomic defects are primary drivers of pathology. Aging is accompanied by progressive accumulation of somatic mutations, telomere attrition, and impaired DNA repair capacity, leading to persistent activation of DNA damage response pathways and increased genomic instability [110]. In aged organoid systems, elevated γH2AX foci, altered ATM/ATR signaling, and reduced repair efficiency have been observed under identical culture conditions, indicating intrinsic age-associated defects rather than acute experimental damage [111]. These comparative young-versus-aged organoid studies uniquely allow separation of chronic genomic decline from transient stress-induced DNA damage, thereby positioning organoids as reductionist platforms to interrogate how accumulated genomic instability drives stem cell exhaustion and tumor initiation.

Mitochondrial dysfunction is another central hallmark of aging, characterized by impaired oxidative phosphorylation, increased reactive oxygen species (ROS) production, and altered metabolic flexibility. Aged pancreatic, hepatic, and intestinal organoids exhibit metabolic rewiring and reduced regenerative fitness consistent with in vivo aging phenotypes. Because organoids isolate epithelial compartments from systemic inflammatory and endocrine confounders, they allow direct testing of whether mitochondrial dysfunction acts as a primary driver of stem cell decline or as a secondary consequence of organismal aging. Such mechanistic dissection is difficult to achieve in whole-animal models where systemic variables obscure cell-intrinsic effects.

Organoid systems also model inflammaging, a persistent inflammatory state that contributes to fibrosis, tumorigenesis, and impaired regeneration. Cytokine exposure or immune co-culture approaches reveal how sustained NF-κB and JAK–STAT activation promotes stem cell dysfunction and senescence accumulation [112]. Importantly, comparative studies of young versus aged organoids show that identical genetic or environmental insults produce divergent outcomes depending on epithelial age, including altered senescence thresholds and proliferative responses. These findings underscore that aging actively reshapes disease trajectories rather than simply increasing disease incidence. By integrating intrinsic stem cell decline, niche deterioration, and inflammatory remodeling, organoid systems position aging as a mechanistic driver of pathology rather than a passive background

https://doi.org/10.3390/targets4020012

factor. An additional systemic mechanism increasingly recognized as a hallmark of aging is clonal hematopoiesis of indeterminate potential (CHIP), characterized by age-associated expansion of hematopoietic stem cell clones harboring mutations in genes such as DNMT3A, TET2, or ASXL1 [113–115]. These mutant clones contribute to chronic low-grade inflammation, altered immune surveillance, and increased risk of cardiovascular disease, metabolic dysfunction, and malignancy. Although organoid systems do not recapitulate hematopoietic clonal expansion directly, they provide a controlled platform to model the downstream consequences of CHIP-associated inflammatory environments [116]. Co-culture approaches incorporating mutant macrophages or exposure to cytokine profiles characteristic of CHIP (e.g., elevated IL-1β, IL-6) enable interrogation of how systemic inflammatory signaling influences epithelial stem cell fitness, senescence accumulation, and tumor initiation. In this context, organoids can serve as reductionist systems to dissect how hematopoietic aging indirectly accelerates dysfunction across non-hematopoietic organs [117].

6.8. Advantages of Organoid Systems for Modeling Aging and Disease

Organoid systems offer several distinct advantages over conventional 2D cell culture and animal models, positioning them as uniquely powerful platforms for studying aging, senescence, and complex human diseases. A primary advantage is the retention of human-specific tissue characteristics, including species-specific transcriptional programs, signaling dependencies, and epigenetic landscapes that are often poorly conserved in animal models. This human fidelity is particularly important for aging research, as many aging-associated pathways, immune responses, and drug sensitivities differ substantially between humans and commonly used model organisms. Unlike traditional 2D cultures, which impose artificial polarity and uniform exposure to nutrients and growth factors, organoids preserve 3D architecture, stem cell hierarchies, and cellular heterogeneity that more closely resemble native tissues. This structural and cellular complexity allows organoids to model lineage dynamics, cell–cell interactions, and niche-dependent regulation of stem cell behavior—features that are essential for understanding age-associated declines in tissue regeneration and the emergence of disease states. Advances in genome editing and scalable culture formats have made it possible to perform targeted and genome-wide perturbation studies directly in human organoids, uncovering functional dependencies that are often missed in 2D systems. Similarly, drug screening in organoids provides more predictive assessments of therapeutic efficacy and toxicity by capturing tissue-specific responses and inter-individual variability.

Another key advantage is the compatibility of organoids with multi-omics and advanced imaging-based approaches, enabling deep mechanistic dissection of complex phenotypes [118]. Single-cell transcriptomics, epigenomics, proteomics, and spatial profiling can be integrated with live-cell imaging and lineage tracing to resolve how aging and disease processes unfold at cellular and subcellular resolution within intact tissue-like structures. While animal models remain indispensable for studying whole-organism physiology, organoids provide a complementary and often superior platform for dissecting human-specific mechanisms underlying multifactorial diseases, including cancer, degenerative disorders, and age-associated pathologies. Together, these advantages make organoids particularly well suited for modeling complex diseases in which human-specific physiology, cellular interactions, and dynamic stem cell behavior play central roles. By bridging the gap between simplified in vitro systems and in vivo models, organoids enable more accurate mechanistic insights and improve the translational relevance of experimental findings.

https://doi.org/10.3390/targets4020012

7. Emerging Technologies Specifically Enabling Mechanistic Dissection of Aging and Senescence in Organoids

Emerging technologies have transformed organoid systems from static tissue mimetics into dynamic platforms for dissecting the mechanisms of aging and cellular senescence. Rather than serving solely as structural replicas of tissues, modern organoid models now enable longitudinal, mechanistic, and age-stratified interrogation of stem cell exhaustion, senescence heterogeneity, inflammaging, and age-modified oncogenic responses. The integration of high-resolution multi-omics, genome engineering, bioengineering approaches, and computational modeling has shifted organoid-based aging research from descriptive phenotyping toward causal and predictive analysis of age-associated decline [119,120].

One major technological advance is the application of single-cell and spatial multi-omics to organoid systems. Single-cell transcriptomics, epigenomics, and chromatin accessibility profiling allow precise mapping of age-associated cellular states, lineage trajectories, and senescence programs within heterogeneous organoid populations [121]. Spatially resolved transcriptomic and proteomic approaches further preserve architectural context, enabling the study of how aging alters cell–cell interactions, niche signaling, and spatial organization within tissues [122]. Together, these approaches dissect how aging manifests at the level of individual cells and tissue microdomains. Advances in live imaging and lineage tracing have also transformed the study of aging in organoids. Long-term time-lapse imaging, combined with fluorescent reporters and inducible genetic labeling, allows direct observation of stem cell dynamics, clonal competition, and senescence onset over extended culture periods [123]. These tools are particularly valuable for distinguishing gradual, age-related functional decline from acute stress responses or culture-induced artifacts.

The incorporation of microfluidic and organoid-on-chip technologies represents another critical innovation. These platforms enable controlled perfusion, mechanical stimulation, and real-time monitoring of tissue responses, addressing key limitations of static organoid cultures such as hypoxia, nutrient gradients, and waste accumulation. For aging research, organoid-on-chip systems facilitate modeling of chronic stress, long-term drug exposure, and inter-organ communication, offering a more faithful approximation of in vivo aging environments. Emerging bioengineering approaches, including synthetic and tunable extracellular matrices, allow precise control of mechanical stiffness, ligand presentation, and matrix remodeling—parameters that are increasingly recognized as key regulators of cellular aging and senescence. By decoupling biochemical and biophysical cues, these systems enable causal testing of how age-associated changes in the tissue microenvironment drive stem cell dysfunction and disease susceptibility.

In parallel, genome engineering and functional genomics tools have been adapted for use in organoids to systematically interrogate aging-associated pathways. CRISPR-based perturbation strategies, including pooled loss-of-function, gain-of-function, and epigenome-editing screens, allow high-throughput identification of regulators of senescence, stress resistance, and regenerative decline in a tissue-specific context [124]. Integration of these approaches with phenotypic and multi-omic readouts is accelerating the discovery of actionable targets in aging biology. Finally, advances in computational modeling and artificial intelligence are beginning to play an important role in organoid-based aging research. Machine learning approaches applied to high-content imaging, single-cell datasets, and drug screening outputs enable pattern recognition, trajectory inference, and predictive modeling of aging phenotypes and therapeutic responses [125]. These computational frameworks are poised to enhance data integration and translational decision-making in complex, multidimensional organoid datasets.

Together, these emerging technologies are transforming organoids into sophisticated platforms for studying aging as a dynamic, multifactorial process (Table 4). By enabling

https://doi.org/10.3390/targets4020012

precise manipulation, longitudinal analysis, and systems-level integration, they substantially enhance the power of organoid models to uncover mechanisms of aging, identify therapeutic targets, and predict intervention outcomes.

Table 4. Emerging technologies advancing aging research in organoid systems. This table summarizes key experimental and computational technologies that enable mechanistic investigation of aging and senescence in organoid models.

Technology	Application in Organoid Systems	Aging-Related Questions Addressed
Single-cell RNA sequencing	Identifies transcriptional heterogeneity in organoids	What cell populations drive aging and senescence?
Spatial transcriptomics	Maps spatial cell–cell interactions within organoids	How does aging alter tissue architecture and niche signaling?
CRISPR functional genomics	Genome-wide perturbation screening	Which genes regulate senescence, DNA repair, and stem cell decline?
Organoid-on-chip/microfluidics	Perfusion, nutrient gradients, inter-organ communication	How do chronic stress and vascular decline influence aging tissues?
Advanced live-cell imaging	Long-term monitoring of stem cell dynamics	How does senescence develop over time in aging tissues?
Metabolomics and mitochondrial assays	Measurement of metabolic activity	How does metabolic decline drive aging phenotypes?
Artificial intelligence and machine learning	Analysis of multi-omics and imaging data	Can aging trajectories and drug responses be predicted?

7.1. *Single-Cell and Spatial Omics*

In the context of aging research, single-cell and spatial omics technologies are particularly powerful for resolving senescence heterogeneity, age-dependent lineage bias, and stem cell exhaustion trajectories within organoid systems [125]. Single-cell RNA sequencing (scRNA-seq), single-cell ATAC-seq, and spatial transcriptomics have revolutionized our understanding of cellular heterogeneity in organoids. These technologies allow dissection of age-associated transcriptional and epigenetic changes at the resolution of individual cells, uncovering subpopulations of stem cells, progenitors, and differentiated cells with distinct aging signatures [126]. For example, scRNA-seq in aged intestinal organoids has identified stem cell subsets with elevated inflammatory signaling and diminished Wnt responsiveness, pinpointing potential targets for rejuvenation [127]. Spatial transcriptomics further enables mapping of microenvironmental interactions and niche-derived signals that influence senescence and tissue decline, providing a more holistic understanding of tissue aging.

7.2. *CRISPR-Based Functional Genomics*

For aging biology, CRISPR-based perturbation strategies enable direct testing of genes that regulate senescence thresholds, DNA repair decline, mitochondrial fitness, and niche responsiveness in aged versus young organoids [128]. CRISPR/Cas9 technologies facilitate targeted genetic manipulation within organoids, enabling high-throughput functional screens to identify regulators of aging, senescence, and stress responses. Pooled or arrayed CRISPR screens allow interrogation of thousands of genes, revealing pathways that control stem cell proliferation, senescence induction, and DNA repair. For instance, CRISPR-mediated knockout of Wnt modulators or DNA damage response genes in aged intestinal organoids can accelerate or reverse senescence, providing mechanistic insight and highlighting potential therapeutic targets [129].

https://doi.org/10.3390/targets4020012

7.3. Microfluidics and Organ-on-Chip Platforms

Because aging is profoundly influenced by chronic stress exposure, altered perfusion, and systemic inflammation, microfluidic organoid-on-chip platforms are uniquely positioned to model long-term aging-associated microenvironmental remodeling. Integration of organoids with microfluidic devices—so-called "organ-on-chip" systems—enhances physiological relevance by introducing dynamic fluid flow, nutrient gradients, mechanical forces, and vascular-like perfusion [130]. These platforms enable modeling of age-associated changes in tissue perfusion, shear stress responses, and nutrient metabolism, which are critical for studying aging in vivo-like contexts. For example, microfluidic intestinal organoids allow investigation of age-dependent epithelial barrier function, stem cell dynamics, and responses to inflammatory or metabolic stressors under controlled flow conditions.

Beyond single-tissue modeling, recent advances in organoid-on-chip engineering have enabled incorporation of endothelial layers, perfusable microvascular networks, and immune components, thereby improving tissue maturation and functional stability. Such systems facilitate more physiologically relevant drug exposure profiles and allow partial modeling of pharmacokinetic and pharmacodynamic parameters that are otherwise absent in static organoid cultures [131]. Multi-organ chip platforms further extend this concept by linking different organoid types (e.g., liver–gut, liver–brain, or immune–epithelium systems) to simulate aspects of inter-organ communication, systemic inflammation, and metabolic crosstalk that contribute to organismal aging. These developments are particularly significant for aging research, where chronic low-grade inflammation, altered nutrient sensing, and vascular decline collectively shape tissue vulnerability and therapeutic response.

Despite these advances, several bottlenecks currently limit the widespread adoption of fluidic organoid platforms. Device fabrication and operation remain technically complex, often requiring specialized engineering expertise and limiting accessibility across laboratories. Variability in microfluidic chip design and materials introduces challenges in reproducibility and standardization, complicating cross-study comparisons. Additionally, scalability and throughput remain lower than conventional organoid cultures, restricting large-scale screening applications [132]. While incorporation of vascular and immune components has improved physiological relevance, fully recapitulating systemic endocrine regulation, circulating immune aging, and whole-body pharmacokinetics remain challenging. Long-term stability of multi-cellular co-culture systems under continuous flow conditions also requires further optimization. Addressing these technical and translational challenges through standardized platforms, modular chip design, and integration with high-throughput workflows will be critical for realizing the full potential of fluidic organoid systems in aging and therapeutic modeling.

7.4. Imaging and Live-Cell Monitoring

Live-cell imaging approaches are especially valuable for aging research, as they enable longitudinal visualization of gradual functional decline, clonal competition, and progressive senescence accumulation in individual organoids. Advanced imaging techniques, including confocal microscopy, light-sheet microscopy, and high-content live-cell imaging, permit real-time monitoring of aging processes in organoids. These approaches enable visualization of senescence markers, DNA damage foci, mitochondrial dynamics, autophagy, and proliferation in three dimensions over time [133]. Time-lapse imaging of aged organoids has revealed dynamic heterogeneity in senescence induction, niche remodeling, and stem cell replacement, highlighting mechanisms of tissue resilience and decline [134].

https://doi.org/10.3390/targets4020012

7.5. Metabolic and Functional Assays

Given that mitochondrial dysfunction and metabolic inflexibility are central hallmarks of aging, integration of organoids with metabolic profiling platforms provides direct functional assessment of age-associated bioenergetic decline [135]. Aging is tightly linked to alterations in metabolism and mitochondrial function. Organoids are increasingly used with metabolomics, mitochondrial flux analysis, and bioenergetic profiling to quantify age-associated changes in energy homeostasis and reactive oxygen species generation. These assays allow correlation of functional decline with molecular changes, facilitating the identification of metabolic vulnerabilities and targets for intervention.

7.6. Artificial Intelligence and Computational Modeling

In aging-focused organoid studies, AI-driven integration of multi-dimensional datasets enables reconstruction of senescence trajectories, prediction of regenerative potential, and identification of age-specific therapeutic vulnerabilities [136]. Integration of organoid data with machine learning and computational modeling enables predictive analysis of aging trajectories, drug responses, and senescence induction. High-dimensional datasets from single-cell omics, imaging, and functional assays can be analyzed to identify biomarkers of aging, predict regenerative potential, and optimize therapeutic interventions. Computational models also allow simulation of organoid behavior under different perturbations, improving experimental design and accelerating discovery. Recent advances in AI and integrative multi-omics are rapidly transforming organoid-based aging research from descriptive profiling toward predictive and mechanistic modeling [137]. Machine-learning frameworks applied to scRNA-seq, single-cell ATAC-seq, spatial transcriptomics, and high-content imaging datasets now enable identification of senescence-associated cellular states, inference of lineage trajectories, and reconstruction of gene regulatory networks at unprecedented resolution [138]. In organoid systems, AI-driven image analysis has been used to quantify morphological aging phenotypes, classify drug responses, and predict functional decline based on multidimensional phenotypic signatures. Integration of transcriptomic, epigenomic, proteomic, and metabolomic layers further allows mapping of age-dependent regulatory circuits and identification of context-specific therapeutic vulnerabilities. Importantly, computational modeling approaches, including deep learning and graph-based network inference, are being leveraged to predict senescence induction, stem cell exhaustion trajectories, and responsiveness to senolytic or senomorphic compounds in PDOs. These integrative strategies position organoids not only as experimental platforms but also as data-rich systems for systems-level modeling of human aging and therapeutic response.

7.7. Multi-Organ and Co-Culture Systems

Multi-organ and co-culture systems are particularly relevant for aging research, as they allow modeling of inflammaging, immune–epithelial crosstalk, and inter-organ metabolic communication that shape organismal aging. Emerging co-culture and multi-organoid platforms facilitate the study of inter-organ interactions and systemic aging effects. For example, co-culturing intestinal organoids with immune cells or fibroblasts enables modeling of age-associated inflammation and niche deterioration, while coupling liver and pancreatic organoids in microfluidic systems allows investigation of metabolic crosstalk during aging [139]. These approaches offer a more integrative perspective on organism-level aging processes. Collectively, these technologies reposition organoids as mechanistic systems for studying aging as a dynamic, multi-layered process rather than a static phenotype. By enabling age-stratified perturbation, longitudinal monitoring, and systems-level integration,

https://doi.org/10.3390/targets4020012

emerging tools enhance the ability of organoid models to uncover causal drivers of stem cell exhaustion, senescence accumulation, and age-modified disease trajectories.

8. Therapeutic and Translational Potential of Organoid Models

Organoid systems extend beyond their utility as experimental models of aging and disease biology to serve as powerful platforms for therapeutic development and translational research. By faithfully recapitulating key aspects of human tissue architecture, cellular heterogeneity, and patient-specific genetic and epigenetic contexts, organoids enable more predictive assessment of drug efficacy, toxicity, and resistance mechanisms than traditional preclinical models [140]. These attributes are particularly valuable for aging-associated diseases, where subtle cell-intrinsic defects and context-dependent signaling critically shape therapeutic outcomes. One of the most impactful translational applications of organoid technology is in preclinical drug discovery and validation. PDOs have been shown to reliably mirror clinical drug responses across multiple disease settings, including colorectal, pancreatic, gastric, and liver cancers, as well as non-malignant disorders. In the context of aging, organoids enable screening for compounds that restore aged stem cell function, modulate senescence-associated pathways, or selectively eliminate senescent cells, thereby facilitating the identification of candidate senotherapeutics with improved human relevance [141]. Organoid models also provide a robust framework for precision medicine, allowing therapeutic strategies to be tailored to individual patients based on organoid-derived functional responses. Biobanked organoids generated from diverse patient populations preserve inter-individual variability and can be used to predict treatment sensitivity, optimize drug combinations, and stratify patients for clinical trials. This approach is particularly promising for age-associated diseases, where heterogeneity in biological aging trajectories often leads to variable treatment responses.

Beyond pharmacological testing, organoids are increasingly being explored in regenerative and cell-based therapies. Advances in long-term expansion, genetic stability, and differentiation control have enabled the generation of transplantable organoid-derived tissues for epithelial repair and regeneration. For aging tissues, organoid-based regenerative strategies offer the potential to replenish dysfunctional stem cell pools, restore tissue integrity, and counteract degenerative decline, although significant challenges related to engraftment, integration, and safety remain. Organoids further serve as valuable platforms for evaluating therapeutic toxicity and off-target effects in a tissue-specific manner. Compared with animal models, human organoids provide more accurate assessments of drug-induced injury, particularly for tissues such as liver, intestine, and kidney that are highly susceptible to age-related functional decline and adverse drug reactions [142]. This capability is especially relevant in elderly populations, where altered pharmacokinetics and increased vulnerability necessitate careful evaluation of therapeutic safety.

Looking forward, the integration of organoid models with genome editing, multi-omics profiling, and computational modeling is expected to further enhance their translational impact. Combining organoid-based functional data with patient clinical information and machine learning approaches may enable more accurate prediction of therapeutic responses and identification of biomarkers linked to aging and senescence [143]. As organoid technologies continue to mature, they are poised to play an increasingly central role in bridging basic aging research and clinical intervention. In summary, organoid systems represent a transformative platform for therapeutic discovery and translational research. By enabling human-relevant drug testing, personalized treatment strategies, and regenerative approaches within a controlled yet physiologically meaningful framework, organoids hold substantial promise for improving outcomes in aging-related diseases and advancing precision medicine.

https://doi.org/10.3390/targets4020012

8.1. Drug Discovery and High-Throughput Screening

Organoids are increasingly employed in high-throughput drug screening to identify small molecules, biologics, and senolytic compounds that selectively target aged, diseased, or oncogene-primed cells. Patient-derived and genetically engineered organoids enable the evaluation of therapeutic efficacy, toxicity, and resistance mechanisms in a human-relevant system [140]. For example, aged intestinal and hepatic organoids have been used to screen senolytic compounds such as BCL-2 inhibitors, which selectively eliminate senescent cells and attenuate SASP-mediated inflammation, highlighting strategies to ameliorate age-related tissue dysfunction [144].

Furthermore, organoids derived from cancer patients allow personalized drug screening, identifying therapeutics that target tumor-specific vulnerabilities while sparing normal tissue. This approach has been particularly successful in colorectal, pancreatic, and liver cancers, where patient-derived organoids faithfully predict responses to chemotherapy, targeted therapy, and immunotherapy. Integrating organoids with automated platforms, robotics, and multi-well assays enhances throughput and reproducibility, accelerating drug discovery pipelines.

8.2. Senolytic and Senomorphic Therapeutics

Senescence contributes to aging-related tissue dysfunction, chronic inflammation, and cancer progression. Organoid models provide a controlled platform for testing senolytic (cell-eliminating) and senomorphic (SASP-modulating) therapies. Genetic or pharmacologic induction of senescence in organoids allows the identification of compounds that specifically target senescent cells without affecting proliferating stem cells [145]. Additionally, organoids allow mechanistic evaluation of how therapeutic interventions modulate secretory phenotypes and tissue microenvironmental interactions, providing insights into potential systemic effects.

8.3. Gene and Cell-Based Therapies

Organoids are ideal platforms for evaluating gene editing and regenerative strategies. CRISPR/Cas9-mediated correction of disease-causing mutations in patient-derived organoids has been successfully demonstrated in cystic fibrosis, metabolic liver disorders, and monogenic intestinal diseases [146,147]. Edited organoids can then be expanded and transplanted into preclinical models to restore tissue function, offering a roadmap for future cell-based regenerative therapies. Additionally, co-culture with immune cells or vascular networks allows testing of gene- and cell-based therapies under physiologically relevant conditions.

8.4. Modeling Therapy Response and Resistance

Organoids provide a predictive platform to study drug resistance mechanisms in aged and diseased tissues. Aging and senescence can alter cellular metabolism, stress responses, and signaling pathways, affecting therapeutic responses. Organoid models allow systematic evaluation of these factors, enabling identification of combinatorial therapies to overcome resistance [148,149]. This is particularly relevant in oncology, where organoids derived from aged or genetically complex tumors can predict patient-specific responses to chemotherapeutics, targeted inhibitors, and immune-modulating agents.

8.5. Integration with Personalized and Precision Medicine

PDOs are at the forefront of personalized medicine, enabling tailoring of therapeutic interventions based on individual patient biology. By capturing donor-specific genetics, epigenetics, and aging phenotypes, PDOs allow evaluation of therapy efficacy, toxicity,

https://doi.org/10.3390/targets4020012

and optimal dosing in vitro before clinical administration. This approach is particularly valuable in elderly patients or those with comorbidities, where standard therapies may be less effective or more toxic.

8.6. Future Directions and Clinical Translation

Despite their promise, several challenges remain for translating organoid-based discoveries into clinical practice, including long-term culture stability, scalability, reproducibility, and integration with immune and vascular systems [150]. Furthermore, combining organoid models with multi-omics, imaging, and computational modeling will accelerate identification of actionable targets and predictive biomarkers for aging-related diseases, cancer, and regenerative interventions (Figure 3).

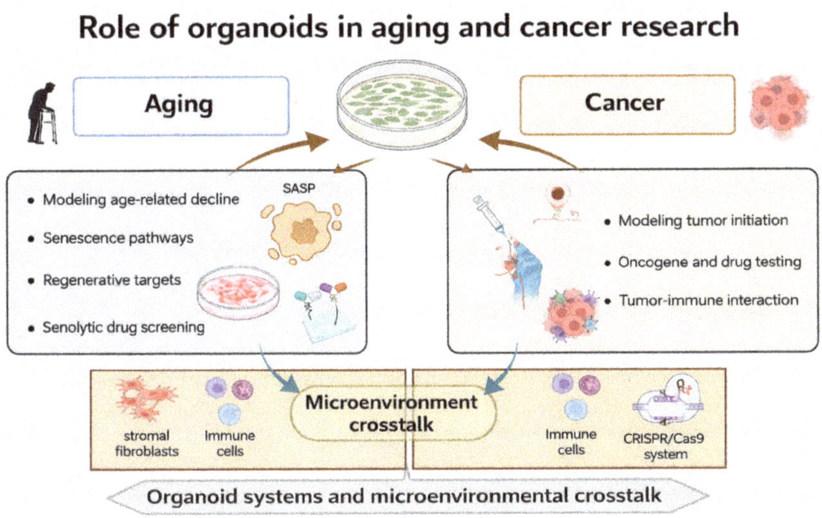

Figure 3. Applications of 3D organoid systems in aging and cancer research. 3D organoid models enable physiologically relevant investigation of aging-associated tissue decline and cancer biology. In aging research, organoids derived from young and aged tissues are used to study stem cell exhaustion, cellular senescence, and regenerative capacity, and to screen senolytic and senomorphic compounds. In cancer research, patient-derived and genetically engineered organoids facilitate modeling of tumor initiation and progression, functional interrogation of oncogenic drivers, and assessment of drug responses and resistance. Across both contexts, integration of stromal and immune components allows analysis of microenvironmental crosstalk that influences tissue homeostasis and disease evolution, supporting translational target discovery and precision medicine applications. https://app.biorender.com/illustrations/canvas-beta/69759fff023658e17ca5bb58 (accessed on 17 March 2026).

In summary, organoid technology offers an unparalleled platform for therapeutic discovery, testing, and translational research. By integrating genetic manipulation, high-throughput screening, personalized modeling, and regenerative approaches, organoids provide a versatile system to bridge basic biological insights with clinical applications. Their use promises to improve treatment strategies for age-associated pathologies, cancer, metabolic disorders, and other human diseases, accelerating the move toward precision medicine.

https://doi.org/10.3390/targets4020012

9. Challenges, Limitations, and Future Perspectives

While organoid technology has revolutionized the study of aging, senescence, and disease biology, several challenges and limitations remain that need to be addressed to fully realize their translational potential. At the same time, ongoing technological innovations provide promising avenues to overcome these barriers and expand the scope of organoid research.

9.1. Technical and Biological Limitations

1. Limited vascularization and immune components: Traditional organoids lack functional vasculature and immune cell populations, which are critical for modeling systemic aging, inflammatory responses, and tissue regeneration [151]. The absence of vascular perfusion limits nutrient and oxygen delivery, impacting long-term culture stability and cellular metabolism.
2. Incomplete tissue maturation: Organoids often resemble fetal or immature tissue rather than fully adult tissue, which can limit their ability to accurately model late-onset diseases or age-associated phenotypes [152]. Strategies to accelerate maturation, such as growth factor modulation, mechanical cues, or co-culture with stromal cells, are under investigation.
3. Donor and culture variability: Inter-donor heterogeneity and differences in culture conditions (e.g., matrix composition, growth factors, oxygen levels) can lead to variable organoid formation efficiency and phenotype reproducibility [153]. Standardized protocols are essential to reduce experimental variability, especially for high-throughput drug screening and translational applications.
4. Limited lifespan and senescence artifacts: Prolonged culture can introduce artificial senescence or genetic drift, complicating interpretation of age-related phenotypes [154]. Careful optimization of passage number and culture duration is crucial for maintaining physiologically relevant aging features.
5. The widespread reliance on animal-derived extracellular matrices such as Matrigel introduces batch-to-batch variability and poorly defined biochemical cues, complicating cross-study comparisons and large-scale translational applications [155,156]. Another key challenge is the incomplete maturation and architectural complexity of many organoid models. While ASC-derived organoids retain aspects of tissue identity, they frequently lack higher-order structures such as vasculature, innervation, and lymphatic components, all of which play critical roles in aging, metabolic regulation, and immune surveillance [157]. The absence of perfusable vascular networks can result in diffusion-limited hypoxia, nutrient gradients, and waste accumulation in long-term cultures, potentially confounding interpretations of senescence and stress-associated phenotypes.
6. A further limitation is the insufficient representation of systemic aging cues. Endocrine signals, circulating metabolites, immune aging (immunosenescence), and neurohumoral regulation profoundly influence tissue aging in vivo but are largely absent in conventional organoid cultures. Consequently, organoids primarily capture cell-intrinsic and local niche–dependent aging mechanisms, while organismal drivers of aging and inter-organ communication remain underrepresented [158].

9.2. Modeling Systemic and Multi-Organ Aging

Aging is a multifactorial process involving interactions between organs, tissues, and systemic factors such as hormones, the immune system, and the microbiome. Current organoid models are primarily tissue-specific and often lack integration with systemic components. Emerging multi-organoid and organ-on-chip platforms, which link organoids

https://doi.org/10.3390/targets4020012

from different tissues under controlled flow conditions, offer a potential solution for studying inter-organ communication and systemic aging effects [159].

9.3. Integration with Microenvironmental and Niche Factors

Organoids typically lack complete representation of extracellular matrix (ECM) complexity, stromal interactions, and mechanical forces present in vivo. These niche signals are critical for stem cell maintenance, senescence regulation, and tissue homeostasis [160]. Advances in synthetic hydrogels, ECM engineering, and co-culture systems are enabling more faithful recapitulation of the tissue microenvironment, which is essential for modeling aging and disease.

9.4. Translational and Clinical Challenges

Despite their predictive potential, clinical translation of organoid-based discoveries faces several hurdles:

- Scalability: Producing organoids at clinical scale remains challenging, particularly for regenerative therapies or personalized medicine applications.
- Regulatory compliance: Safety, reproducibility, and standardization must meet regulatory standards for therapeutic use.
- Cost and accessibility: Organoid culture is resource-intensive, limiting broad adoption in some research or clinical settings.

9.5. Future Directions

1. Vascularized and immune-competent organoids: Incorporation of vasculature, immune cells, and other niche components will improve physiological relevance and allow modeling of inflammation, immune surveillance, and systemic aging.
2. Advanced multi-omics integration: Combining single-cell transcriptomics, proteomics, epigenomics, and metabolomics will enable comprehensive mapping of aging networks and identification of robust therapeutic targets.
3. Personalized aging models: Patient-derived organoids from different age groups and genetic backgrounds will facilitate individualized studies of aging, senescence, and drug response, advancing precision medicine approaches.
4. Automation and high-throughput platforms: Robotics, microfluidics, and organ-on-chip integration will allow large-scale screening of drugs, senolytic compounds, and regenerative therapies in physiologically relevant contexts.
5. Computational modeling and AI integration: Machine learning algorithms can analyze complex organoid datasets to predict aging trajectories, therapy responses, and optimal intervention strategies, accelerating translational research [161,162]. AI-driven frameworks—including deep learning, trajectory inference algorithms, and graph-based network modeling—enable integration of transcriptomic, epigenomic, proteomic, and metabolic layers to reconstruct aging trajectories and identify regulatory circuits that drive stem cell exhaustion or senescence induction. In organoid-based model systems, machine learning–driven image analysis has increasingly been applied to extract quantitative information from complex morphological features. These computational approaches enable the classification of aging-associated phenotypes, such as changes in organoid size, shape, structural organization, and cellular heterogeneity. By analyzing large image datasets, machine learning models can also quantify drug-induced phenotypic responses and identify subtle morphological changes that may not be easily detected through manual observation. Importantly, these image-derived phenotypic signatures can be integrated with predictive algo-

https://doi.org/10.3390/targets4020012

rithms to forecast therapeutic sensitivity, thereby providing a powerful platform for high-throughput drug screening and precision medicine applications.

10. Conclusions

Organoid technology offers unprecedented opportunities to study aging, senescence, and disease in human-relevant systems. It has emerged as a transformative paradigm in biomedical research, offering unprecedented opportunities to model human aging, cellular senescence, and disease progression within physiologically relevant 3D systems. By recapitulating native tissue architecture, cellular heterogeneity, and stem cell hierarchies, organoids bridge the long-standing gap between simplified in vitro cultures and complex in vivo models, enabling more precise dissection of epithelial-intrinsic aging mechanisms across multiple organ systems [163]. This human-relevant fidelity is particularly critical for aging research, where species-specific differences and systemic confounders often limit the translational value of traditional models. Organoids derived from aged tissues, as well as genetically or environmentally engineered to model senescence, have yielded important insights into age-associated alterations in signaling pathways, stem cell competence, and microenvironmental regulation of tissue homeostasis. These systems have clarified how aging reshapes proliferative capacity, lineage commitment, and stress responses, while revealing context-dependent roles for OIS and SASP in tissue decline and disease initiation [5,85]. Importantly, organoid-based models allow aging-related phenotypes to be studied in isolation from systemic influences, thereby enabling causal mechanistic interrogation that is difficult to achieve in vivo.

The integration of organoid platforms with emerging technologies—including single-cell and spatial multi-omics, CRISPR-based functional genomics, organoid-on-chip systems, advanced live imaging, and computational modeling—has further expanded their analytical power [164]. These approaches enable longitudinal, high-resolution, and systems-level analysis of aging processes, transforming organoids from descriptive models into predictive platforms for hypothesis testing and therapeutic discovery [165]. As a result, organoids now provide a versatile framework to uncover novel regulators of senescence, identify age-specific vulnerabilities, and interrogate dynamic stem cell–niche interactions. Looking ahead, organoid systems hold exceptional promises for the development and translation of senescence-modulating therapies, including senolytic and senomorphic agents, regenerative strategies, and patient-tailored interventions. Their compatibility with high-throughput drug screening, functional genomics, and patient-derived biobanks positions organoids as key drivers of precision medicine approaches for age-associated diseases. Furthermore, organoid-based modeling of therapy resistance, toxicity, and tissue-specific responses is poised to improve clinical decision-making, particularly in elderly populations where treatment efficacy and safety remain major challenges.

Despite ongoing limitations related to standardization, maturation, and systemic integration, rapid advances in bioengineering, multicellular co-culture, and microphysiological systems are steadily enhancing the physiological relevance and translational utility of organoid models. Continued efforts to harmonize protocols, develop synthetic matrices, and incorporate immune and vascular components will be essential to fully realize the potential of organoids as comprehensive aging models [166,167]. In summary, organoid systems represent a paradigm shift in aging and disease research. They provide human-relevant, experimentally tractable platforms capable of elucidating fundamental mechanisms of cellular aging, uncovering novel therapeutic targets, and guiding the development of interventions aimed at mitigating senescence and preserving tissue resilience across the lifespan. As these technologies continue to evolve, organoids are poised to play a

https://doi.org/10.3390/targets4020012

central role in translating insights from aging biology into effective, personalized therapies that address the growing burden of age-related disease.

Author Contributions: Conceptualization, S.B. and A.G.; writing—original draft preparation, S.B., P.V. and A.G.; writing—review and editing, S.B., P.V. and A.G.; visualization, S.B. and A.G.; supervision, S.B. and A.G.; project administration, S.B. and A.G. All authors have read and agreed to the published version of the manuscript.

Funding: This research received no external funding.

Institutional Review Board Statement: Not applicable.

Informed Consent Statement: Not applicable.

Data Availability Statement: No new data were created or analyzed in this study. Data sharing is not applicable to this article.

Conflicts of Interest: The authors declare no conflicts of interest.

Abbreviations

The following abbreviations are used in this manuscript:

2D	Two-dimensional
3D	Three-dimensional
SASP	Senescence-associated secretory phenotype
OIS	Oncogene-induced senescence
ASCs	Adult stem cells
PSCs	Pluripotent stem cells (such as ESCs)
ESCs	Embryonic stem cells
iPSCs	Induced pluripotent stem cells
RNAi	RNA interference
PDOs	Patient-derived organoids
scRNA-seq	Single-cell RNA sequencing
ECM	Extracellular matrix

References

1. López-Otín, C.; Blasco, M.A.; Partridge, L.; Serrano, M.; Kroemer, G. The Hallmarks of Aging. *Cell* **2013**, *153*, 1194–1217. [CrossRef]
2. Partridge, L. The New Biology of Ageing. *Philos. Trans. R. Soc. B Biol. Sci.* **2010**, *365*, 147–154. [CrossRef] [PubMed]
3. Goodell, M.A.; Rando, T.A. Stem Cells and Healthy Aging. *Science* **2015**, *350*, 1199–1204. [CrossRef]
4. Kirkland, J.L.; Tchkonia, T. Cellular Senescence: A Translational Perspective. *eBioMedicine* **2017**, *21*, 21–28. [CrossRef]
5. Campisi, J. Aging, Cellular Senescence, and Cancer. *Annu. Rev. Physiol.* **2013**, *75*, 685–705. [CrossRef]
6. Di Micco, R.; Krizhanovsky, V.; Baker, D.; d'Adda di Fagagna, F. Cellular Senescence in Ageing: From Mechanisms to Therapeutic Opportunities. *Nat. Rev. Mol. Cell Biol.* **2021**, *22*, 75–95. [CrossRef] [PubMed]
7. Campisi, J.; d'Adda di Fagagna, F. Cellular Senescence: When Bad Things Happen to Good Cells. *Nat. Rev. Mol. Cell Biol.* **2007**, *8*, 729–740. [CrossRef]
8. Baker, D.J.; Childs, B.G.; Durik, M.; Wijers, M.E.; Sieben, C.J.; Zhong, J.; Saltness, R.A.; Jeganathan, K.B.; Verzosa, G.C.; Pezeshki, A. Naturally Occurring p16Ink4a-Positive Cells Shorten Healthy Lifespan. *Nature* **2016**, *530*, 184–189. [CrossRef]
9. Edmondson, R.; Broglie, J.J.; Adcock, A.F.; Yang, L. Three-Dimensional Cell Culture Systems and Their Applications in Drug Discovery and Cell-Based Biosensors. *Assay Drug Dev. Technol.* **2014**, *12*, 207–218. [CrossRef]
10. Dutta, S.; Sengupta, P. Men and Mice: Relating Their Ages. *Life Sci.* **2016**, *152*, 244–248. [CrossRef] [PubMed]
11. Sato, T.; Vries, R.G.; Snippert, H.J.; Van De Wetering, M.; Barker, N.; Stange, D.E.; Van Es, J.H.; Abo, A.; Kujala, P.; Peters, P.J. Single Lgr5 Stem Cells Build Crypt-Villus Structures In Vitro without a Mesenchymal Niche. *Nature* **2009**, *459*, 262–265. [CrossRef]
12. Lancaster, M.A.; Knoblich, J.A. Organogenesis in a Dish: Modeling Development and Disease Using Organoid Technologies. *Science* **2014**, *345*, 1247125. [CrossRef] [PubMed]
13. Pentinmikko, N.; Iqbal, S.; Mana, M.; Andersson, S.; Cognetta, A.B., III; Suciu, R.M.; Roper, J.; Luopajärvi, K.; Markelin, E.; Gopalakrishnan, S. Notum Produced by Paneth Cells Attenuates Regeneration of Aged Intestinal Epithelium. *Nature* **2019**, *571*, 398–402. [CrossRef]

https://doi.org/10.3390/targets4020012

14. Jothi, D.; Kulka, L.A.M. Strategies for Modeling Aging and Age-Related Diseases. *npj Aging* **2024**, *10*, 32. [CrossRef]
15. Chao, C.-C.; Shen, P.-W.; Tzeng, T.-Y.; Kung, H.-J.; Tsai, T.-F.; Wong, Y.-H. Human iPSC-Derived Neurons as a Platform for Deciphering the Mechanisms behind Brain Aging. *Biomedicines* **2021**, *9*, 1635. [CrossRef]
16. Richter, K.M.; Wrage, M.; Krekeler, C.; De Oliveira, T.; Conradi, L.-C.; Menck, K.; Bleckmann, A. Model Systems to Study Tumor-Microbiome Interactions in Early-Onset Colorectal Cancer. *EMBO Mol. Med.* **2025**, *17*, 395–413. [CrossRef] [PubMed]
17. Fujii, M.; Shimokawa, M.; Date, S.; Takano, A.; Matano, M.; Nanki, K.; Ohta, Y.; Toshimitsu, K.; Nakazato, Y.; Kawasaki, K. A Colorectal Tumor Organoid Library Demonstrates Progressive Loss of Niche Factor Requirements during Tumorigenesis. *Cell Stem Cell* **2016**, *18*, 827–838. [CrossRef]
18. Lu, X.; Yang, J.; Xiang, Y. Modeling Human Neurodevelopmental Diseases with Brain Organoids. *Cell Regen.* **2022**, *11*, 1. [CrossRef] [PubMed]
19. Ogrodnik, M.; Miwa, S.; Tchkonia, T.; Tiniakos, D.; Wilson, C.L.; Lahat, A.; Day, C.P.; Burt, A.; Palmer, A.; Anstee, Q.M. Cellular Senescence Drives Age-Dependent Hepatic Steatosis. *Nat. Commun.* **2017**, *8*, 15691. [CrossRef]
20. Beumer, J.; Artegiani, B.; Post, Y.; Reimann, F.; Gribble, F.; Nguyen, T.N.; Zeng, H.; Van den Born, M.; Van Es, J.H.; Clevers, H. Enteroendocrine Cells Switch Hormone Expression along the Crypt-to-Villus BMP Signalling Gradient. *Nat. Cell Biol.* **2018**, *20*, 909–916. [CrossRef]
21. Ritschka, B.; Storer, M.; Mas, A.; Heinzmann, F.; Ortells, M.C.; Morton, J.P.; Sansom, O.J.; Zender, L.; Keyes, W.M. The Senescence-Associated Secretory Phenotype Induces Cellular Plasticity and Tissue Regeneration. *Genes Dev.* **2017**, *31*, 172–183. [CrossRef]
22. González-Gualda, E.; Baker, A.G.; Fruk, L.; Muñoz-Espín, D. A Guide to Assessing Cellular Senescence In Vitro and In Vivo. *FEBS J.* **2021**, *288*, 56–80. [CrossRef]
23. Suhito, I.R.; Sunil, C.; Tay, A. Engineering Human Immune Organoids for Translational Immunology. *Bioact. Mater.* **2025**, *44*, 164–183. [CrossRef]
24. Zhou, L.; Huang, J.; Li, C.; Gu, Q.; Li, G.; Li, Z.A.; Xu, J.; Zhou, J.; Tuan, R.S. Organoids and Organs-on-Chips: Recent Advances, Applications in Drug Development, and Regulatory Challenges. *Med* **2025**, *6*, 100667. [CrossRef]
25. Clevers, H. Modeling Development and Disease with Organoids. *Cell* **2016**, *165*, 1586–1597. [CrossRef]
26. Sato, T.; Stange, D.E.; Ferrante, M.; Vries, R.G.; Van Es, J.H.; Van Den Brink, S.; Van Houdt, W.J.; Pronk, A.; Van Gorp, J.; Siersema, P.D. Long-Term Expansion of Epithelial Organoids from Human Colon, Adenoma, Adenocarcinoma, and Barrett's Epithelium. *Gastroenterology* **2011**, *141*, 1762–1772. [CrossRef] [PubMed]
27. Lancaster, M.A.; Renner, M.; Martin, C.-A.; Wenzel, D.; Bicknell, L.S.; Hurles, M.E.; Homfray, T.; Penninger, J.M.; Jackson, A.P.; Knoblich, J.A. Cerebral Organoids Model Human Brain Development and Microcephaly. *Nature* **2013**, *501*, 373–379. [CrossRef]
28. Takebe, T.; Sekine, K.; Enomura, M.; Koike, H.; Kimura, M.; Ogaeri, T.; Zhang, R.-R.; Ueno, Y.; Zheng, Y.-W.; Koike, N. Vascularized and Functional Human Liver from an iPSC-Derived Organ Bud Transplant. *Nature* **2013**, *499*, 481–484. [CrossRef] [PubMed]
29. Barker, N.; Huch, M.; Kujala, P.; van de Wetering, M.; Snippert, H.J.; van Es, J.H.; Sato, T.; Stange, D.E.; Begthel, H.; van den Born, M. Lgr5+ ve Stem Cells Drive Self-Renewal in the Stomach and Build Long-Lived Gastric Units In Vitro. *Cell Stem Cell* **2010**, *6*, 25–36. [CrossRef] [PubMed]
30. Takasato, M.; Er, P.X.; Chiu, H.S.; Maier, B.; Baillie, G.J.; Ferguson, C.; Parton, R.G.; Wolvetang, E.J.; Roost, M.S.; Chuva de Sousa Lopes, S.M. Kidney Organoids from Human iPS Cells Contain Multiple Lineages and Model Human Nephrogenesis. *Nature* **2015**, *526*, 564–568. [CrossRef]
31. Sato, T.; Clevers, H. Primary Mouse Small Intestinal Epithelial Cell Cultures. In *Epithelial Cell Culture Protocols*, 2nd ed.; Springer: Berlin/Heidelberg, Germany, 2012; pp. 319–328.
32. Yin, X.; Farin, H.F.; Van Es, J.H.; Clevers, H.; Langer, R.; Karp, J.M. Niche-Independent High-Purity Cultures of Lgr5+ Intestinal Stem Cells and Their Progeny. *Nat. Methods* **2014**, *11*, 106–112. [CrossRef] [PubMed]
33. Eiraku, M.; Takata, N.; Ishibashi, H.; Kawada, M.; Sakakura, E.; Okuda, S.; Sekiguchi, K.; Adachi, T.; Sasai, Y. Self-Organizing Optic-Cup Morphogenesis in Three-Dimensional Culture. *Nature* **2011**, *472*, 51–56. [CrossRef] [PubMed]
34. McCracken, K.W.; Catá, E.M.; Crawford, C.M.; Sinagoga, K.L.; Schumacher, M.; Rockich, B.E.; Tsai, Y.-H.; Mayhew, C.N.; Spence, J.R.; Zavros, Y. Modelling Human Development and Disease in Pluripotent Stem-Cell-Derived Gastric Organoids. *Nature* **2014**, *516*, 400–404. [CrossRef] [PubMed]
35. Khorraminejad-Shirazi, M.; Dorvash, M.; Estedlal, A.; Hoveidaei, A.H.; Mazloomrezaei, M.; Mosaddeghi, P. Aging: A Cell Source Limiting Factor in Tissue Engineering. *World J. Stem Cells* **2019**, *11*, 787. [CrossRef]
36. Aspal, M.; Pushlar, N.; Melameka, M.; McVicar, R.N.; Smith, E.; Ogunyamoju, T.; Kumar, M.; Marth, J.D.; Chun, J.; Snyder, E.Y. Modeling the Aging Human Lung: Generation of a Senescent Human Lung Organoid Culture System. *bioRxiv* **2025**. [CrossRef]
37. Broutier, L.; Mastrogiovanni, G.; Verstegen, M.M.; Francies, H.E.; Gavarró, L.M.; Bradshaw, C.R.; Allen, G.E.; Arnes-Benito, R.; Sidorova, O.; Gaspersz, M.P. Human Primary Liver Cancer–Derived Organoid Cultures for Disease Modeling and Drug Screening. *Nat. Med.* **2017**, *23*, 1424–1435. [CrossRef]

https://doi.org/10.3390/targets4020012

38. Nozaki, K.; Mochizuki, W.; Matsumoto, Y.; Matsumoto, T.; Fukuda, M.; Mizutani, T.; Watanabe, M.; Nakamura, T. Co-Culture with Intestinal Epithelial Organoids Allows Efficient Expansion and Motility Analysis of Intraepithelial Lymphocytes. *J. Gastroenterol.* **2016**, *51*, 206–213. [CrossRef]
39. Dave, R.; Pandey, K.; Patel, R.; Gour, N.; Bhatia, D. Leveraging 3D Cell Culture and AI Technologies for Next-Generation Drug Discovery. *Cell Biomater.* **2025**, *1*, 100050. [CrossRef]
40. Poornima, K.; Francis, A.P.; Hoda, M.; Eladl, M.A.; Subramanian, S.; Veeraraghavan, V.P.; El-Sherbiny, M.; Asseri, S.M.; Hussamuldin, A.B.A.; Surapaneni, K.M. Implications of Three-Dimensional Cell Culture in Cancer Therapeutic Research. *Front. Oncol.* **2022**, *12*, 891673. [CrossRef]
41. Liu, X.; Zhou, Z.; Zhang, Y.; Zhong, H.; Cai, X.; Guan, R. Recent Progress on the Organoids: Techniques, Advantages and Applications. *Biomed. Pharmacother.* **2025**, *185*, 117942. [CrossRef]
42. Yang, S.; Hu, H.; Kung, H.; Zou, R.; Dai, Y.; Hu, Y.; Wang, T.; Lv, T.; Yu, J.; Li, F. Organoids: The Current Status and Biomedical Applications. *MedComm* **2023**, *4*, e274. [CrossRef] [PubMed]
43. Wang, Q.; Yuan, F.; Zuo, X.; Li, M. Breakthroughs and challenges of organoid models for assessing cancer immunotherapy: A cutting-edge tool for advancing personalised treatments. *Cell Death Discov.* **2025**, *11*, 222. [CrossRef]
44. Kretzschmar, K.; Clevers, H. Organoids: Modeling Development and the Stem Cell Niche in a Dish. *Dev. Cell* **2016**, *38*, 590–600. [CrossRef] [PubMed]
45. Bhaduri, A.; Andrews, M.G.; Mancia Leon, W.; Jung, D.; Shin, D.; Allen, D.; Jung, D.; Schmunk, G.; Haeussler, M.; Salma, J. Cell Stress in Cortical Organoids Impairs Molecular Subtype Specification. *Nature* **2020**, *578*, 142–148. [CrossRef]
46. Kapetanos, K.; Asimakopoulos, D.; Christodoulou, N.; Vogt, A.; Khan, W. Chronological Age Affects MSC Senescence in Vitro—A Systematic Review. *Int. J. Mol. Sci.* **2021**, *22*, 7945. [CrossRef]
47. Sun, X.; Sun, F.; Zhang, Y.; Qu, J.; Zhang, W.; Liu, G.-H. A Narrative Review of Organoids for Investigating Organ Aging: Opportunities and Challenges. *J. Bio-X Res.* **2023**, *6*, 3–14. [CrossRef]
48. Llorente, C. The Imperative for Innovative Enteric Nervous System–Intestinal Organoid Co-Culture Models: Transforming GI Disease Modeling and Treatment. *Cells* **2024**, *13*, 820. [CrossRef]
49. Neal, J.T.; Li, X.; Zhu, J.; Giangarra, V.; Grzeskowiak, C.L.; Ju, J.; Liu, I.H.; Chiou, S.-H.; Salahudeen, A.A.; Smith, A.R. Organoid Modeling of the Tumor Immune Microenvironment. *Cell* **2018**, *175*, 1972–1988. [CrossRef]
50. Magré, L.; Verstegen, M.M.; Buschow, S.; Van Der Laan, L.J.; Peppelenbosch, M.; Desai, J. Emerging Organoid-Immune Co-Culture Models for Cancer Research: From Oncoimmunology to Personalized Immunotherapies. *J. Immunother. Cancer* **2023**, *11*, e006290. [CrossRef]
51. Kamroo, A.; Kakroudi, M.H.; Sarmadian, A.J.; Firouzabadi, A.; Mousavi, S.; Yazdanpanah, N.; Saleki, K.; Rezaei, N. Immunosenescence and Organoids: Pathophysiology and Therapeutic Opportunities. *Immun. Ageing* **2025**, *22*, 46. [CrossRef]
52. Chauhdari, T.; Zaidi, S.A.; Su, J.; Ding, Y. Organoids Meet Microfluidics: Recent Advancements, Challenges, and Future of Organoids-on-Chip. *Vitr. Models* **2025**, *4*, 71–88. [CrossRef]
53. Gjorevski, N.; Sachs, N.; Manfrin, A.; Giger, S.; Bragina, M.E.; Ordóñez-Morán, P.; Clevers, H.; Lutolf, M.P. Designer Matrices for Intestinal Stem Cell and Organoid Culture. *Nature* **2016**, *539*, 560–564. [CrossRef]
54. López-Otín, C.; Blasco, M.A.; Partridge, L.; Serrano, M.; Kroemer, G. Hallmarks of Aging: An Expanding Universe. *Cell* **2023**, *186*, 243–278. [CrossRef] [PubMed]
55. Azar, J.; Bahmad, H.F.; Daher, D.; Moubarak, M.M.; Hadadeh, O.; Monzer, A.; Al Bitar, S.; Jamal, M.; Al-Sayegh, M.; Abou-Kheir, W. The Use of Stem Cell-Derived Organoids in Disease Modeling: An Update. *Int. J. Mol. Sci.* **2021**, *22*, 7667. [CrossRef] [PubMed]
56. Zhu, Y.; Lin, N.; Li, J.; Zhang, H.; Zhang, P.; Cheng, X.; Yang, Q.; Liu, L. Organoids in Genetic Disorders: From Disease Modeling to Translational Applications. *Stem Cell Rev. Rep.* **2025**, *21*, 2578–2596. [CrossRef] [PubMed]
57. Fane, M.; Weeraratna, A.T. How the Ageing Microenvironment Influences Tumour Progression. *Nat. Rev. Cancer* **2020**, *20*, 89–106. [CrossRef]
58. Volkmer, E.; Drosse, I.; Otto, S.; Stangelmayer, A.; Stengele, M.; Kallukalam, B.C.; Mutschler, W.; Schieker, M. Hypoxia in Static and Dynamic 3D Culture Systems for Tissue Engineering of Bone. *Tissue Eng. Part A* **2008**, *14*, 1331–1340. [CrossRef]
59. Beerman, I.; Bhattacharya, D.; Zandi, S.; Sigvardsson, M.; Weissman, I.L.; Bryder, D.; Rossi, D.J. Functionally Distinct Hematopoietic Stem Cells Modulate Hematopoietic Lineage Potential during Aging by a Mechanism of Clonal Expansion. *Proc. Natl. Acad. Sci. USA* **2010**, *107*, 5465–5470. [CrossRef]
60. Ermolaeva, M.; Neri, F.; Ori, A.; Rudolph, K.L. Cellular and Epigenetic Drivers of Stem Cell Ageing. *Nat. Rev. Mol. Cell Biol.* **2018**, *19*, 594–610. [CrossRef]
61. Sharpless, N.E.; Sherr, C.J. Forging a Signature of In Vivo Senescence. *Nat. Rev. Cancer* **2015**, *15*, 397–408. [CrossRef]
62. Scesa, G.; Adami, R.; Bottai, D. iPSC Preparation and Epigenetic Memory: Does the Tissue Origin Matter? *Cells* **2021**, *10*, 1470. [CrossRef]
63. Sprangers, J.; Zaalberg, I.C.; Maurice, M.M. Organoid-based modeling of intestinal development, regeneration, and repair. *Cell Death Differ.* **2021**, *28*, 95–107. [CrossRef] [PubMed]

64. Shay, J.E.; Yilmaz, Ö.H. Dietary and Metabolic Effects on Intestinal Stem Cells in Health and Disease. *Nat. Rev. Gastroenterol. Hepatol.* **2025**, *22*, 23–38. [CrossRef]
65. Peng, W.C.; Logan, C.Y.; Fish, M.; Anbarchian, T.; Aguisanda, F.; Álvarez-Varela, A.; Wu, P.; Jin, Y.; Zhu, J.; Li, B. Inflammatory Cytokine TNFα Promotes the Long-Term Expansion of Primary Hepatocytes in 3D Culture. *Cell* **2018**, *175*, 1607–1619. [CrossRef]
66. Kuehnemann, C.; Wiley, C.D. Senescent cells at the crossroads of aging, disease, and tissue homeostasis. *Aging Cell* **2024**, *23*, e13988. [CrossRef] [PubMed]
67. Hu, J.L.; Todhunter, M.E.; LaBarge, M.A.; Gartner, Z.J. Opportunities for Organoids as New Models of Aging. *J. Cell Biol.* **2018**, *217*, 39–50. [CrossRef]
68. Yao, Q.; Cheng, S.; Pan, Q.; Yu, J.; Cao, G.; Li, L.; Cao, H. Organoids: Development and applications in disease models, drug discovery, precision medicine, and regenerative medicine. *MedComm* **2024**, *5*, e735. [CrossRef]
69. Merenda, A.; Fenderico, N.; Maurice, M.M. Wnt signaling in 3D: Recent advances in the applications of intestinal organoids. *Trends Cell Biol.* **2020**, *30*, 60–73. [CrossRef]
70. Igarashi, R.; Oda, M.; Okada, R.; Yano, T.; Takahashi, S.; Pastuhov, S.; Matano, M.; Masuda, N.; Togasaki, K.; Ohta, Y.; et al. Generation of human adult hepatocyte organoids with metabolic functions. *Nature* **2025**, *641*, 1248–1257. [CrossRef]
71. Yu, P.; Liu, B.; Dong, C.; Chang, Y. Induced Pluripotent Stem Cells-Based Regenerative Therapies in Treating Human Aging-Related Functional Decline and Diseases. *Cells* **2025**, *14*, 619. [CrossRef] [PubMed]
72. Miller, J.D.; Ganat, Y.M.; Kishinevsky, S.; Bowman, R.L.; Liu, B.; Tu, E.Y.; Mandal, P.K.; Vera, E.; Shim, J.; Kriks, S. Human iPSC-Based Modeling of Late-Onset Disease via Progerin-Induced Aging. *Cell Stem Cell* **2013**, *13*, 691–705. [CrossRef]
73. Capeling, M.M.; Chen, B.; Aliar, K.; Penna, E.; Lopez, V.I.; Foo, C.; Rost, S.; Holokai, L.; Tong, X.; Phillips, D.; et al. Dictionary of human intestinal organoid responses to secreted niche factors at single cell resolution. *Nat. Commun.* **2026**, *17*, 1527. [CrossRef] [PubMed]
74. Krauß, D.; Moreno-Viedma, V.; Adachi-Fernandez, E.; de Sá Fernandes, C.; Genger, J.W.; Fari, O.; Blauensteiner, B.; Kirchhofer, D.; Bradaric, N.; Gushchina, V.; et al. EGFR controls transcriptional and metabolic rewiring in KRASG12D colorectal cancer. *EMBO Mol. Med.* **2025**, *17*, 1355. [CrossRef] [PubMed]
75. Chatsirisupachai, K.; Lagger, C.; de Magalhães, J.P. Age-associated differences in the cancer molecular landscape. *Trends Cancer* **2022**, *8*, 962–971. [CrossRef] [PubMed]
76. Kumari, R.; Jat, P. Mechanisms of cellular senescence: Cell cycle arrest and senescence associated secretory phenotype. *Front. Cell Dev. Biol.* **2021**, *9*, 645593. [CrossRef]
77. Yang, H.; Liu, D.; Qiu, L.; Wang, R.; Zhang, C.; Yu, D.; Zhong, Q.; Yuki, N.; Song, Z.; Zhu, T. Reprogramming Cellular Senescence and Aging Clocks for Advanced Cancer Immunotherapy. *Mol. Cancer* **2025**, *24*, 237. [CrossRef]
78. Safwan-Zaiter, H.; Wagner, N.; Wagner, K.-D. P16INK4A—More than a Senescence Marker. *Life* **2022**, *12*, 1332. [CrossRef]
79. Sorrentino, V.; Romani, M.; Mouchiroud, L.; Beck, J.S.; Zhang, H.; D'Amico, D.; Moullan, N.; Potenza, F.; Schmid, A.W.; Rietsch, S. Enhancing Mitochondrial Proteostasis Reduces Amyloid-β Proteotoxicity. *Nature* **2017**, *552*, 187–193. [CrossRef]
80. Wang, B.; Han, J.; Elisseeff, J.H.; Demaria, M. The senescence-associated secretory phenotype and its physiological and pathological implications. *Nat. Rev. Mol. Cell Biol.* **2024**, *25*, 958–978. [CrossRef]
81. Takasugi, M.; Yoshida, Y.; Ohtani, N. Cellular senescence and the tumour microenvironment. *Mol. Oncol.* **2022**, *16*, 3333–3351. [CrossRef]
82. Zuo, J.; Fang, Y.; Wang, R.; Liang, S. High-throughput solutions in tumor organoids: From culture to drug screening. *Stem Cells* **2025**, *43*, sxae070. [CrossRef]
83. Mukhare, R.; Gandhi, K.A.; Kadam, A.; Raja, A.; Singh, A.; Madhav, M.; Chaubal, R.; Pandey, S.; Gupta, S. Integration of organoids with CRISPR screens: A narrative review. *Biol. Cell* **2025**, *117*, e70006. [CrossRef]
84. Geurts, M.H.; Clevers, H. CRISPR engineering in organoids for gene repair and disease modelling. *Nat. Rev. Bioeng.* **2023**, *1*, 32–45. [CrossRef]
85. Torrens-Mas, M.; Perelló-Reus, C.; Navas-Enamorado, C.; Ibargüen-González, L.; Sanchez-Polo, A.; Segura-Sampedro, J.J.; Masmiquel, L.; Barcelo, C.; Gonzalez-Freire, M. Organoids: An emerging tool to study aging signature across human tissues. modeling aging with patient-derived organoids. *Int. J. Mol. Sci.* **2021**, *22*, 10547. [CrossRef] [PubMed]
86. Li, S.; Agudelo Garcia, P.A.; Aliferis, C.; Becich, M.J.; Calyeca, J.; Cosgrove, B.D.; Elisseeff, J.; Farzad, N.; Fertig, E.J.; Glass, C. Advancing Biological Understanding of Cellular Senescence with Computational Multiomics. *Nat. Genet.* **2025**, *57*, 2381–2394. [CrossRef]
87. Xue, C.; Sun, F.; Zhu, H.; Shi, J.; Wang, J.; Sun, Q.; Tong, M.; Gu, X.; Han, Q.; Shi, W. Advancements, strategies, and challenges in organoid-based drug evaluation for tissue engineering and regenerative medicine. *Engineering*, 2025, *in press*. [CrossRef]
88. Su, Y.; Chen, Y.; Wei, G.; Shi, Y. Organoid-based novel technology for antitumor drug screening. *View* **2025**, *6*, 20250107. [CrossRef]
89. Glaviano, A.; Foo, A.S.; Lam, H.Y.; Yap, K.C.; Jacot, W.; Jones, R.H.; Eng, H.; Nair, M.G.; Makvandi, P.; Geoerger, B.; et al. PI3K/AKT/mTOR signaling transduction pathway and targeted therapies in cancer. *Mol. Cancer* **2023**, *22*, 138. [CrossRef]

https://doi.org/10.3390/targets4020012

90. Wu, Y.; Zhang, F.; Du, F.; Huang, J.; Wei, S. Combination of tumor organoids with advanced technologies: A powerful platform for tumor evolution and treatment response. *Mol. Med. Rep.* **2025**, *31*, 140. [CrossRef]
91. Li, T.; Yin, J.; Hao, Y.; Gao, W.; Li, Q.; Feng, Q.; Tao, B.; Hao, M.; Liu, Y.; Lin, C.; et al. Single-cell sequencing and organoids: Applications in organ development and disease. *Mol. Biomed.* **2025**, *6*, 134. [CrossRef]
92. Sun, J.; Huang, S.; Zhang, Q.; Lei, L.; Nie, H. Development of organoids for drug research and tissue regeneration. *Precis. Med. Eng.* **2025**, *2*, 100045. [CrossRef]
93. Cui, H.; Tang, D.; Garside, G.B.; Zeng, T.; Wang, Y.; Tao, Z.; Zhang, L.; Tao, S.I. Wnt signaling mediates the aging-induced differentiation impairment of intestinal stem cells. *Stem Cell Rev. Rep.* **2019**, *15*, 448–455. [CrossRef] [PubMed]
94. Trastus, L.A.; d'Adda di Fagagna, F. The complex interplay between aging and cancer. *Nat. Aging* **2025**, *5*, 350–365. [CrossRef] [PubMed]
95. Xie, Z.; Zhan, Q. The complex interplay between aging and cancer: Unraveling the clues. *Med. Rev.* **2024**, *4*, 2–4. [CrossRef]
96. Silverman, L.I.; Flanagan, F.; Rodriguez-Granrose, D.; Simpson, K.; Saxon, L.H.; Foley, K.T. Identifying and managing sources of variability in cell therapy manufacturing and clinical trials. *Regen. Eng. Transl. Med.* **2019**, *5*, 354–361. [CrossRef]
97. Wang, M.; Yu, H.; Zhang, T.; Cao, L.; Du, Y.; Xie, Y.; Ji, J.; Wu, J. In-Depth Comparison of Matrigel Dissolving Methods on Proteomic Profiling of Organoids. *Mol. Cell. Proteom.* **2022**, *21*, 100181. [CrossRef] [PubMed]
98. Minciullo, P.L.; Catalano, A.; Mandraffino, G.; Casciaro, M.; Crucitti, A.; Maltese, G.; Morabito, N.; Lasco, A.; Gangemi, S.; Basile, G. Inflammaging and Anti-Inflammaging: The Role of Cytokines in Extreme Longevity. *Arch. Immunol. Ther. Exp.* **2016**, *64*, 111–126. [CrossRef]
99. Liu, K.; Chen, X.; Fan, Z.; Ren, F.; Liu, J.; Hu, B. From Organoids to Organoids-on-a-Chip: Current Applications and Challenges in Biomedical Research. *Chin. Med. J.* **2025**, *138*, 792–807. [CrossRef]
100. Zhang, S.; Wan, Z.; Kamm, R.D. Vascularized Organoids on a Chip: Strategies for Engineering Organoids with Functional Vasculature. *Lab A Chip* **2021**, *21*, 473–488. [CrossRef]
101. Cuenca, M.V.; Bulut, M.; Mummery, C.L.; Orlova, V.V. Vascularization of Organoid Microenvironments: Perfusable Networks for Organoid Growth and Maturation. *Curr. Opin. Biomed. Eng.* **2025**, *34*, 100586. [CrossRef]
102. LeSavage, B.L.; Suhar, R.A.; Broguiere, N.; Lutolf, M.P.; Heilshorn, S.C. Next-Generation Cancer Organoids. *Nat. Mater.* **2022**, *21*, 143–159. [CrossRef]
103. Zhou, Z.; Cong, L.; Cong, X. Patient-Derived Organoids in Precision Medicine: Drug Screening, Organoid-on-a-Chip and Living Organoid Biobank. *Front. Oncol.* **2021**, *11*, 762184. [CrossRef] [PubMed]
104. Schutgens, F.; Clevers, H. Human Organoids: Tools for Understanding Biology and Treating Diseases. *Annu. Rev. Pathol. Mech. Dis.* **2020**, *15*, 211–234. [CrossRef] [PubMed]
105. Tian, C.-M.; Yang, M.-F.; Xu, H.-M.; Zhu, M.-Z.; Yue, N.-N.; Zhang, Y.; Shi, R.-Y.; Yao, J.; Wang, L.-S.; Liang, Y.-J.; et al. Stem cell-derived intestinal organoids: A novel modality for IBD. *Cell Death Discov.* **2023**, *9*, 255. [CrossRef]
106. Meirelles, L.A.; Persat, A. Decoding host-microbe interactions with engineered human organoids. *EMBO J.* **2025**, *44*, 1569. [CrossRef] [PubMed]
107. Gebert, J.T.; Scribano, F.; Engevik, K.A.; Perry, J.L.; Hyser, J.M. Gastrointestinal organoids in the study of viral infections. *Am. J. Physiol. Gastrointest. Liver Physiol.* **2023**, *324*, G51–G59. [CrossRef]
108. Wang, J.; Tao, X.; Zhu, J.; Dai, Z.; Du, Y.; Xie, Y.; Chu, X.; Fu, G.; Lei, Z. Tumor organoid-immune co-culture models: Exploring a new perspective of tumor immunity. *Cell Death Discov.* **2025**, *11*, 195. [CrossRef]
109. Harter, M.F.; Recaldin, T.; Gjorevski, N. Organoids as Models of Immune-Organ Interaction. *Cell Rep.* **2025**, *44*, 116214. [CrossRef]
110. Vijg, J. From DNA damage to mutations: All roads lead to aging. *Ageing Res. Rev.* **2021**, *68*, 101316. [CrossRef]
111. Anglada, T.; Repullés, J.; Espinal, A.; LaBarge, M.A.; Stampfer, M.R.; Genescà, A.; Martín, M. Delayed γH2AX foci disappearance in mammary epithelial cells from aged women reveals an age-associated DNA repair defect. *Aging* **2019**, *11*, 1510. [CrossRef]
112. Sarapultsev, A.; Gusev, E.; Komelkova, M.; Utepova, I.; Luo, S.; Hu, D. JAK-STAT Signaling in Inflammation and Stress-Related Diseases: Implications for Therapeutic Interventions. *Mol. Biomed.* **2023**, *4*, 40. [CrossRef]
113. Evans, M.A.; Walsh, K. Clonal Hematopoiesis, Somatic Mosaicism, and Age-Associated Disease. *Physiol. Rev.* **2023**, *103*, 649–716. [CrossRef] [PubMed]
114. Mejia-Ramirez, E.; Florian, M.C. Understanding Intrinsic Hematopoietic Stem Cell Aging. *Haematologica* **2019**, *105*, 22. [CrossRef]
115. Fuster, J.J.; MacLauchlan, S.; Zuriaga, M.A.; Polackal, M.N.; Ostriker, A.C.; Chakraborty, R.; Wu, C.-L.; Sano, S.; Muralidharan, S.; Rius, C. Clonal Hematopoiesis Associated with TET2 Deficiency Accelerates Atherosclerosis Development in Mice. *Science* **2017**, *355*, 842–847. [CrossRef] [PubMed]
116. Avagyan, S.; Zon, L.I. Clonal Hematopoiesis and Inflammation–the Perpetual Cycle. *Trends Cell Biol.* **2023**, *33*, 695–707. [CrossRef]
117. Cobo, I.; Tanaka, T.; Glass, C.K.; Yeang, C. Clonal Hematopoiesis Driven by DNMT3A and TET2 Mutations: Role in Monocyte and Macrophage Biology and Atherosclerotic Cardiovascular Disease. *Curr. Opin. Hematol.* **2022**, *29*, 1–7. [CrossRef]
118. Artegiani, B.; Hendriks, D. Organoids from Pluripotent Stem Cells and Human Tissues: When Two Cultures Meet Each Other. *Dev. Cell* **2025**, *60*, 493–511. [CrossRef]

119. Sibilio, P.; De Smaele, E.; Paci, P.; Conte, F. Integrating multi-omics data: Methods and applications in human complex diseases. *Biotechnol. Rep.* **2025**, *48*, e00938. [CrossRef]
120. Baião, A.R.; Cai, Z.; Poulos, R.C.; Robinson, P.J.; Reddel, R.R.; Zhong, Q.; Vinga, S.; Gonçalves, E. A technical review of multi-omics data integration methods: From classical statistical to deep generative approaches. *Brief. Bioinform.* **2025**, *26*, bbaf355. [CrossRef]
121. Kester, L.; Van Oudenaarden, A. Single-Cell Transcriptomics Meets Lineage Tracing. *Cell Stem Cell* **2018**, *23*, 166–179. [CrossRef]
122. Kim, D.; Youn, J.; Kim, J.; Lee, J.; Yoon, J.; Kim, D.S. From Organoid Culture to Manufacturing: Technologies for Reproducible and Scalable Organoid Production. *npj Biomed. Innov.* **2026**, *3*, 12. [CrossRef]
123. Stuart, T.; Satija, R. Integrative Single-Cell Analysis. *Nat. Rev. Genet.* **2019**, *20*, 257–272. [CrossRef] [PubMed]
124. Larsson, L.; Frisén, J.; Lundeberg, J. Spatially Resolved Transcriptomics Adds a New Dimension to Genomics. *Nat. Methods* **2021**, *18*, 15–18. [CrossRef]
125. Longo, S.K.; Guo, M.G.; Ji, A.L.; Khavari, P.A. Integrating Single-Cell and Spatial Transcriptomics to Elucidate Intercellular Tissue Dynamics. *Nat. Rev. Genet.* **2021**, *22*, 627–644. [CrossRef]
126. Ritsma, L.; Ellenbroek, S.I.; Zomer, A.; Snippert, H.J.; De Sauvage, F.J.; Simons, B.D.; Clevers, H.; Van Rheenen, J. Intestinal Crypt Homeostasis Revealed at Single-Stem-Cell Level by In Vivo Live Imaging. *Nature* **2014**, *507*, 362–365. [CrossRef]
127. Fan, R. Single-Cell and Spatial Omics for Mapping Cellular Senescence in Health, Aging And Disease. *Innov. Aging* **2023**, *7*, 473. [CrossRef]
128. Azani, A.; Sharafi, M.; Doachi, R.; Akbarzadeh, S.; Lorestani, P.; Haji Kamanaj Olia, A.; Zahed, Z.; Gharedaghi, H.; Ghasrsaz, H.; Foroozand, H.; et al. Applications of CRISPR-Cas9 in Mitigating Cellular Senescence and Age-Related Disease Progression. *Clin. Exp. Med.* **2025**, *25*, 237. [CrossRef]
129. Lamprecht, J.; Sanad, S.A.; Kusumbe, A.P.; Habib, S.J. The Wnt–NAD+ Axis in Cancer, Aging, and Tissue Regeneration. *Trends Cell Biol.* **2025**, *36*, 154–174. [CrossRef]
130. Surina, S.; Chmielewska, A.; Pratscher, B.; Freund, P.; Rodríguez-Rojas, A.; Burgener, I.A. Organ-on-a-Chip: A Roadmap for Translational Research in Human and Veterinary Medicine. *Int. J. Mol. Sci.* **2025**, *26*, 10753. [CrossRef]
131. Cai, H.; Wang, D.; Zhao, Y.; Yang, C. Recent Advances in Microfluidic Chip Technology for Laboratory Medicine: Innovations and Artificial Intelligence Integration. *Biosensors* **2026**, *16*, 104. [CrossRef] [PubMed]
132. Fan, X.; Hou, K.; Liu, G.; Shi, R.; Wang, W.; Liang, G. Strategies to overcome the limitations of current organoid technology-engineered organoids. *J. Tissue Eng.* **2025**, *16*, 20417314251319475. [CrossRef] [PubMed]
133. Shroff, H.; Testa, I.; Jug, F.; Manley, S. Live-cell imaging powered by computation. *Nat. Rev. Mol. Cell Biol.* **2024**, *25*, 443–463. [CrossRef]
134. Rabinowitz, Z.M.; Cui, L. Detecting Cellular Senescence In Vivo: Imagining Imaging Better. *Aging Cancer* **2023**, *4*, 97–110. [CrossRef]
135. Zhang, X.; Gao, Y.; Zhang, S.; Wang, Y.; Pei, X.; Chen, Y.; Zhang, J.; Zhang, Y.; Du, Y.; Hao, S. Mitochondrial Dysfunction in the Regulation of Aging and Aging-Related Diseases. *Cell Commun. Signal.* **2025**, *23*, 290. [CrossRef]
136. Bai, L.; Wu, Y.; Li, G.; Zhang, W.; Zhang, H.; Su, J. AI-Enabled Organoids: Construction, Analysis, and Application. *Bioact. Mater.* **2024**, *31*, 525–548. [CrossRef] [PubMed]
137. Wen, J. Refining the generation, interpretation and application of multi-organ, multi-omics biological aging clocks. *Nat. Aging* **2025**, *5*, 1897–1913. [CrossRef]
138. Zhu, S.; Hua, H.; Chen, S. Rigorous integration of single-cell ATAC-seq data using regularized barycentric mapping. *Nat. Mach. Intell.* **2025**, *7*, 1461–1477. [CrossRef]
139. Bogoslowski, A.; An, M.; Penninger, J.M. Incorporating immune cells into organoid models: Essential for studying human disease. *Organoids* **2023**, *2*, 140–155. [CrossRef]
140. Abbasian, M.H.; Sobhani, N.; Sisakht, M.M.; D'Angelo, A.; Sirico, M.; Roudi, R. Patient-derived organoids: A game-changer in personalized cancer medicine. *Stem Cell Rev. Rep.* **2025**, *21*, 211–225. [CrossRef]
141. Zhu, Z.; Cheng, Y.; Liu, X.; Ding, W.; Liu, J.; Ling, Z.; Wu, L. Advances in the development and application of human organoids: Techniques, applications, and future perspectives. *Cell Transplant.* **2025**, *34*, 09636897241303271. [CrossRef] [PubMed]
142. Wang, D.; Villenave, R.; Stokar-Regenscheit, N.; Clevers, H. Human organoids as 3D in vitro platforms for drug discovery: Opportunities and challenges. *Nat. Rev. Drug Discov.* **2026**, *25*, 204–226. [CrossRef] [PubMed]
143. Balkhair, O.; Albalushi, H. Artificial Intelligence in Organoid-Based Disease Modeling: A New Frontier in Precision Medicine. *Biomimetics* **2025**, *10*, 845. [CrossRef]
144. Kolesnichenko, M. Rejuvenating the aging gut by targeting senescence: Organ rejuvenation. *Nat. Aging* **2026**, *6*, 32–33. [CrossRef] [PubMed]
145. Zhu, Y.; Tchkonia, T.; Pirtskhalava, T.; Gower, A.C.; Ding, H.; Giorgadze, N.; Palmer, A.K.; Ikeno, Y.; Hubbard, G.B.; Lenburg, M. The Achilles' Heel of Senescent Cells: From Transcriptome to Senolytic Drugs. *Aging Cell* **2015**, *14*, 644–658. [CrossRef]
146. Takebe, T.; Wells, J.M. Organoids by Design. *Science* **2019**, *364*, 956–959. [CrossRef]

https://doi.org/10.3390/targets4020012

147. Bulcaen, M.; Kortleven, P.; Liu, R.B.; Maule, G.; Dreano, E.; Kelly, M.; Ensinck, M.M.; Thierie, S.; Smits, M.; Ciciani, M.; et al. Prime editing functionally corrects cystic fibrosis-causing CFTR mutations in human organoids and airway epithelial cells. *Cell Rep. Med.* **2024**, *5*, 101544. [CrossRef]
148. Kumar, A.; Thirumurugan, K. Understanding cellular senescence: Pathways involved, therapeutics and longevity aiding. *Cell Cycle* **2023**, *22*, 2324–2345. [CrossRef]
149. Hautefort, I.; Poletti, M.; Papp, D.; Korcsmaros, T. Everything you always wanted to know about organoid-based models (and never dared to ask). *Cell. Mol. Gastroenterol. Hepatol.* **2022**, *14*, 311–331. [CrossRef]
150. Takebe, T.; Wells, J.M.; Helmrath, M.A.; Zorn, A.M. Organoid center strategies for accelerating clinical translation. *Cell Stem Cell* **2018**, *22*, 806–809. [CrossRef] [PubMed]
151. Papp, D.; Korcsmaros, T.; Hautefort, I. Revolutionizing immune research with organoid-based co-culture and chip systems. *Clin. Exp. Immunol.* **2024**, *218*, 40–54. [CrossRef] [PubMed]
152. Tenreiro, M.F.; Muotri, A.R. Reconstructing human corticogenesis: Insights from cerebral organoids into neurodevelopment and disease modeling. *Dev. Cell*, 2026; *in press*. [CrossRef]
153. Mohammadi, S.; Morell-Perez, C.; Wright, C.W.; Wyche, T.P.; White, C.H.; Sana, T.R.; Lieberman, L.A. Assessing donor-to-donor variability in human intestinal organoid cultures. *Stem Cell Rep.* **2021**, *16*, 2364–2378. [CrossRef]
154. Yamauchi, S.; Takahashi, A. Cellular Senescence: Mechanisms and Relevance to Cancer and Aging. *J. Biochem.* **2025**, *177*, 163–169. [CrossRef] [PubMed]
155. Wolff, L.; Hendrix, S. Rethinking Matrigel: The Complex Journey to Matrix Alternatives in Organoid Culture. *Adv. Sci.* **2025**, *12*, e08734. [CrossRef]
156. Aisenbrey, E.A.; Murphy, W.L. Synthetic alternatives to Matrigel. *Nat. Rev. Mater.* **2020**, *5*, 539–551. [CrossRef]
157. Heinzelmann, E.; Piraino, F.; Costa, M.; Roch, A.; Norkin, M.; Garnier, V.; Homicsko, K.; Brandenberg, N. iPSC-derived and Patient-Derived Organoids: Applications and challenges in scalability and reproducibility as pre-clinical models. *Curr. Res. Toxicol.* **2024**, *7*, 100197. [CrossRef]
158. Murrow, L.M.; Weber, R.J.; Gartner, Z.J. Dissecting the stem cell niche with organoid models: An engineering-based approach. *Development* **2017**, *144*, 998–1007. [CrossRef]
159. Zhao, Y.; Landau, S.; Okhovatian, S.; Liu, C.; Lu, R.X.Z.; Lai, B.F.L.; Wu, Q.; Kieda, J.; Cheung, K.; Rajasekar, S.; et al. Integrating organoids and organ-on-a-chip devices. *Nat. Rev. Bioeng.* **2024**, *2*, 588–608. [CrossRef]
160. Hofer, M.; Lutolf, M.P. Engineering Organoids. *Nat. Rev. Mater.* **2021**, *6*, 402–420. [CrossRef]
161. Ramesan, S.P.; Poonacha, J.B.; Pathirana, D.; Merkt, S.; Maass, C.; Hasenauer, J.; Reckzeh, E.S. Next-generation discovery: Empowering organoid research with machine learning, artificial intelligence, and mathematical modeling. *Trends Biotechnol.* **2026**. [CrossRef]
162. Yang, J.; Fischer, N.G.; Ye, Z. Revolutionising oral organoids with artificial intelligence. *Biomater. Transl.* **2024**, *5*, 372.
163. Suarez-Martinez, E.; Suazo-Sanchez, I.; Celis-Romero, M.; Carnero, A. 3D and organoid culture in research: Physiology, hereditary genetic diseases and cancer. *Cell Biosci.* **2022**, *12*, 39. [CrossRef] [PubMed]
164. Quintard, C.; Tubbs, E.; Jonsson, G.; Jiao, J.; Wang, J.; Werschler, N.; Laporte, C.; Pitaval, A.; Bah, T.-S.; Pomeranz, G. A Microfluidic Platform Integrating Functional Vascularized Organoids-on-Chip. *Nat. Commun.* **2024**, *15*, 1452. [CrossRef] [PubMed]
165. Ong, H.T.; Karatas, E.; Poquillon, T.; Grenci, G.; Furlan, A.; Dilasser, F.; Raffi, S.B.M.; Blanc, D.; Drimaracci, E.; Mikec, D.; et al. Digitalized organoids: Integrated pipeline for high-speed 3D analysis of organoid structures using multilevel segmentation and cellular topology. *Nat. Methods* **2025**, *22*, 1343–1354. [CrossRef] [PubMed]
166. Liu, H.; Zhang, X.; Liu, J.; Qin, J. Vascularization of engineered organoids. *BMEMat* **2023**, *1*, e12031. [CrossRef]
167. Shirure, V.S.; Hughes, C.C.; George, S.C. Engineering vascularized organoid-on-a-chip models. *Annu. Rev. Biomed. Eng.* **2021**, *23*, 141–167. [CrossRef] [PubMed]

https://doi.org/10.3390/targets4020012

targets

MDPI

Review

Comprehensive Roles of ZIP and ZnT Zinc Transporters in Metabolic Inflammation

Susmita Barman [1,2,3,*,†], Seetur R. Pradeep [1,3,†] and Krishnapura Srinivasan [1]

1 Department of Biochemistry, CSIR—Central Food Technological Research Institute, Mysore 570020, Karnataka, India; pradeep.sr23@gmail.com (S.R.P.)
2 Department of Obstetrics and Gynecology, University of Nebraska Medical Center, Omaha, NE 68198, USA
3 Division of Yoga & Life Sciences, Swami Vivekananda Yoga Anusandhana Samsthana (S-VYASA), Swami Vivekananda Rd., Jigani, Bingipura, Bangalore 560105, Karnataka, India
* Correspondence: susmita.barman1987@gmail.com
† These authors contributed equally to this work.

Abstract

Zinc homeostasis is fundamental to metabolic health, orchestrated by the coordinated actions of two major zinc transporter families: ZIP (Zrt- and Irt-like proteins) and ZnT (zinc transporters). ZIP transporters facilitate zinc influx into the cytosol from the extracellular space or from the lumen of intracellular organelles, whereas ZnT transporters control zinc efflux from the cytosol to the extracellular space or facilitate its sequestration into intracellular vesicles and organelles, concurrently harboring the meticulous intracellular zinc homeostasis. This equilibrium is essential for all critical functions like cellular response, metabolic control, and immune pathway alteration. Disruption of this homeostasis is a driver of different pathological alterations like metabolic inflammation, a chronic low-grade inflammatory state underlying obesity; type 2 diabetes; and nonalcoholic fatty liver disease. Recent studies revealed that ZIP and ZnT transporters dynamically regulate metabolic and inflammatory cues, with their tissue-specific expression varying by tissue and acclimating to different physiological and pathological conditions. Recent advanced research in molecular and genetic understanding has helped to deepen our knowledge of the interplay of activity between ZIP and ZnT transporters and their crosstalk in metabolic tissues, underscoring the potential therapeutic prospect for restoring zinc balance and ameliorating metabolic inflammation. This review provides a comprehensive overview that covers the function, regulation, and interactive crosstalk of ZIP and ZnT zinc transporters in metabolic tissues and their pathological conditions.

Keywords: zinc transporters; ZIP family; ZnT family; metabolic inflammation; obesity; diabetes; immune response; zinc homeostasis

1. Introduction

Zinc is an essential trace element intricately interwoven into the foundation of human physiology. It protects cellular homeostasis, controls gene transcription, supports metabolic reactions, and intervenes in intercellular communication and many more. Computational research indicates that approximately one tenth of the human proteome possesses zinc-binding capability, while over three percent of human genes encode zinc finger-containing proteins, underscoring the profound functional significance of zinc in regulating diverse biological systems [1]. A tightly organized orchestration of membrane-bound transport mechanisms controls zinc's intracellular equilibrium, essential for its biological activity.

Two primary families of zinc transporters are chiefly responsible for maintaining zinc homeostasis: the SLC30 family (ZnT), which facilitates zinc efflux from the cytoplasm to the extracellular space or into intracellular compartments, and the SLC39 family (ZIP), which mediates zinc influx into the cytoplasm from either extracellular sources or intracellular organelles [1]. Complementing these transporters is another indispensable class of proteins, metallothioneins (MTs), cysteine-rich, zinc-binding proteins that function as intracellular zinc reservoirs which also bind to excess free zinc ions to avoid cytotoxicity [2].

Researchers have identified an intricate association between zinc equilibrium and metabolic inflammation, a phenomenon also known as metaflammation, which is a chronic, low-grade inflammatory state arising from metabolic disorders rather than classical infectious triggers. Metaflammation is a defining feature of several prevalent metabolic diseases, including type 2 diabetes (T2D), nonalcoholic fatty liver disease (NAFLD), obesity, and cardiovascular disease. Persistent metabolic stress, especially developed from dietary habits which are rich in saturated fats and refined carbohydrates, initiates inflammatory cascades in different metabolic organs. These insults result in the secretion of pro-inflammatory cytokines like interleukin-1β (IL-1β) and tumor necrosis factor-alpha (TNF-α), which subsequently interfere with insulin signaling, impair glucose metabolism, and promote metabolic dysfunction [3].

Emerging research evidence indicates that zinc transporter dysfunction plays a crucial role in the pathophysiology of metaflammation. To maintain the intracellular zinc pool, several zinc transporters exhibit tissue-specific critical roles that significantly contribute to immune and metabolic regulation. In light of this, ZIP14 is increased in hepatocytes during inflammation and affects insulin sensitivity. In pancreatic cells, ZnT8 performs a crucial function in insulin formation and secretion, and its malfunction leads to decreased insulin production or β-cell stress [4,5]. Meanwhile, ZIP4 and -8 and ZnT2 are involved in immune cell activation and gut epithelial barrier function, respectively, both of which are pivotal for maintaining systemic inflammatory balance [6,7].

Importantly, besides maintaining zinc flux, zinc transporters are active participants in immune and metabolic signaling pathways. Transcriptional regulation of zinc transporter genes involves key inflammatory transcription factors such as nuclear factor-kappa B (NF-κB) and metal-regulatory transcription factor 1 (MTF-1) [8]. Furthermore, the innate immune system components, including Toll-like receptors (TLRs) and the NLRP3 (NOD-like receptor family, pyrin domain-containing 3) inflammasome, are known to modulate the expression of various transporters in response to metabolic stressors. Recent insights have further identified zinc as a second messenger in intracellular signaling, which is influenced by this stimulus [9].

Few commendable prior reviews have addressed zinc biology in immunity or metabolic illness independently, although they have often concentrated on either systemic zinc status or particular transporters in isolation. In this review, we endeavored to consolidate existing information about the unique and interrelated roles of ZIP and ZnT transporters in metabolic inflammation. We tried to present an overview of their molecular processes, tissue-specific functions, and intersection with metaflammation in primary metabolic disorders (T2D, NAFLD, and obesity). We also accentuated their potential as therapeutic targets in metabolic disorders and inflammation by elucidating the zinc-dependent regulatory webs that govern these functions. Overall, we wanted to highlight the translational prospect of zinc signaling ordinance in reforming metabolic health and mitigating chronic inflammation as a cohesive framework, which has not been comprehensively addressed in previous research. To systematically identify and evaluate studies examining ZIP and ZnT transporters in the context of metabolic inflammation, we performed a structured literature search across multiple electronic databases [e.g., PubMed,

https://doi.org/10.3390/targets4010005

Scopus, and Web of Science]. The selection process, including the specific counts of records identified, screened, and excluded based on predefined eligibility criteria, is detailed in the PRISMA flow diagram (Figure S1).

2. Key Machinery for Zinc Homeostasis

Maintaining tightly regulated cellular and tissue zinc concentrations is essential for the function of hundreds of enzymes, transcription factors, and signaling molecules. Zinc homeostasis requires the coordinated action of membrane-bound transporter families, metallothioneins, and organelle-specific trafficking mechanisms enabling dynamic control over import, export, and intracellular zinc distribution.

2.1. ZnT and ZIP Families: The Core Zinc Transporters

The ZIP (Zrt/Irt-like protein) transporter family (1–14 members) is essential for preserving intercellular zinc status by facilitating zinc transfer from the extracellular milieu to the cytoplasm or from intracellular organelle reserves. The transport system is crucial for maintaining zinc homeostasis within the cells or the whole system, particularly in zinc deficiency or cellular stress, to ensure adequate zinc supply to support the variety of biochemical processes that are essential for cellular function [10].

On the other hand, the ZnT (zinc transporter) transporter family (ZnT1–ZnT10) is responsible for facilitating the efflux of zinc from the cytoplasm, either into the extracellular space or into intracellular organelles such as the Golgi apparatus, vesicles, and mitochondria. These transporters are essential for maintaining cellular zinc homeostasis by excess zinc-induced cytotoxicity and ensuring proper compartmentalization of zinc pools, critical for specific signaling pathways within the cell [11].

The "yin–yang" regulation mechanism of these two families maintains cellular zinc homeostasis in opposition. ZIP transporters deliver cytosolic zinc for cellular processes, whereas ZnT transporters export excess zinc to ensure optimal storage for optimal cellular performance [12]. While numerous ZIP and ZnT members exhibit overlapping localizations and can partially compensate for the absence of a single transporter, such as ZnT4, ZnT5/6, and ZnT7 collectively regulating Golgi zinc, or the upregulation of ZIP4/ZIP14 in the liver in response to ZnT8 deficiency, genetic and experimental investigations also demonstrate significant non-redundant, tissue-specific functions, indicating that the loss of critical transporters like ZnT8 in β-cells or ZIP14 in hepatocytes cannot be entirely compensated by other family members. Comprehensive details are discussed in the following sections.

2.2. Metallothioneins: Crucial Intracellular Zinc Buffers

Metallothioneins (MTs), a group of small, cysteine-rich proteins, function as the crucial intracellular zinc buffers and dynamic reservoirs, which help in tightly regulating the concentration of free/labile zinc ions [Zn2+] in the picomolar (pM) to low nanomolar (nM) range [13]. This dynamic binding allows MTs to buffer both steady-state and transient levels of free zinc ions (which are typically in the picomolar range), thereby controlling zinc availability for a wide range of biological processes, including enzyme activity, gene expression, boosting immunity, and fighting against oxidative stress via zinc-finger transcription factors. MTs are not exclusively zinc-specific; they can also chelate other divalent and certain monovalent metal ions, including copper, cadmium, mercury, and silver. MTs play dual roles in essential metal homeostasis and detoxification of toxic metals [14]. Considering the complex role of metallothioneins, the partly zinc-depleted variant MT2 functions not only as a zinc buffer but also dynamically regulates cytosolic zinc concentrations by both donating and accepting zinc ions to maintain the zinc pool, releasing them when they are required for enzymatic or structural activities inside the cell [13,15]. Moreover, MTs

https://doi.org/10.3390/targets4010005

are essential for eliminating surplus zinc and thereby evading the cytotoxic consequences of zinc aggregation. Of note, MTs are involved in redox regulation, as their zinc–thiolate clusters are highly sensitive to oxidative and reductive changes, allowing them to modulate cellular redox status and mitigate oxidative stress as well as cellular defense [6,16]. MTF-1 meticulously controls MT synthesis. MTF-1 recognizes an increase in free zinc and activates MTs to control the surplus zinc [17].

The synchronized and antagonistic function of ZIP and ZnT transporters, in conjunction with zinc-binding protein MTs, is crucial for maintaining the cytosolic free zinc content within a confined, low range (pM to low nM values). By shuttling zinc across designated membranes, with precise orchestrated mechanisms they maintain the intracellular zinc gradients essential for the finest cellular signaling and general physiological function.

2.3. Key Cellular Zinc Compartments and Organelle-Specific Zinc Trafficking

Zinc has a markedly heterogeneous distribution across different cellular compartments, each of which maintains specific zinc concentrations targeted for certain enzymatic, structural, or signaling roles. The complementary functions of ZnT and ZIP transporters coordinate zinc flow across organellar membranes to dynamically modulate subcellular zinc distribution. Precise regulation of the zinc pool is critical for basic cell survival, redox equilibrium, and intracellular signaling.

Endoplasmic Reticulum (ER): In the endoplasmic reticulum, zinc transporters such as ZnT5 and ZnT7 play a critical role in facilitating zinc influx into the ER lumen, where this transported zinc is essential for the proper folding and structural maturation of nascent proteins, including several crucial zinc-requiring ectoenzymes like Phosphatidylinositol-glycan anchor biosynthesis (PIG) proteins, Sphingomyelin phosphodiesterase 1 (SMPD1), chaperones, etc. [18,19]. Zinc present in the ER promotes the activity of certain folding enzymes and chaperones, and it is crucial for managing unfolded protein response (UPR) pathways for cellular stress adaptation, hence aiding cellular responses to metabolic or oxidative stressors [20].

Golgi Apparatus: Zinc is crucial in the Golgi apparatus as a cofactor for specific enzymes involved in protein post-translational modification and transport. ERp44 (Endoplasmic Reticulum protein 44), an early secretory pathway chaperone, binds Zn2+ to regulate client binding and release for protein flow and homeostasis. In return for protons, ZnT4, ZnT5/ZnT6, and ZnT7 membrane transporter complexes import Zn2+ into the Golgi lumen. The Amagai research group, in their study, showed that systematic ZnT-knockdowns exhibited distinct control of labile Zn2+ within the Golgi stack. ZnT4, ZnT5/ZnT6, and ZnT7 control labile Zn2+ content in the distal, middle, and proximal Golgi, respectively, consistent with their local ZnT-mediated Zn2+ fluxes regulating ERp44 localization, trafficking, and client-retrieval activity, as shown by time-course imaging of cells undergoing synchronized secretory protein traffic and functional experiments [21]. Additionally, key zinc-dependent processes and enzymes include Golgi alpha-mannosidase II (GMII or MAN2A1/MAN2A2), Beta-4-galactosyltransferase, and alkaline phosphatases (ALPs), which mostly depend on ZnT4–7 to maintain zinc homeostasis within the Golgi lumen [22–24]. Zinc supports these enzymatic activities related to glycosylation, protein packaging, and vesicle trafficking. Disruption in these transporters, particularly ZnT4 mutations, has been shown to impair secretory pathway efficiency and compromise protein maturation, potentially leading to broader cellular dysfunction in secretory tissues.

Mitochondria: Mitochondrial zinc homeostasis, albeit less well characterized, is increasingly acknowledged as a vital regulatory axis. Recent research emphasizes the importance of precisely regulated mitochondrial zinc homeostasis, as both deficiency and excess can affect redox homeostasis, increase oxidative stress, and cause metabolic inflammation

https://doi.org/10.3390/targets4010005

and mitochondrial dysfunction by targeting specific electron transport chain (ETC) complexes [25]. Several ongoing studies have shown interest in defining the involvement of transporters in mitochondrial zinc transportation, highlighting a promising novel approach for targeted therapies in different mitochondria-dependent pathological conditions, including metabolic disorders and cancer.

Vesicles and Insulin Granules: Vesicles accumulate high concentrations of zinc within their lumen through specific transporters of the ZnT (SLC30) family, such as ZnT2, ZnT3, and ZnT8, which pump zinc from the cytosol into secretory and synaptic vesicles. According to a cryo-EM and immunolocalization study, ZnT transporters are required for vesicular zinc storage and controlled zinc secretion in neuronal, endocrine, and epithelial cells [26]. ZnT2 (SLC30A2) is a vesicular zinc transporter that directs zinc into lysosomes and secretory vesicles, especially in mammary epithelial cells, enabling zinc secretion into breast milk [27]. The ZnT2 mutation degrades this process, which in turn leads to transient newborn zinc deficit (TNZD) in breastfed babies due to low milk zinc levels [28]. ZnT8 transports zinc into insulin granules in pancreatic β-cells, facilitating insulin crystallization and further maturation and secretion. Furthermore, zinc-mediated crystallization is crucial for maintaining proper glucose homeostasis in the body, and dysfunction of this transporter contributes to β-cell stress and T2D [29].

Lysosomes and Endosomes: Zinc trafficking within lysosomes and endosomes is systematically regulated by the ZnT and ZIP family transporters, the TRPML1 (transient receptor potential mucolipin 1) ion channel, and lysosomal exocytosis. ZnT2 and ZnT4 localize to the lysosomal membrane and mediate zinc import into the lumen for storage or detoxification, as documented by recent experimental knockdown (TRPML1KD) and colocalization studies. TRPML1 (MCOLN1) serves as a key channel for zinc efflux from lysosomal stores to the cytoplasm, with loss of function (LOF) causing pathological zinc accumulation and lysosomal enlargement, a phenotype seen in mucolipidosis type IV, a rare genetic disorder. The function of these transporters is modulated by lysosomal acidification (via Vascular-ATPase) and the transcription factor MTF-1, which upregulate the expression of ZnT2 and ZnT4 in response to excess zinc [30]. ZIP transporters like ZIP8 and ZIP13 can also regulate lysosomal–cytosolic zinc exchange and immune signaling [31]. Proper zinc balance in these organelles is crucial for protein degradation, autophagy, immune activation, and receptor recycling, and dysregulation is implicated in immune disorders, metabolic disease, and lysosomal storage pathologies.

This intricate and compartmentalized zinc trafficking system ensures that each organelle maintains the optimal zinc concentration for its specialized functions. Alterations in the zinc concentration within organelles, either due to mutations in transporter genes, inflammation, or metabolite disturbances, can compromise their integrity and proper functioning. This may lead to disruption of fundamental operations, including autophagy, protein folding, and immune signaling, which ultimately leads to cellular dysfunction and associated pathological conditions. Thus, precise regulation of zinc distribution within organelles is a critical mechanism underpinning cellular physiology and pathology, going beyond basic maintenance to serve as a foundation for organelle and cell health. Key components of Compartments and Organelle-Specific zinc homeostasis are summarized in Table 1.

https://doi.org/10.3390/targets4010005

Table 1. Zinc homeostasis components.

Organelle	Transporters Involved	Function and Physiological Significance	Key Experimental Study Details and Insights	Reference
Cytosol	ZIP family (influx), ZnT family (efflux into organelles/extracellular space)	Maintains low free Zn2+ (picomolar) to support signaling and prevent toxicity; main hub for zinc-sensitive enzymes and kinases.	ZIP1/ZIP3 found in intracellular organelles (HEK293, mouse), localize dynamically based on zinc status. Cytosolic zinc waves observed in mast cells, dependent on ZIP/ZnT activity.	[32]
Secretory Vesicles (e.g., synaptic, insulin granules)	ZnT3, ZnT8, ZnT2, ZnT4	Accumulate high vesicular zinc (millimolar range); ZnT3 loads synaptic vesicles (neurons); ZnT8 loads insulin granules (pancreatic β-cells); ZnT2 regulates glandular vesicles; ZnT4 traffics vesicles in secretory tissues.	ZnT8 knockout and variant studies confirm granule-specific insulin packaging/diabetes risk. ZnT3 is involved in heterogeneous synaptic vesicle assembly and neurotransmission. ZnT2 is critical for zinc vesicle formation and stress protection.	[29,33–38]
Golgi Apparatus and ER	ZnT5, ZnT6, ZnT7 (influx into lumen); ZIP7, ZIP13 (efflux into cytosol)	Zinc is required for folding and activation of secreted/membrane proteins, e.g., tissue-nonspecific alkaline phosphatase (TNAP), ERp44; ZIP exports zinc in "zinc wave" for cytosolic signaling; ZnT5/6/7 localized to Golgi and ER, essential for ALP activation.	ZnT5 variant B localizes to ER, colocalizes with ZIP7, forming zinc efflux pathway; ZIP7 essential for cytosolic zinc signaling. ZnT7 localizes to proximal Golgi and regulates ERp44-dependent homeostasis. ZnT5/6/7 activate tissue-nonspecific alkaline phosphatase in two-step mechanism.	[15,18,21,22,39,40]
Mitochondria	ZnT2 (suggested), ZnT9 (SLC30A9), ZIP family (potential roles)	Mitochondrial zinc pools regulate oxidative metabolism, apoptosis, mitophagy; zinc influx/efflux affects mitochondrial stress resilience and cytochrome c release.	SLC30A9 (ZnT9) loss causes zinc mishandling and mitochondrial overload in HeLa cells, shown by live dye tracking and ERC coevolution analysis. Zn-induced mitochondrial swelling triggers mPTP opening, mediates apoptosis.	[25,41–44]

https://doi.org/10.3390/targets4010005

Table 1. *Cont.*

Organelle	Transporters Involved	Function and Physiological Significance	Key Experimental Study Details and Insights	Reference
Lysosome-Related Organelles (LROs)	CDF-2 (ZnT family), ZIPT-2.3 (ZIP family, C. elegans), ZnT4	Dynamic zinc storage and release; maintain organellar zinc pools for protein degradation/homeostasis, critical in stress adaptation.	In C. elegans, CDF-2 stores zinc in LROs during excess, ZIPT-2.3 releases zinc during deficiency; co-regulation and colocalization confirmed by super-resolution microscopy and transgenics. Morphological changes reflect zinc status and transporter levels.	[45]
Nucleus	Metallothioneins (MTs), potential transporters	Zinc primarily bound to transcription factors (zinc fingers), essential for gene expression/DNA replication; labile pool controls TF binding/dynamics.	Single-molecule microscopy reveals that zinc availability modulates DNA binding of zinc finger TFs (MTF-1, CTCF, GR) in live mammalian cells; zinc depletion shortens TF dwell time. MTs buffer nuclear zinc and protect against oxidative injury.	[2,46]

3. Functional Links Between Zinc and Metaflammation

Metaflammation is classified as persistent, low-grade systemic low-grade inflammation that originates in metabolic tissues in response to different factors. Chronic overnutrition and sedentary lifestyle drive metabolic dysfunction that lays the foundation for metaflammation by overloading key metabolic pathways in the liver, adipose tissue, skeletal muscle, and pancreatic β-cells. This condition is distinct from acute inflammation, which is transitory and subsides upon rejuvenation. Metaflammation, which affects metabolically active tissues such as adipose tissue, the liver, the pancreas, and skeletal muscle, stays constant, though it is less intense [3].

The hallmark of metaflammation is the sustained production of pro-inflammatory cytokines, such as IL-1β, TNF-α, and interleukin-6 (IL-6). Metaflammation is intricately associated with metabolic overload, a condition resulting from uncontrolled intake of food components, which triggers complex and crucial crosstalk between the systemic immune and metabolic pathways. The core mechanism involves specific immune cells that reside within metabolic tissues, most notably tissue-resident macrophages and other components of the innate immune system [47]. These interactions can lead to insulin resistance, glucose intolerance, and disturbances in lipid metabolism, thereby contributing to the development of chronic health conditions such as type 2 diabetes (T2D), nonalcoholic fatty liver disease (NAFLD), atherosclerosis, and hypertension associated with obesity. Collectively, alterations in insulin signaling, mitochondrial–ER–Golgi functions, and PRR (pattern recognition receptor) activation represent the onset of metaflammation from metabolic dysfunction, transforming nutrient-sensing pathways into chronic inflammatory signaling hubs in different metabolic tissues. Zinc dyshomeostasis and altered activity of ZIP/ZnT

https://doi.org/10.3390/targets4010005

transporters further modulate and amplify this response by modulating NF-κB, NLRP3, and oxidative stress pathways. Interestingly, zinc maintains a bidirectional relationship with inflammation: both deficiency and excess can promote metaflammation through overlapping mechanisms. Inadequate zinc disrupts metallothionein buffering, increases NF-κB and NLRP3 action, and favors immunity towards pro-inflammatory conditions. On the other hand, zinc overload, commonly adjudicated by transporters, disrupts mitochondrial redox equilibrium, lysosomal function, and ER homeostasis, while also triggering NLRP3 and oxidative stress cascade. While zinc deficiency is clinically more prevalent, pathological zinc accumulation in specific compartments (such as ZIP14-mediated hepatic zinc influx and ZnT lysosomal sequestration) reflects these pro-inflammatory influences. Thus, maintaining optimal intracellular zinc mandates careful coordination between ZIP and ZnT to control both extremes.

Therefore, the choice of dietary intake plays a major role in establishing and aggravating metaflammation. As a prominent example, Western diets high in saturated fatty acids (SFAs) and processed carbs fuel metabolic organ inflammatory signals and lead to ER stress, mitochondrial dysfunction, and ROS [48–50]. Lipotoxicity, ER stress, and NLRP3 inflammasome activation arise from saturated fatty acid stimulation of TLR4. This cascade increases IL-1β production and promotes pro-inflammatory macrophage polarization in adipose tissue [51,52]. Importantly, overconsuming glucose and fructose facilitates the production of advanced glycation end-products (AGEs), which worsen liver inflammation and lipid metabolism via the receptor for AGEs (RAGE) signaling pathway [53–55]. The chronic state of overnutrition also drives the production of ROS due to mitochondrial overload and the activity of NADPH oxidase [56,57]. This redox can displace zinc from specific zinc-binding proteins like cysteine thiols, MTs, which can further interfere in signaling dependent on cytosolic zinc status [58]. Thus, metabolic tissues can convert into immunological-active organs, addressing the close relationship between food signaling and immune responses in metaflammation. These relationships are summarized schematically in Figure 1, which illustrates how metabolic stressors, zinc fluxes, and zinc transporters converge on inflammatory signaling pathways to drive metaflammation.

3.1. The Influence of Zinc Transporter Imbalance in Metaflammation

Zinc plays a dual role in modulating metaflammation. An imbalance in zinc equilibrium, either through deficiency or mis-localization within the cell's compartments, disrupts immune homeostasis and increases both oxidative stress and inflammatory signaling pathways [59,60]. Importantly, the zinc-deprived condition influences cytokine production, affecting pro-inflammatory profile activation, primarily associated with increased levels of TNF-α, IL-1β, and IL-6, coupled with a decline in anti-inflammatory cytokines such as IL-10 [61]. This imbalance fosters the polarization of macrophages towards the M1 phenotype, impairs T-cell growth, and limits the efficacy of regulatory T-cells, consequently intensifying insulin resistance [62,63].

The expression of zinc transporters is tightly regulated by inflammatory signals. For instance, in the liver, ZIP14 is upregulated by IL-6. This upregulation leads to enhanced zinc accumulation in the liver and subsequent changes in how zinc is distributed throughout the body, ultimately affecting insulin sensitivity [64]. Dysfunctional ZnT8 can impair the crystallization and secretion of insulin in pancreatic β-cells, while ZIP8 influences NF-κB signaling in immune cells. Disruptions in the function or expression of specific zinc transporters are indeed linked to impaired metabolic function and sustained inflammatory responses involving several mechanisms. Specific levels of zinc are also critical in modulating inflammatory responses within the system, primarily by inhibiting the NF-κB signaling pathway [65,66]. This inhibition occurs through the blockage of IκB kinase

https://doi.org/10.3390/targets4010005

activity, a key enzyme responsible for the phosphorylation and subsequent degradation of IκB proteins. Under normal, zinc-sufficient conditions, NF-κB is held in an inactive state in the cytoplasm by IκB proteins. In response to inflammatory stimuli (like infection or cytokines), IκB is phosphorylated by the IκB kinase (IKK) complex, ubiquitinated, and degraded by the proteasome. This degradation allows NF-κB to translocate into the nucleus and activate pro-inflammatory gene expression [66]. In contrast, zinc deficiency impairs inflammation regulation by enhancing NLRP3 inflammasome activation, a multiprotein complex central to innate immune responses that promotes IL-1β secretion, a key pro-inflammatory cytokine. This occurs through zinc-dependent mechanisms, including lysosomal stress, reduced metallothionein buffering, and increased reactive oxygen species that prime NLRP3 assembly across immune and metabolic cells [67]. While causality varies by tissue, these changes contribute to the sustained low-grade inflammation characteristic of metaflammation, underscoring the importance of maintaining optimal zinc levels to support immune–metabolic balance.

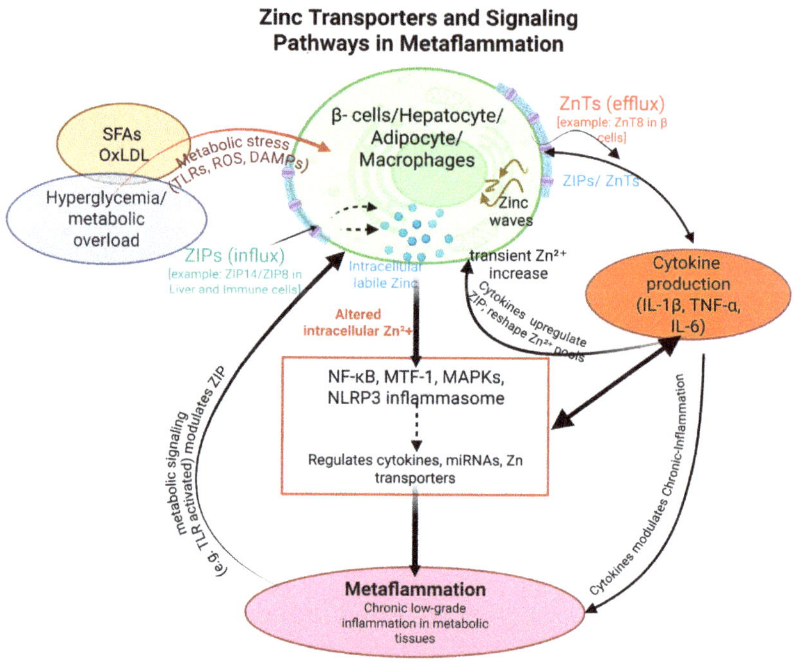

Figure 1. Zinc transporters and signaling pathways in metaflammation. Hyperglycemia, metabolic overload, saturated fatty acids (SFAs), oxidized LDL (OxLDL), and damage-associated molecular patterns (DAMPs) stimulate pattern recognition receptors, including Toll-like receptors (TLRs), inducing metabolic stress in β-cells, hepatocytes, adipocytes, and macrophages. The stress revises the activity and expression of ZIP (influx) and ZnT (efflux) zinc transporters, resulting in transient elevation and redistribution of intracellular labile Zn2+ ("zinc waves"). Modifications in zinc pools, characterized by shifts in intracellular free zinc concentration, act as signaling mechanisms that control immune cell activity and inflammatory responses. Pro-inflammatory cytokines upregulate ZIPs and modulate zinc homeostasis in metabolic and immunological organs, creating a positive feedback loop involving cytokine signaling, TLR activation, zinc transporters, and zinc reservoirs. Metaflammation/persistent low-grade inflammation in metabolic tissues is the foundation for several metabolism-associated diseases, including obesity, insulin resistance, and type 2 diabetes.

https://doi.org/10.3390/targets4010005

3.2. Zinc Transporters and MicroRNA-Mediated Regulation of Metaflammation

Zinc transporters regulate the biogenesis and expression of specific microRNAs (miRNAs) involved in immune and metabolic pathways, in part by altering intracellular zinc concentrations that affect zinc-sensitive transcription factors and signaling nodes. Rather than acting as generic regulators, zinc-modulated miRNAs influence defined targets that connect zinc status to cytokine production, immune cell behavior, and metabolic programming. Recent studies demonstrated that transporters such as ZIP14 modulate the expression of zinc-sensitive miRNAs such as miR-675 and others within intestinal and hepatic tissues [68]. In enterocytes, ZIP14 deficiency has been shown to impact intracellular zinc balance, resulting in modified miRNA profiles, which in turn influence the transcriptional regulation of inflammatory pathways, including NF-κB and STAT3 signaling, thereby fostering a pro-inflammatory condition and undermining gut barrier integrity. Moreover, aberrant zinc homeostasis resulting from transporter dysfunction in different metabolic tissues has been shown to regulate the expression of miRNAs by the "zinc–miRNA axis" linked to insulin resistance and lipid metabolism, further contributing to metaflammation [69]. Another study has shown that zinc deficiency upregulates miR-21 and promotes inflammation in adipose tissue by targeting the tumor suppressor PDCD4, whereas zinc repletion can normalize miR-21 levels and attenuate these inflammatory effects [69]. Similarly, miR-34a is increased in states of zinc deficiency and obesity, exacerbating insulin resistance and driving pro-inflammatory M1 macrophage polarization through suppression of SIRT1 and Klf4, as confirmed in both murine and human adipose tissue models.

In obesity and NAFLD, miR-122 influences systemic cholesterol and lipid metabolism by regulating key enzyme expressions and signaling pathways in hepatocytes. Studies show that miR-122 generally acts to promote hepatic lipogenesis and cholesterol synthesis in a healthy liver, and its dysregulation in disease states can either aggravate steatosis or, when markedly suppressed, alter lipid export and circulate lipid profiles in divergent ways. Different studies have evidenced a complex, dose-dependent relationship between zinc status and the expression of specific miRNAs, specifically involving miR-122 and miR-144-3p in metabolic tissues [70]. Moderate zinc deficiency has been shown to reduce miR-122 expression, leading to altered hepatic lipid handling and changes in serum lipid levels, while zinc supplementation can partially restore miR-122 and normalize aspects of lipid metabolism. Alternatively, excessive zinc exposure upregulated miR-144-3p, which has been exhibited to directly impair the Nrf2 antioxidant defense pathway and exacerbates oxidative stress and insulin resistance, as demonstrated in in vitro hepatocyte and adipocyte models. Together, these findings suggest that zinc status, acting through transporter-regulated miRNAs, such as miR-122 and miR-144-3p, forms a functional molecular axis that links zinc homeostasis to inflammatory, metabolic, and redox signaling pathways, with different zinc concentrations driving distinct miRNA-mediated outcomes rather than a simple linear effect. Of note, while these studies collectively establish a zinc–miRNA axis linking transporter status to inflammation, important limitations remain. Most evidence derives from cell culture or rodent models, with human data limited to correlative expression changes rather than causal transporter–miRNA relationships. Conflicting reports on miR-122 directionality (up- vs. downregulation) in NAFLD highlight tissue- and context-specific effects that require clarification through targeted genetic and longitudinal human studies. Key zinc transporter–regulated inflammatory and metabolic pathways in different tissues are summarized in Table 2.

https://doi.org/10.3390/targets4010005

Table 2. Zinc-related impacts of pathogenic triggers on inflammatory outcomes.

Pathogenic Trigger	Zinc-Related Impact	Inflammatory Outcome/Mechanism	Mechanistic/Clinical Notes	
SFAs, ROS, glucose overload	Disruption of zinc transporter expression: ZIP14 ↑ (in hepatocytes, adipocytes), ZnT8 ↓ (in pancreatic β-cells)	NLRP3 inflammasome activation, increased IL-1β/IL-18 release, chronic metabolic inflammation	ZIP14 upregulation in response to TLR4 activation and IL-6 drives hepatic/adipose zinc accumulation and insulin resistance. ZnT8 downregulation impairs insulin granule formation and β-cell function. High glucose and ROS amplify IL-1β/IL-18 via NLRP3.	[51,64,65]
Zinc deficiency	Reduced Treg (regulatory T) cell numbers, increased NF-κB activation, impaired metallothionein buffering	Chronic low-grade inflammation, heightened NLRP3 activation, increased cytokine output	Zn deficiency leads to lysosomal stress, ROS generation, and NLRP3 inflammasome activation/secretion of IL-1β. Zinc supplementation inhibits NLRP3 and supports immune balance.	[56,67,71]
Zinc transporter dysfunction	Alters zinc distribution in pancreas (ZnT8), liver/adipose (ZIP14), gut (ZnT2/ZIP8)	Insulin resistance, gut barrier leakiness, cytokine imbalance	Genetic or acquired dysfunction in ZnT8/ZIP14 impairs insulin packaging/secretion and hepatic/adipose zinc homeostasis. ZnT2/ZIP8 regulate intestinal barrier integrity; dysfunction increases permeability and systemic inflammation.	[64,65,72,73]
Oxidative stress	Displaces zinc from protein binding sites, impairs antioxidant function	Amplifies ROS, triggers NLRP3 activation, further immune cell recruitment and cytokine release	Oxidative stress displaces zinc, activates stress kinases, and amplifies pro-inflammatory signaling. Zinc repletion reduces ROS and NLRP3 activity, supporting antioxidant defenses.	[56,74]

4. Tissue-Specific Roles of Zinc Transporters

Zinc transporters exhibit specific, non-redundant roles depending on the tissue and the subcellular compartment where they are expressed, allowing for fine-tuned control of zinc homeostasis for particular physiological functions. The following discussion delves into the evidence highlighting the functions of key zinc transporters in specific organs, particularly their implications for metaflammation and metabolic disorders.

4.1. Liver

The Role of ZIP and ZnT Transporters in inflammation and insulin resistance: The liver serves as a pivotal regulator of metabolism and acts as a primary site for the redistribution of zinc during inflammatory responses. Among the several zinc transporters, ZIP14 is a pivotal zinc importer, with its expression markedly increased in hepatocytes during both acute and chronic inflammation. This induction is mostly initiated by pro-inflammatory

https://doi.org/10.3390/targets4010005

cytokines, including IL-6 and TNF-α [75]. The overexpression of ZIP14 enhances the inflow of zinc into hepatocytes, which is closely associated with the acute-phase response and marked by a fast decrease in blood zinc levels while hepatic zinc reserves simultaneously increase. ZIP14 is located in the cytosol and the mitochondria of hepatocytes, where it conjunctively works with ZIP8 to maintain zinc balance within mitochondria [76]. Controlling pathways like glycolysis, gluconeogenesis, glycogenesis, and glycogenolysis helps liver cells conserve energy and adapt to insulin and glucose fluctuations. ZIP14 regulates insulin signaling pathways via affecting PTP1B (protein tyrosine phosphatase 1B) and c-Met phosphorylation [75]. ZIP14KO animals demonstrated hepatic insulin resistance, glucose intolerance, and lipid metabolic issues. These mice also exhibited increased body fat and altered expression of glucose and lipid metabolism genes, such as GLUT2, SREBP-1c, and FASN [77]. ZIP4/14 and ZnT9 depletion or dysregulation impairs the liver's response to inflammatory stimuli and regeneration signals, increasing the likelihood of NAFLD and metabolic syndrome [78]. ZnT transporters export zinc from hepatocytes or store it intracellularly to lower cytosolic zinc levels. ZnT8 is an interesting exception to the steady expression of most ZnT transporters during inflammation. Studies have shown that ZnT8 deficiency in ZnT8 KO mice can protect the liver from injury due to acetaminophen (APAP)-induced oxidative stress and damage [79]. This protection is linked to an increase in hepatic zinc and MT levels, which enhances the liver's antioxidant defense system, and it is also associated with reduced liver cell death and inflammation. This protective strategy involves compensatory ZIP4 and ZIP14 overexpression, which increases hepatic zinc absorption, antioxidant protein production, and hepatocyte proliferation [78,79]. Overall, ZIP14 and ZnT8 work in concert to control the metabolic pathways and response to inflammation in the liver for proper maintenance of zinc homeostasis. Understanding these mechanisms is essential for elucidating the pathways that contribute to metabolic diseases and developing targeted therapeutic strategies.

4.2. Pancreas

ZIP and ZnT Transporters in Insulin Granule Biogenesis and Inflammation: Zinc is indispensable for insulin production, storage, and controlled secretion in pancreatic β-cells. The ZIP and ZnT transporters perform distinct, non-overlapping roles necessary for insulin granule production and cellular responses to metabolic and inflammatory stressors. ZnT8 is the main zinc exporter in insulin-producing β-cell membranes, where it transports cytosolic zinc into the lumen for insulin crystallization [80]. This transport is essential for insulin crystallization, which allows dense storage, proteolytic degradation protection, and structural stability. ZnT8-KO mice showed a significant reduction in crystalline insulin granules. Moreover, in ZnT8−/−-knockout mice, β-cells form immature, less dense granules packed with water-soluble insulin instead of solid zinc–insulin crystals [81]. These mice can still process and secrete basal insulin, but they struggle to maintain glucose homeostasis under metabolic stress, such as a high-fat diet. During high metabolic demand, crystalline insulin produced from ZnT8 is essential for reserve and packaging. Human genome-wide association studies (GWASs) have identified risk variants in the SLC30A8 gene, which encodes ZnT8, that are associated with a higher risk of T2D [82,83]. This gene is strongly linked to T2D pathogenesis, with specific common polymorphisms and rare LOF mutations impacting the risk of developing the disease. The finding underscores ZnT8's role in T2D, suggesting it as a potential therapeutic target. In addition to ZnT8, islet cells have ZnT5 and ZnT7 in their ER and Golgi apparatus [84]. These transporters collaborate to give zinc for proinsulin folding, hexamerization, and maturation before granulation. Proper insulin production and storage granule integrity are orchestrated by the functioning of these transporters. The ZIP family of zinc importers helps maintain β-cell zinc homeostasis, especially

https://doi.org/10.3390/targets4010005

under high insulin demand. Glucose stimulation upregulates transporters such ZIP6, ZIP7, ZIP8, and ZIP14 in mouse and human islets, increasing zinc absorption to replace granule reserves and fulfill cell biosynthetic needs. A recent study highlighted that ZIP14 acts as a negative regulator of glucose-stimulated insulin secretion (GSIS) by regulating the zinc trafficking and proper compartmentalization of intracellular zinc, particularly involving the ER [85]. This signifies the dynamic interplay between ZIP and ZnT transporters in regulating zinc trafficking during insulin production, storage, and secretion. On the other hand, inflammatory states associated with exposure to cytokines may drastically alter the expression and function of zinc transporters in β-cells during inflammatory conditions [86]. This modification may cause β-cell malfunction and death, which is frequently evinced in metabolic disorders like T2D. Pro-inflammatory stimuli can disrupt the expression of ZnT8 and ZIP transporters, thereby impairing zinc-dependent processes and linking immunometabolic stress to β-cell failure. In overview, zinc transporters, particularly ZnT8 and the various members of the ZIP family, are indispensable for orchestrating the biogenesis, structural integrity, and functionality of insulin granules in pancreatic β-cells. Dysregulation of these transporters results in decreased insulin processing, secretion, and β-cell resilience under inflammatory and metabolic stress. Understanding zinc homeostasis in pancreas functioning is crucial since such abnormalities are the key cause of diabetes and other metabolic diseases.

4.3. *Adipose Tissue*

As is well known, ZIP and ZnT transporters serve as crucial regulators of zinc homeostasis in adipose tissue, and their dysregulation profoundly affects adipose inflammation, insulin resistance, and overall metabolic health in this tissue. Within this complex tissue, zinc transporters, particularly those belonging to the ZIP and ZnT families, are essential regulators of immune responses, metabolic functions, inflammation, and insulin sensitivity across various adipose depots. Among these transporters, the most important is ZIP14 due to its significant increased expression in white adipose tissue (WAT) during systemic inflammation and metabolic stress [87]. This upregulation is especially prominent in adipocytes in conditions like lipopolysaccharide-induced endotoxemia, where inflammatory signals concomitantly stimulate a rise in ZIP14 expression. ZIP14 overexpression facilitates a surge in zinc influx into cells, effectively redistributing zinc in response to pro-inflammatory cytokines: IL-6 and TNF-α. Studies have revealed striking consequences of ZIP14 deficiency in the ZIP14KO murine model. ZIP14-deficient mice exhibited significant adipocyte hypertrophy, increased leptin production, and raised concentrations of pro-inflammatory cytokines such as IL-6, TNF-α, and IL-1β [88]. Additionally, ZIP14 deletion disrupts normal adipocyte differentiation and alters the fat mass-to-lean mass ratio, resulting in a phenotype marked by metabolic endotoxemia, excessive adipose tissue growth, and systemic insulin resistance [89]. This diseased condition closely resembles the characteristics of human metabolic syndrome and T2D, highlighting the significance of ZIP14 in metabolic health. Moreover, in mice, deficiency in ZIP14 leads to extensive systemic dysfunctions beyond adipocyte impacts, particularly chronic, low-grade systemic inflammation (metabolic endotoxemia), substantial manganese (Mn) accumulation in the brain resulting in neurotoxicity and motor impairments, as well as disrupted glucose and iron metabolism [77,90].

Another study showed that ZIP14 inhibition resulted in reduced bioavailability of cytosolic zinc, leading to dysregulation of pivotal inflammatory pathways, including the JAK/STAT and NF-κB signaling pathways [87]. Other members of the family, such as ZIP13, ZIP8, and ZnT7, have also been shown to play significant roles in modulating adipose tissue functionality and inflammatory responses [91]. In one study, ZIP13-null

https://doi.org/10.3390/targets4010005

mice displayed enhanced thermogenesis and improved insulin sensitivity, suggesting a complex interplay between zinc transport and adipose tissue browning [92]. On the other hand, apart from zinc efflux, the other family, i.e., ZnT transporters, significantly influence adipokine secretion, fat metabolism, and local immune responses. In chronic inflammatory conditions, such as obesity, studies have documented associations with altered expression and functionality of both ZIP and ZnT transporters, perpetuating a vicious cycle of adipose inflammation and metabolic dysfunction. Overall, the ZIP and ZnT transporter families' delicate balance of zinc influx and efflux determines adipose tissue's immunological and metabolic characteristics. The findings on ZIP disruption and inflammation, abnormal fat accumulation, and insulin resistance highlight the mechanistic importance of zinc homeostasis in adipose health and illness, making zinc transporters interesting targets for obesity-related metabolic diseases. The investigation into these transporters may lead to new treatments for metabolic illnesses associated with malfunctioning adipose tissue.

4.4. Gut Barrier

Firstly, it is well documented that zinc homeostasis is a critical player in maintaining gut health and its permeability. Zinc is essential for maintaining the integrity of the intestinal mucosal barrier and protecting against pathogens and toxins [93]. In a complex network, all the zinc transporters are intricately regulated to maintain the intestinal barrier and regulate immunological responses. The gut epithelium and immunological cells need these transporters to precisely absorb and distribute zinc. Proper zinc homeostasis maintains tight junctions, the mucosal layer, antimicrobial peptide production, and resident and recruited immune cell activity. ZIP4, the ZIP family's major zinc transporter, is located on enterocytes' apical membranes [6,94]. Of note, studies have shown that ZIP4 mutation results in acrodermatitis enteropathica, a condition that highlights the transporter's critical role in zinc homeostasis and overall gut health [95]. Other ZIP transporter family members, including ZIP2, ZIP6, ZIP7, and ZIP10, are also transcriptionally regulated in response to dietary zinc levels and activation of the aryl hydrocarbon receptor (AHR), which collectively boosts zinc import and fortifies intestinal barrier function [96]. Zinc influx is a critical player in maintaining the integrity of tight junction proteins, including claudins and occludins, which is again mediated by specific zinc transporters, primarily ZIP4 for dietary absorption and ZIP14 for systemic zinc uptake into enterocytes [73]. In ZIP14-knockout (ZIP14KO) mice, the observed increase in intestinal permeability, local inflammatory responses, and reduction in major histocompatibility complex class II (MHCII) gene expression are a consequence of intracellular zinc deficiency in intestinal epithelial cells (IECs) and the resulting epigenetic modifications and altered signaling pathways [97]. At the same time, loss of the ZIP14 transporter in Zip14ΔIEC experimental mice notably impacted intestinal integrity by disrupting chromatin accessibility and altering gene expression, which compromised the gut barrier and immune response through mechanisms involving zinc-dependent histone deacetylases (HDACs) and key transcription factors like CIITA. Furthermore, ZIP14 has been shown to influence gut microbiome composition, important for epithelial zinc homeostasis, microbial diversity, and immune functionality [89]. ZnT2 is predominantly found in the secretory granules of Paneth and goblet cells, where it plays a critical role in the secretion of antimicrobial peptides and mucins [98]. In the ZnT2-null mouse model, Paneth cells exhibited impaired function due to zinc-deficient secretory granules, leading to a compromised chemical barrier. Though ZnT1 and other ZnT members have been studied less extensively in the context of the intestine, they are believed to play a protective role by preventing cytosolic zinc overload and modulating zinc signaling during episodes of intestinal stress.

https://doi.org/10.3390/targets4010005

On the other hand, zinc is essential for tight junction protein production and maintenance, and deficits due to transporter failure or diet may cause intestinal permeability or "leaky gut." A weakened mucosal layer, decreased antimicrobial peptide synthesis, and dysregulated innate and adaptive immune responses are linked to this disease. In past decades, research showed the significance of zinc acting as a signaling cofactor via GPR39, activating PKCζ to improve epithelial barrier function [99]. Zinc supplementation and plant-derived AHR ligands increase zinc importer transcription, notably ZIP4, which has been shown to upregulate tight junction and mucin genes, strengthening barrier defense mechanisms [96]. Moreover, zinc transporter dysfunction is shown to weaken gut physical and chemical barriers and intestinal immune cell antigen presentation and cytokine transmission. This disruption may enhance microbial product translocation into systemic circulation, causing inflammation and metabolic abnormalities which may further lead to different pathological conditions such as metaflammation, IBD, obesity, and T2D [6,7,72,100]. Thus, ZIP and ZnT transporters are essential for zinc-dependent gut barrier function, immunological homeostasis, and microbiota composition. Their malfunction might significantly impact gut integrity and immunological competency, exposing people to chronic inflammatory diseases and health concerns, which underscore the need for better understanding of transporters' complex functions in zinc homeostasis and metabolic illness.

The tissue-specific results indicate both convergent (dysregulation of NF-κB/NLRP3) and divergent (protective versus pathogenic roles of ZnT/ZIPs) functions of zinc transporters; however, cross-study comparisons are hindered by variations in models (genetic versus pharmacological) and endpoints (insulin sensitivity versus inflammation). Human genetic correlations provide corroborative but inconclusive data since population studies are unable to differentiate causation from confounding variables. Extending this analysis to signaling mechanisms reveals similar patterns of convergence and complexity. Critical gaps include tissue-specific "optimal zinc set points" and the potential for single transporter manipulation to provide therapeutic advantages without off-target consequences. On the other hand, the apparent convergence of zinc effects on NF-κB, NLRP3, and TLR pathways across studies supports zinc as a central immune–metabolic regulator, yet mechanistic discrepancies persist. Some reports emphasize transcriptional regulation via MTF-1, while others highlight rapid zinc wave signaling, suggesting context-dependent mechanisms that remain poorly integrated. Clinical translation is limited by a lack of zinc compartment-specific biomarkers and uncertainty about whether systemic zinc supplementation can recapitulate tissue-specific transporter effects. Representative experimental and clinical evidence linking zinc transporter dysregulation to metabolic inflammation across tissues is summarized in Table 3.

Table 3. Tissue-specific zinc transporters, functions, and pathological implications and supporting studies.

Tissue	Zinc Transporters	Functions	Pathological Implications and Supporting Studies	References
Liver	ZIP14, ZIP8, ZIP1, ZIP10, ZnT1, ZnT5, ZnT6	ZIP14/ZIP8: hepatic zinc uptake; ZIP14 acute-phase-responsive; ZIP1/ZIP10 vesicular influx; ZnT1 zinc efflux.	ZIP14: hepatic inflammation, NAFLD, insulin resistance. ZnT1: systemic zinc balance. ZIP1: membrane/vesicle shuttling.	[76,78,101–103]

Table 3. *Cont.*

Tissue	Zinc Transporters	Functions	Pathological Implications and Supporting Studies	References
Pancreas	ZnT8, ZnT5, ZnT7, ZIP6, ZIP7, ZIP8, ZIP1	ZnT8: insulin granule zinc loading (T2D-linked); ZnT5/7: ER/Golgi zinc for hormone biosynthesis; ZIPs (ZIP1/ZIP3): cytosolic and organelle zinc homeostasis.	ZnT8: β-cell failure, impaired insulin secretion, diabetes risk. ZIP6/7/8: proinsulin processing, stress responses. ZIP1: vesicular localization.	[37,84,104]
Adipose Tissue	ZIP14, ZIP13, ZIP8, ZIP1, ZnT7, ZnT5	ZIP14: inflammatory zinc influx and immune signaling; ZIP13: secretory pathway and adipocyte differentiation/BMP–TGF-β; ZnT7: fat metabolism.	ZIP13: adipose inflammation, altered fat mass. ZIP14: metabolic syndrome, obesity-related inflammation. ZnT7: insulin sensitivity, adiposity. ZIP1/ZIP8: adipocyte zinc and cytokine regulation.	[88,91,92,105,106]
Gut (Intestine)	ZIP4, ZIP8, ZIP1, ZIP10, ZnT1, ZnT2, ZnT4	ZIP4: dietary zinc uptake (mutations → acrodermatitis enteropathica); ZIP8: immune cell zinc; ZIP1/ZIP10: epithelial zinc balance; ZnT1: basolateral export; ZnT2: zinc secretion from Paneth/goblet cells.	ZIP4: intestinal integrity, systemic zinc. ZnT1: gut zinc export, serum zinc. ZnT4: vesicle trafficking in enterocytes. ZIP8/ZIP1/ZIP10: intestinal immunity and barrier function.	[94–96,98,107,108]
Kidney	ZIP8, ZIP1, ZnT3, ZIP10, ZnT1, ZnT4, ZnT8	ZIPs/ZnTs support renal zinc reabsorption, homeostasis, and excretion; ZnT1/2 mRNA unique in kidneys; ZnT4 and ZnT6 traffic in vesicular compartment.	ZnT3, ZIP8, ZIP1: nephropathy risk, renal zinc handling. Transporter regulation in acute-phase zinc redistribution.	[57,109–112]
Brain	ZIP3, ZIP8, ZIP1, ZIP6, ZIP7, ZnT1, ZnT3, ZnT4, ZnT6	ZnT3: synaptic vesicle zinc; ZIP3/8/1/6: neuronal zinc influx and homeostasis; ZIP7: Golgi/ER zinc in neurons and glia; ZnT4/ZnT6: neural vesicular zinc.	ZnT3: synaptic zinc, plasticity, neurodegeneration. ZIP7: ER/Golgi zinc signaling in neurons/glia. Imbalance: cognition and neuroinflammation.	[34,113,114]

https://doi.org/10.3390/targets4010005

Roles of ZIP and ZnT Zinc Transporters in Key Metabolic Diseases:

To summarize the disease-specific roles of ZIP and ZnT transporters in the aforementioned pathological conditions, the following paragraphs highlight their key contributions. Zinc transporters play a significant role in the pathophysiology of T2D mainly by disrupting insulin synthesis and overall insulin sensitivity (as elaborated in Sections 4.2 and 5). As mentioned before, ZnT8 dysfunction hinders zinc loading and release of insulin granules in β-cells, elevating the risk of T2D, as shown by human genetic variations and knockout models [81,115]. Additionally, modifications in ZIP14 and ZnT7 within the liver disturb glucose homeostasis and PTP1B/c-Met signaling [77]. Additionally, in NAFLD, the overexpression of ZIP14 facilitates hepatic zinc accumulation, insulin resistance, and steatosis via the deregulation of SREBP-1c/FASN, whereas animals defective in ZIP14 exhibit protection against metabolic syndrome (refer to Sections 4.1 and 5). The deletion of ZnT8 paradoxically provides hepatoprotection against oxidative damage by inducing compensatory ZIP expression and upregulating metallothionein. Of note, obesity implicates ZIP14 and ZIP13 in white adipose tissue, facilitating inflammation, adipocyte hypertrophy, and inhibited browning, in conjunction with ZnT7's influence on adipokine release and insulin sensitivity (detailed in Sections 4.3 and 5). These transporters perpetuate metaflammation and macrophage polarization, exacerbating insulin resistance and comorbidities associated with obesity.

5. Zinc-Modulated Signaling Pathways

Zinc is known as a dynamic signal transducer and serves as a second messenger by regulating signaling cascades, affecting protein activities, and enabling cellular communication between external stimuli and intracellular responses. Recent studies have shown that rapid intracellular zinc release concomitant with calcium signaling influences cellular responses [116]. Intracellular zinc signaling is divided into two categories. One involves the relatively swift, stimulus-driven "zinc waves," and the other involves more gradual, transcription-mediated changes depending on the expression of zinc transporters [1]. Zinc waves occur within minutes following receptor activation, such as the FcεRI in mast cells, requiring both calcium influx and the activation of the MEK signaling cascade. This phenomenon leads to a transient increase in free cytosolic zinc levels, which directly modulates critical protein phosphatase and kinase activity. Notably, zinc inhibits the activity of phosphatases, thus prolonging the activation of mitogen-activated protein kinases (MAPKs), including ERK, JNK, and p38. Beyond these rapid signaling events, zinc also influences long-term cellular responses by affecting the expression and function level of different transcription factors, most notably MTF-1. Conversely, MTF-1 senses cytosolic zinc fluctuations and subsequently regulates the expression of MTs and zinc transporters involved in zinc homeostasis [17]. By activating these genes, MTF-1 helps maintain intracellular zinc balance and cellular adaptation to stress or excess metal exposure.

Zinc's Crosstalk with Canonical Pathways: NF-κB, MAPKs, NLRP3, and TLRs

Zinc plays a crucial role in suppressing the nuclear translocation of NF-κB by stabilizing its inhibitor, IκB, and directly inhibiting the activity of upstream kinases [117]. Conversely, when zinc levels are depleted, there is an enhancement in NF-κB-driven inflammatory gene expression, which can activate NLRP3 inflammasomes, a key component in metaflammation and immune responses [51]. The interplay between zinc waves and extracellular zinc influx activates key signaling pathways such as ERK and Akt. This happens through the phosphatase inhibition and direct stimulation of Ras and MEK, resulting in the subsequent phosphorylation and activation of CREB and Fosl1 [1,118]. Depleted zinc levels also activate stress-responsive kinases like JNK and p38, which are integral to the cellular

https://doi.org/10.3390/targets4010005

response to inflammatory stimuli and stress [119]. Adequate intracellular zinc levels serve to restrict the activation of the NLRP3 inflammasome, thereby mitigating the release of the pro-inflammatory cytokine IL-1β [67]. In contrast, zinc chelation or deficiency removes this inhibitory effect, resulting in an exaggerated inflammatory response.

Crosstalk Between Zinc, TLRs, and Metabolic Sensors: Zinc is essential for the immune system, playing a dual role as a cofactor and a signaling molecule (secondary messenger) that connects innate immunity and metabolic regulation. As a cofactor, it regulates thousands of proteins involved in basic cell functions like DNA replication and cell division. Interestingly, zinc ions can also function as a signaling molecule, modulating intracellular signaling pathways in immune cells and serving as second messengers in response to stimuli. This dual function allows zinc to markedly affect cellular responses to diverse danger signals, especially via its interaction with TLR and metabolic sensor pathways. On the other hand, TLRs are crucial pattern recognition receptors (PRRs) that serve as a first line of defense in the immune system by detecting molecules from microbes (like LPS) and also from the body (damage-associated molecular patterns or DAMPs) [120]. Recent research shows that TLRs can also be activated by metabolic stresses, such as saturated fatty acids, and that this activation is linked to chronic inflammation in metabolic diseases like obesity. Zinc's involvement in TLR signaling can be categorized into several critical functions which have been discussed briefly here. After being stimulated by LPS or palmitate, it directs rapid activation of NF-κB, with subsequent enhancement of pro-inflammatory cytokines [121], though this initial response is tightly regulated by negative feedback mechanisms to prevent excessive inflammation and potential tissue damage. As a feedback mechanism, this activation also triggers the upregulation of zinc importers, particularly ZIP14, facilitating the sequestration of zinc in the liver, a commonly observed response during acute inflammation [77], wherein an adequate zinc level is critical for negative regulation to curb excessive inflammation driven by TLRs.

In addition, zinc pilots MyD88-dependent pathways to decrease pro-inflammatory cytokine production by direct inhibition of IKK, preventing the activation and nuclear translocation of NF-κB. By blocking NF-κB, zinc restricts the downstream inflammatory response [122]. Zinc does indeed influence the functions of precise regulatory proteins like A-20 and the other zinc-finger protein, which again underscores its crucial role in preserving the critical equilibrium or the fine line between immune activation and its tolerance [123]. Research from past decades consistently highlights the importance of zinc as a trace element essential for the immune system, often modulating immune responses by serving as a critical cofactor for various enzymes and regulatory molecules.

Zinc transporters, particularly ZIP8 and ZIP14, therefore serve as critical modulators that restrain NF-κB activity and adjust protein tyrosine phosphatase function, thereby fine-tuning cellular responses to TLR activation in metabolic tissues [66,77,86]. In this way, zinc and its transporters act at the interface of immune and metabolic signaling, linking TLR-driven inflammatory responses with nutrient- and energy-sensing pathways summarized in the following sections, rather than functioning as independent or universal metabolic sensors.

Zinc is known to modulate AMPK activity, which serves as a master regulator of cellular energy homeostasis and fatty acid oxidation. It is known that AMPK integrates metabolic stress signals to maintain cellular energy equilibrium and that it has anti-inflammatory characteristics. Moreover, zinc may affect mTOR signaling pathways, which are vital for controlling autophagy, immune cell differentiation, and anabolic activities necessary for cell growth and metabolism [124]. Of note, zinc has been shown as a critical regulator in experimental tissues for activating insulin receptors and the ensuing IRS-1/PI3K/Akt signaling pathway, facilitating effective glucose absorption and

https://doi.org/10.3390/targets4010005

metabolism [125]. The zinc transporter ZnT8 is particularly important for insulin crystallization and secretion in β-cells, while zinc in peripheral tissues plays a role in modulating insulin sensitivity and downstream metabolic pathways. As mentioned earlier, in pathological conditions such as obesity and type 2 diabetes, metabolic stressors like SFA and ROS trigger chronic, low-grade inflammation via activation of TLRs and inflammasomes, which in turn leads to altered expression of zinc transporters like increased ZIP14 and decreased ZnT8 [77]. Disruption of intracellular zinc levels initiates a malicious cycle of metabolic and inflammatory signaling pathways by impairing the body's innate anti-inflammatory and insulin-sensitizing mechanisms, which in turn further dysregulates zinc equilibrium. Thus, extensive experimental and clinical research underscores zinc's critical role as an integrator of immune and metabolic signaling pathways. As has been shown, zinc modulates TLR responses and metabolic sensors, hence influencing cytokine production, metabolic adaptability, and the course of many illnesses. Consequently, focusing on zinc homeostasis and the related transporter pathways offers a viable therapeutic strategy for treating inflammatory and metabolic disorders. Figure 1 provides an overview of these zinc-modulated signaling events, highlighting ZIP/ZnT-dependent changes in labile Zn^{2+} and their impact on NF-κB, MAPK, MTF-1, and NLRP3 activation in metabolic tissues. Zinc-modulated signaling pathways, their functional roles, and associated physiological impacts, are summarized in Table 4.

Table 4. Zinc-modulated signaling pathways, roles, and physiological impacts (with literature support).

Signaling Pathway	Zinc's Role	Physiological Impact	Study Details	References
Zinc Waves	Acts as a second messenger; rapid release from ER/perinuclear stores after receptor stimulation (FcεRI, TLR, cAMP/PKA)	Modulates protein tyrosine phosphatase activity, prolongs MAPK activation, amplifies/controls cytokine (IL-6, TNF-α) production	Zinc waves occur within minutes after FcεRI crosslinking, dependent on Ca^{2+} and MEK signals. Inhibits phosphatases and sustains MAPKs; first described in mast cells.	[1]
NF-κB	Inhibits IκB kinase (IKK), stabilizes IκB, directly and indirectly restricts NF-κB nuclear translocation	Suppresses pro-inflammatory gene expression (e.g., TNF-α, IL-1β); zinc deficiency or transporter dysfunction relieves this suppression	Zinc wave enhances cytokine gene induction via prolonged MAPK and potentially NF-κB activation after FcεRI stimulation. Zinc essentially gates the amplitude/duration of the NF-κB response.	[1,9,51]
MTF-1	Direct zinc sensor: zinc binding activates metal response elements, upregulating metallothioneins and select ZnT genes	Promotes cellular defense against oxidative stress; increases zinc buffering capacity; adapts transporter profile to stress	Zinc exposure or cytosolic elevation leads to MTF-1 nuclear translocation and oxidative stress protection, well documented in immune and liver cells.	[17]

https://doi.org/10.3390/targets4010005

Table 4. *Cont.*

Signaling Pathway	Zinc's Role	Physiological Impact	Study Details	References
MAPKs (ERK, JNK, p38)	Zinc waves/influx modulate phosphorylation, inhibiting protein phosphatases, sustaining MAPK signaling	Controls cell proliferation, inflammation, cytokine output, and survival/differentiation signals	Zinc ionophores mimic zinc waves by prolonging MAPK activation, increasing late-phase IL-6/TNF-α expression in mast cells.	[1]
NLRP3 Inflammasome	Zinc deficiency or oxidation-driven displacement of zinc from proteins activates NLRP3 inflammasome, increases IL-1β	Promotes metaflammation, insulin resistance, and chronic inflammatory disease	Zinc supplementation inhibits NLRP3 activation; deficiency/oxidative stress enhances it. Linked to response in macrophages, adipose tissue.	[51]
TLRs	Zinc suppresses MyD88 and canonical NF-κB pathway activation in TLR4/2 signaling, modulates inflammatory threshold	Prevents excessive cytokine release on microbial/metabolic stimulation; restricts prolonged inflammation	TLR activation results in rapid transporter regulation and a decrease in free zinc as an early signal for dendritic cell activation. ZIP14 and ZIP8 up-/downregulation tightly couple TLR activity to zinc homeostasis.	[77,122]
Insulin Signaling	Zinc enhances Akt activation, supports phosphorylation cascade; ZnT8 ensures proper insulin packaging/release	Promotes glucose uptake, insulin secretion, and β-cell function; deficiency linked to impaired glycemic control	ZnT8 mutations disrupt insulin granule biogenesis and secretion, increasing T2D risk. Zinc signaling also influences IRS-1/PI3K/Akt sensitivity in target tissues.	[125,126]

6. Therapeutic Implications and Translational Potential of Zinc Transporters

Recent advancements have established zinc transporters as promising therapeutic targets for metabolic disorders like T2D, NAFLD, and obesity. The therapeutic landscape has been traversed via principal intervention strategies, which include employing promising clinically approved small-molecule modulators to affect zinc transporter activity, the implementation of nutritional zinc supplementation, and the direct targeting of zinc transporter proteins as viable drug candidates in clinical applications.

Small-Molecule Modulators of Zinc Transporters: The development of small compounds that may regulate the activity or expression of zinc transporters has considerable promise for reestablishing zinc homeostasis and improving cellular signaling pathways affected in several disease conditions. Currently there are no specific, approved small-molecule drugs that directly target the zinc transporter ZIP5 for clinical use in T2D, NAFLD, or obesity. Existing research in the form of genetic studies and experimental models suggests that pharmacological inhibition of SLC39A5 could be a viable therapeutic strategy in the future. This effect has been observed in human subjects harboring LOF mutations as well as in genetically modified mouse models, even when subjected to high-fat, high-fructose diets [127]. Furthermore, targeting the transporter ZIP13 has shown promise in modulating adipocyte browning, which could be beneficial in the context of obesity and T2D management [92,128]. ZnT8 in pancreatic β- cells supports insulin synthesis,

https://doi.org/10.3390/targets4010005

storage, and release, positioning it as a key target for diabetes therapies. Strategy-based therapies for insulin resistance, glucose intolerance, and dysregulated lipid metabolism may include zinc transporter augmentation or inhibition [129] for targeted therapeutics for metabolic illnesses such as T2D, NAFLD, and obesity. Research shows that addressing zinc imbalances in metabolic organs, including the pancreas, liver, adipose tissue, and muscle, may restore cellular function.

Nutritional interventions involving zinc supplementation have been linked to favorable outcomes in glycemic control and overall metabolic health, as evidenced by various animal studies and clinical trials [71,130,131]. Supplementation studies carried out by different researchers have shown that it has the ability to improve insulin sensitivity in individuals, decrease fasting glucose levels, lower triglyceride concentrations, and alleviate systemic inflammation, especially in those with zinc insufficiency [56,71]. In populations such as obese mice and humans suffering from metabolic syndrome or T2D, zinc supplementation has been associated with improved insulin response and beneficial alterations in metabolic profiles. Personalized zinc therapy is an emerging field, and while routine clinical practice does not yet widely offer specific zinc dosing based on an individual's full genetic profile, certain genetic tests and specialized medical consultation are available. To pursue personalized zinc therapy based on genetic variations, you would typically need to undergo genetic testing ordered by a medical professional or a specialized nutritionist, followed by consultation with a dietitian or genetic counselor to tailor a plan. This field is a developing area of nutritional science, and the clinical application of specific genetic variants for general zinc supplementation recommendations is still evolving.

Zinc Transporters as Druggable Targets in T2D, NAFLD, and Obesity: Several zinc transporters have emerged as promising candidates for future therapeutic targeting in metabolic disease, although most supporting data are currently preclinical or genetic in nature. Research indicates that genetic ablation or pharmacological inhibition of the ZIP5 transporter can lower the risk of diabetes and provide protection against hepatic metabolic dysfunction, supported by clinical evidence from large population studies and direct animal experimentation [132]. Variants of the ZnT8 transporter, whether gain- or loss-of-function variants, have been associated with altered susceptibility to T2D, primarily through their impact on insulin storage and release. Current investigations are exploring the modulation of ZnT8 as a therapeutic strategy for diabetes management primarily in experimental systems [29,126]. On the other hand, the ZIP13 and ZIP14 transporters have been documented as essential controllers of adipogenesis, insulin sensitivity, and hepatic glucose metabolism, as discussed earlier. Their modification may target concerns such as insulin resistance, adipose tissue inflammation, and steatosis, with considerable implications for the treatment of obesity and NAFLD. Though there are no currently FDA-approved drugs that specifically target ZIP13 or ZIP14 for the treatment of metabolic disorders, research in animal models has identified potential therapeutic strategies involving the modulation of zinc signaling pathways associated with these transporters. For example, proof-of-concept studies showed that bortezomib, used to treat the genetic disorder spondylocheirodysplastic Ehlers–Danlos syndrome (SCD-EDS), caused by mutations in the ZIP13 gene, helped to restore normal ZIP13 protein levels and intracellular zinc homeostasis in cellular models, such that it could be a promising repurposed or adapted drug option for common metabolic diseases in the future [133]. Therefore, targeting the zinc signaling cascade might reveal a potential restorative approach. From the above discussion, we know that zinc transporters not only orchestrate systemic zinc disbandment but also engage in cellular signaling mechanisms that control metabolic operations and inflammatory responses. Continued advancements in our understanding of the structure, regulation, and functions of these transporters will accelerate the development of innovative small molecules and biologics,

https://doi.org/10.3390/targets4010005

paving the way for new therapeutic options in the management of diabetes, fatty liver disease, and obesity.

Pharmaceutical Development and Contemporary Case Analyses

In the realm of drug development, significant advancements have emerged, particularly with regard to targeting different zinc transporters considering different pathological conditions. These efforts are mainly focused on two primary strategies: the direct modulation of transporter activity and the chemical inhibition of excessive zinc trafficking. A recent study, published in April 2025, used gene-edited human embryonic stem cell-derived β-cells (SC-β-cells) and human primary islet models to specifically examine the role of zinc and demonstrated that an increased influx of zinc causes ER stress, eventually resulting in β-cell death, a mechanism of considerable importance in diabetes pathogenesis [134]. Another study by Batta et al. aligns closely with current efforts to pharmacologically modulate zinc homeostasis by identifying small-molecule ligands that selectively target members of the ZIP and ZnT transporter families [135]. Using in silico docking, Batta and a different research group screened multiple phytochemical and drug-like compounds against modeled structures of ZIP4, ZIP8, ZIP10, and ZnT8, identifying several candidates with notable predicted binding affinity, most prominently quercetin, kaempferol, genistein, resveratrol, and curcumin [136]. The recognized compounds had promising potential and interactions with the transporters' active or regulatory regions, suggesting possible altering influences on zinc translocation. This study sought to realize the ultimate goal of developing isoform-specific modulators for zinc transporters. It established the first computational framework to facilitate the selection of potent compounds, thereby encouraging further biochemical confirmation and therapeutic outcomes.

On the other hand, utilizing a sophisticated drug-screening platform centered on isogenic SC-β-cells, researchers made a groundbreaking discovery that low doses of anisomycin (at a concentration of 25 nM), which is conventionally recognized as a protein synthesis inhibitor at higher doses, showed selective inhibition of the over-transportation of zinc [134]. Remarkably, treatment with a low dose (25 nM) of anisomycin was found to protect human pancreatic beta cells from stress-induced cell death and prevent the onset of T2D in mice fed a high-fat diet. This low-dose anisomycin not only safeguarded human β-cells from stress-induced cell death but also demonstrated efficacy in vivo. Rui et al. showed that when administered to primary human islets and diabetic mice, it successfully prevented the onset of type 2 diabetes induced by a high-fat diet. This unique pharmaceutical technique targeting zinc transporters is a novel therapeutic approach to maintaining β-cell function in the diabetic population [134]. A significant milestone in the field of translational zinc biology has been achieved with the identification of LOF mutations in the zinc transporter ZIP5. Comprehensive large-scale exome and cohort studies across many ethnic groups have shown that people with uncommon LOF mutations in ZIP5 have significantly increased blood zinc levels and a much decreased risk of developing T2D [127]. Functional assays corroborated that these mutations lead to a complete loss of transporter function, resulting in systemic zinc elevation. Similar to human genetic findings, ZIP5/Slc39a5-null mice had elevated tissue zinc levels, decreased fasting glucose, improved hepatic insulin sensitivity, and reduced liver inflammation and fibrosis, even with congenital or diet-induced obesity [127]. These strong in vivo data suggest that ZIP5 may be a viable metabolic illness treatment target, especially for individuals who do not benefit from zinc supplementation alone. Clinical investigations of zinc supplementation in metabolic syndrome, obesity, and diabetes patients have consistently shown improvements in insulin resistance and metabolic biomarkers. However, recent approaches aim to prioritize the modulation of intracellular zinc trafficking.

https://doi.org/10.3390/targets4010005

To circumvent the limits of "one-size-fits-all" supplementing techniques, small compounds or transporter proteins are used to address the molecular causes of metabolic disease dysregulations. Due to both gain- and loss-of-function variants of ZnT8 having been shown to affect β-cell function and insulin granule biogenesis, it has drawn attention in diabetes research. This duality in genetic variation justifies the exploration of both pharmacological enhancement and inhibition as viable strategies, tailored to individual patient genotypes.

Recent advancements in zinc-centric pharmacological research have highlighted zinc transporters as potential therapeutic targets for inflammatory disorders, such as rheumatoid arthritis (RA). Case studies in experimental arthritis models, namely, collagen-induced arthritis (CIA) mice and ex vivo synovial fibroblast systems, show that the modulation of zinc flux may influence critical inflammatory pathways [137,138]. ZIP8 has been shown to be highly elevated in the arthritic synovium and has been proposed as a druggable target due to its zinc-dependent stimulation of MMP-9 and MMP-13, which is directly responsible for cartilage degradation [139]. Small-molecule ZIP8 inhibitors and siRNA-mediated suppression methods have shown effectiveness in mitigating joint inflammation and matrix deterioration in CIA models [138]. Moreover, targeting the ZIP14 transporter, which moves zinc in response to cytokines in tissues like the liver and joints or other synovial tissue, can reduce systemic inflammation and the acute-phase response [64,87,140]. Compounds that regulate ZnT1 and ZnT5 activity are being studied for their ability to restore cytosolic zinc buffering capacity and inhibit NF-κB-mediated cytokine production [141]. These case studies together highlight the potential of zinc transporter-focused therapies, establishing them as promising candidates for the development of next-generation anti-inflammatory drugs. Key therapeutic interventions and experimental strategies targeting zinc transporters and zinc-modulated pathways are summarized in Table 5.

Table 5. Zinc transporter proteins as therapeutic drug targets: clinical trials and translational studies.

Zinc Transporter/ Target	Mechanism/ Rationale	Disease/ Condition	Drug/ Intervention Type	Study Details	References
ZnT8	Zinc transport into insulin granules (β-cell specific); impacts insulin maturation and secretion	Type 2 diabetes (T2D)	Targeted modulator/precision therapy (in development)	GWASs and rare-variant studies: loss-of-function alleles reduce T2D risk. Ongoing drug development focused on enhancing or mimicking protective variants. ZnT8-KO mouse studies confirm islet-specific function.	[142,143]
ZIP5	Regulates glucose sensing and insulin secretion in β-cells; impacts gut/pancreas zinc handling	Diabetes, metabolic diseases	Small-molecule/genetic modulation (preclinical)	Mouse knockout protects against glucose dysregulation and pancreatic zinc toxicity. SLC39A5 variants studied in large cohorts; shown to modulate serum zinc and glucose homeostasis in humans and animals.	[127]
ZIP8	Modulates zinc uptake in gut/liver/adipose; influences innate immunity, metabolism, Crohn's disease risk	Crohn's disease, gut/liver inflammation	Genetic and pharmacological modulation (early translational phase)	Functional variant linked to Crohn's disease and microbiome composition. Modifiers of ZIP8 studied in immune/inflammatory disease animal models.	[144,145]
ZIP10	Controls B-cell receptor signaling, humoral immunity, anti-apoptotic signaling	Hematologic malignancy, immunodeficiency	Genetic targeting/therapeutic antibodies (preclinical)	ZIP10 critical for B cell survival; mouse genetic studies. ZIP10 inhibitors/enhancers are under investigation for immune modulation; drug development in early preclinical phase.	[146]

https://doi.org/10.3390/targets4010005

Table 5. *Cont.*

Zinc Transporter/ Target	Mechanism/ Rationale	Disease/ Condition	Drug/ Intervention Type	Study Details	References
ZIP13	Regulates vascular and cardiac/skin function; upregulation linked to fibrosis and inflammation	Cardiovascular disease, fibrosis	Small-molecule inhibitor/antisense RNA (in development)	Mouse ZIP13 downregulation reduces ischemia/reperfusion injury via CaMKII regulation. Pharmacologic inhibition as a therapeutic strategy is under study.	[147]
ZIP4	Dietary zinc absorption/homeostasis; overexpression in cancer	Pancreatic and GI cancers, acrodermatitis enteropathica	Antibody drugs/antisense oligonucleotides	Anti-ZIP4 therapies in preclinical cancer studies. Genetic therapies for acrodermatitis enteropathica are under development.	[94,148]
ZnT1	Exports zinc from cells; affects systemic and tissue zinc levels	Zinc deficiency/excess, intestinal disorders	Dietary/ pharmacological/ translational biomarker	Plays a role in dietary and supplemental zinc absorption; involved in biomarker trial (NCT01062347).	[149,150]
SLC transporters (class)	General therapeutic target class: several subtypes, including SLC30A, SLC39A, individually druggable	T2D, metabolic, cancer, inflammation	Small-molecule/ monoclonal antibody/combo therapy	SLC30A8 and related SLCs identified as most promising for metabolic indications from human genetic and animal studies; some SLCs targeted in marketed and experimental cancer/metabolic drugs.	[151–153]

7. Future Perspectives and Open Questions

Zinc's role as a dynamic immunometabolic regulator has changed our knowledge of chronic illness development and opened new treatment avenues. However, zinc signaling's molecular details, tissue-specific control, and translational promise for precision metabolic therapy are still little understood. One major challenge lies in the context-specific nature of zinc transporter regulation. To achieve this, single-cell mapping of transporter expression in physiological and pathological states and tissue-specific delivery strategies like nanoparticle carriers or gene promoters are needed to precisely modulate transporter function. To achieve therapeutic selectivity, ligand-specific modulators that utilize transporter isoform structural differences might be designed. We also need biomarkers that can track intracellular zinc dynamics and transporter function in real time. Circulating miRNAs (e.g., miR-34a in NAFLD), metallothionein levels as proxies for cytosolic zinc buffering and redox balance, and zinc-sensitive contrast agents or genetically encoded biosensors for noninvasive PET or MRI zinc flux imaging seem promising. Genetic variants in the SLC30 and SLC39 families may predict zinc-targeted therapy response, allowing patient classification and customized intervention. T2D and NAFLD are complicated; hence, zinc-based monotherapies may not be enough. Instead, integrative combination techniques can work better. Zinc modulation may boost the metabolic advantages of GLP-1 receptor agonists, synergize with anti-inflammatory medicines like IL-1β inhibitors or NLRP3 antagonists, or improve gut–immune interaction with microbiome-directed therapy. Co-targeting AMPK, mTOR, and zinc homeostasis may improve cellular energy balance and immune–metabolic integration. To maximize its therapeutic potential, future translational research should focus on multi-axis techniques and incorporate zinc signaling into metabolic and inflammatory control.

8. Conclusions

Both preclinical and human genetic studies have underscored the critical role of zinc homeostasis in metabolic health, showing that alterations in transporter expression, genetic mutations, or disease-induced mis-localization can all lead to the pathogeneses of T2D, NFLD, obesity, and/or related metabolic disturbances. Breakthrough studies in imaging and biosensor advances have revealed zinc as a second messenger, as well as its signaling

https://doi.org/10.3390/targets4010005

mechanisms. Zinc transporter mutations protect metabolic illness models, proving that targeting transporter activity when zinc supplementation is inadequate is therapeutic. Though precise regulation of zinc flux in disease-relevant tissues is a critical emerging area in biomedical research, future therapeutic success is highly dependent on addressing tissue specificity and context-driven treatment methods. Altogether, modulation of zinc transporter functions either by small molecules, gene targeting, or precision nutritional strategies holds significant translational potential to restore immunometabolic balance and reduce the burden of chronic metabolic diseases. By integrating organelle, tissue, and signaling perspectives on ZIP and ZnT transporters in metaflammation, this review provides a unified framework that extends beyond prior zinc-centered overviews and highlights specific transporters and axes (zinc–miRNA and NF-κB/NLRP3) with translational potential in metabolic disease. Translating these insights into clinical practice for metabolic inflammation remains an active research area facing substantial challenges and opportunities in biomarker development, drug delivery, and combinatorial treatment strategies. Continued multidisciplinary efforts will be essential to advance these findings toward next-generation therapies for T2D, NAFLD, and obesity.

Supplementary Materials: The following supporting information can be downloaded at: https://www.mdpi.com/article/10.3390/targets4010005/s1, Figure S1: PRISMA flow diagram of study selection.

Author Contributions: Conceptualization, S.B. and K.S.; writing—original draft preparation, S.B. and S.R.P.; writing—review and editing, S.B.; visualization, S.B.; supervision, K.S.; project administration, K.S. All authors have read and agreed to the published version of the manuscript.

Funding: This research received no external funding.

Institutional Review Board Statement: Not Applicable.

Informed Consent Statement: Not Applicable.

Data Availability Statement: No new data were created or analyzed in this study.

Conflicts of Interest: The authors declare no conflict of interest.

References

1. Chen, B.; Yu, P.; Chan, W.N.; Xie, F.; Zhang, Y.; Liang, L.; Leung, K.T.; Lo, K.W.; Yu, J.; Tse, G.M.K.; et al. Cellular Zinc Metabolism and Zinc Signaling: From Biological Functions to Diseases and Therapeutic Targets. *Signal Transduct. Target. Ther.* **2024**, *9*, 6. [CrossRef]
2. Kimura, T.; Kambe, T. The Functions of Metallothionein and ZIP and ZnT Transporters: An Overview and Perspective. *Int. J. Mol. Sci.* **2016**, *17*, 336. [CrossRef]
3. Charles-Messance, H.; Mitchelson, K.A.J.; Castro, E.D.M.; Sheedy, F.J.; Roche, H.M. Regulating Metabolic Inflammation by Nutritional Modulation. *J. Allergy Clin. Immunol.* **2020**, *146*, 706–720. [CrossRef] [PubMed]
4. Aydemir, T.B.; Troche, C.; Kim, M.-H.; Cousins, R.J. Hepatic ZIP14-Mediated Zinc Transport Contributes to Endosomal Insulin Receptor Trafficking and Glucose Metabolism. *J. Biol. Chem.* **2016**, *291*, 23939–23951. [CrossRef]
5. Barman, S.; Srinivasan, K. Diabetes and Zinc Dyshomeostasis: Can Zinc Supplementation Mitigate Diabetic Complications? *Crit. Rev. Food Sci. Nutr.* **2022**, *62*, 1046–1061. [CrossRef] [PubMed]
6. Barman, S.; Pradeep, S.R.; Srinivasan, K. Zinc Supplementation Mitigates Its Dyshomeostasis in Experimental Diabetic Rats by Regulating the Expression of Zinc Transporters and Metallothionein. *Metallomics* **2017**, *9*, 1765–1777. [CrossRef]
7. Barman, S.; Srinivasan, K. Ameliorative Effect of Zinc Supplementation on Compromised Small Intestinal Health in Streptozotocin-Induced Diabetic Rats. *Chem. Biol. Interact.* **2019**, *307*, 37–50. [CrossRef] [PubMed]
8. Fan, Y.-G.; Wu, T.-Y.; Zhao, L.-X.; Jia, R.-J.; Ren, H.; Hou, W.-J.; Wang, Z.-Y. From Zinc Homeostasis to Disease Progression: Unveiling the Neurodegenerative Puzzle. *Pharmacol. Res.* **2024**, *199*, 107039. [CrossRef]
9. Paik, S.; Kim, J.K.; Silwal, P.; Sasakawa, C.; Jo, E.-K. An Update on the Regulatory Mechanisms of NLRP3 Inflammasome Activation. *Cell. Mol. Immunol.* **2021**, *18*, 1141–1160. [CrossRef]
10. Jeong, J.; Eide, D.J. The SLC39 Family of Zinc Transporters. *Mol. Asp. Med.* **2013**, *34*, 612–619. [CrossRef]

11. Kambe, T.; Taylor, K.M.; Fu, D. Zinc Transporters and Their Functional Integration in Mammalian Cells. *J. Biol. Chem.* **2021**, *296*, 100320. [CrossRef]
12. Roca-Umbert, A.; Garcia-Calleja, J.; Vogel-González, M.; Fierro-Villegas, A.; Ill-Raga, G.; Herrera-Fernández, V.; Bosnjak, A.; Muntané, G.; Gutiérrez, E.; Campelo, F.; et al. Human Genetic Adaptation Related to Cellular Zinc Homeostasis. *PLoS Genet.* **2023**, *19*, e1010950. [CrossRef]
13. Krężel, A.; Maret, W. The Bioinorganic Chemistry of Mammalian Metallothioneins. *Chem. Rev.* **2021**, *121*, 14594–14648. [CrossRef]
14. Thirumoorthy, N.; Manisenthil Kumar, K.; Shyam Sundar, A.; Panayappan, L.; Chatterjee, M. Metallothionein: An Overview. *World J. Gastroenterol.* **2007**, *13*, 993–996. [CrossRef] [PubMed]
15. Bafaro, E.; Liu, Y.; Xu, Y.; Dempski, R.E. The Emerging Role of Zinc Transporters in Cellular Homeostasis and Cancer. *Signal Transduct. Target. Ther.* **2017**, *2*, 17029. [CrossRef] [PubMed]
16. Oteiza, P.I. Zinc and the Modulation of Redox Homeostasis. *Free Radic. Biol. Med.* **2012**, *53*, 1748–1759. [CrossRef]
17. Laity, J.H.; Andrews, G.K. Understanding the Mechanisms of Zinc-Sensing by Metal-Response Element Binding Transcription Factor-1 (MTF-1). *Arch. Biochem. Biophys.* **2007**, *463*, 201–210. [CrossRef] [PubMed]
18. Watanabe, S.; Amagai, Y.; Sannino, S.; Tempio, T.; Anelli, T.; Harayama, M.; Masui, S.; Sorrentino, I.; Yamada, M.; Sitia, R.; et al. Zinc Regulates ERp44-Dependent Protein Quality Control in the Early Secretory Pathway. *Nat. Commun.* **2019**, *10*, 603. [CrossRef]
19. Wagatsuma, T.; Shimotsuma, K.; Sogo, A.; Sato, R.; Kubo, N.; Ueda, S.; Uchida, Y.; Kinoshita, M.; Kambe, T. Zinc Transport via ZNT5-6 and ZNT7 Is Critical for Cell Surface Glycosylphosphatidylinositol-Anchored Protein Expression. *J. Biol. Chem.* **2022**, *298*, 102011. [CrossRef]
20. Kim, M.-H.; Aydemir, T.B.; Kim, J.; Cousins, R.J. Hepatic ZIP14-Mediated Zinc Transport Is Required for Adaptation to Endoplasmic Reticulum Stress. *Proc. Natl. Acad. Sci. USA* **2017**, *114*, E5805–E5814. [CrossRef]
21. Amagai, Y.; Yamada, M.; Kowada, T.; Watanabe, T.; Du, Y.; Liu, R.; Naramoto, S.; Watanabe, S.; Kyozuka, J.; Anelli, T.; et al. Zinc Homeostasis Governed by Golgi-Resident ZnT Family Members Regulates ERp44-Mediated Proteostasis at the ER-Golgi Interface. *Nat. Commun.* **2023**, *14*, 2683. [CrossRef]
22. Suzuki, T.; Ishihara, K.; Migaki, H.; Ishihara, K.; Nagao, M.; Yamaguchi-Iwai, Y.; Kambe, T. Two Different Zinc Transport Complexes of Cation Diffusion Facilitator Proteins Localized in the Secretory Pathway Operate to Activate Alkaline Phosphatases in Vertebrate Cells. *J. Biol. Chem.* **2005**, *280*, 30956–30962. [CrossRef]
23. Yuasa, H.; Morino, N.; Wagatsuma, T.; Munekane, M.; Ueda, S.; Matsunaga, M.; Uchida, Y.; Katayama, T.; Katoh, T.; Kambe, T. ZNT5-6 and ZNT7 Play an Integral Role in Protein N-Glycosylation by Supplying Zn^{2+} to Golgi α-Mannosidase II. *J. Biol. Chem.* **2024**, *300*, 107378. [CrossRef]
24. McCormick, N.H.; Kelleher, S.L. ZnT4 Provides Zinc to Zinc-Dependent Proteins in the Trans-Golgi Network Critical for Cell Function and Zn Export in Mammary Epithelial Cells. *Am. J. Physiol.-Cell Physiol.* **2012**, *303*, C291–C297. [CrossRef]
25. Liu, H.Y.; Gale, J.R.; Reynolds, I.J.; Weiss, J.H.; Aizenman, E. The Multifaceted Roles of Zinc in Neuronal Mitochondrial Dysfunction. *Biomedicines* **2021**, *9*, 489. [CrossRef]
26. Ishida, H.; Yo, R.; Zhang, Z.; Shimizu, T.; Ohto, U. Cryo-EM Structures of the Zinc Transporters ZnT3 and ZnT4 Provide Insights into Their Transport Mechanisms. *FEBS Lett.* **2025**, *599*, 41–52. [CrossRef] [PubMed]
27. Rivera, O.C.; Geddes, D.T.; Barber-Zucker, S.; Zarivach, R.; Gagnon, A.; Soybel, D.I.; Kelleher, S.L. A Common Genetic Variant in Zinc Transporter ZnT2 (Thr288Ser) Is Present in Women with Low Milk Volume and Alters Lysosome Function and Cell Energetics. *Am. J. Physiol.-Cell Physiol.* **2020**, *318*, C1166–C1177. [CrossRef] [PubMed]
28. Chowanadisai, W.; Lönnerdal, B.; Kelleher, S.L. Identification of a Mutation in SLC30A2 (ZnT-2) in Women with Low Milk Zinc Concentration That Results in Transient Neonatal Zinc Deficiency. *J. Biol. Chem.* **2006**, *281*, 39699–39707. [CrossRef] [PubMed]
29. Davidson, H.W.; Wenzlau, J.M.; O'Brien, R.M. Zinc Transporter 8 (ZnT8) and β Cell Function. *Trends Endocrinol. Metab.* **2014**, *25*, 415–424. [CrossRef]
30. Kukic, I.; Lee, J.K.; Coblentz, J.; Kelleher, S.L.; Kiselyov, K. Zinc-Dependent Lysosomal Enlargement in TRPML1-Deficient Cells Involves MTF-1 Transcription Factor and ZnT4 (Slc30a4) Transporter. *Biochem. J.* **2013**, *451*, 155–163. [CrossRef]
31. Aydemir, T.B.; Liuzzi, J.P.; McClellan, S.; Cousins, R.J. Zinc Transporter ZIP8 (SLC39A8) and Zinc Influence IFN-γ Expression in Activated Human T Cells. *J. Leukoc. Biol.* **2009**, *86*, 337–348. [CrossRef]
32. Wang, F.; Dufner-Beattie, J.; Kim, B.-E.; Petris, M.J.; Andrews, G.; Eide, D.J. Zinc-Stimulated Endocytosis Controls Activity of the Mouse ZIP1 and ZIP3 Zinc Uptake Transporters. *J. Biol. Chem.* **2004**, *279*, 24631–24639. [CrossRef]
33. Xian, Y.; Zhou, M.; Hu, Y.; Liu, J.; Zhu, W.; Wang, Y. Free Zinc Determines the Formability of the Vesicular Dense Core in Diabetic Beta Cells. *Cell Insight* **2022**, *1*, 100020. [CrossRef] [PubMed]
34. McAllister, B.B.; Dyck, R.H. Zinc Transporter 3 (ZnT3) and Vesicular Zinc in Central Nervous System Function. *Neurosci. Biobehav. Rev.* **2017**, *80*, 329–350. [CrossRef]

https://doi.org/10.3390/targets4010005

35. Lee, S.; Rivera, O.C.; Kelleher, S.L. Zinc Transporter 2 Interacts with Vacuolar ATPase and Is Required for Polarization, Vesicle Acidification, and Secretion in Mammary Epithelial Cells. *J. Biol. Chem.* **2017**, *292*, 21598–21613. [CrossRef] [PubMed]
36. Kambe, T.; Matsunaga, M.; Takeda, T. Understanding the Contribution of Zinc Transporters in the Function of the Early Secretory Pathway. *Int. J. Mol. Sci.* **2017**, *18*, 2179. [CrossRef]
37. Wijesekara, N.; Dai, F.F.; Hardy, A.B.; Giglou, P.R.; Bhattacharjee, A.; Koshkin, V.; Chimienti, F.; Gaisano, H.Y.; Rutter, G.A.; Wheeler, M.B. Beta Cell Specific ZnT8 Deletion in Mice Causes Marked Defects in Insulin Processing, Crystallization and Secretion. *Diabetologia* **2010**, *53*, 1656–1668. [CrossRef]
38. Palmiter, R.D.; Cole, T.B.; Quaife, C.J.; Findley, S.D. ZnT-3, a Putative Transporter of Zinc into Synaptic Vesicles. *Proc. Natl. Acad. Sci. USA* **1996**, *93*, 14934–14939. [CrossRef]
39. Fukunaka, A.; Kurokawa, Y.; Teranishi, F.; Sekler, I.; Oda, K.; Ackland, M.L.; Faundez, V.; Hiromura, M.; Masuda, S.; Nagao, M.; et al. Tissue Nonspecific Alkaline Phosphatase Is Activated via a Two-Step Mechanism by Zinc Transport Complexes in the Early Secretory Pathway. *J. Biol. Chem.* **2011**, *286*, 16363–16373. [CrossRef]
40. Thornton, J.K.; Taylor, K.M.; Ford, D.; Valentine, R.A. Differential Subcellular Localization of the Splice Variants of the Zinc Transporter ZnT5 Is Dictated by the Different C-Terminal Regions. *PLoS ONE* **2011**, *6*, e23878. [CrossRef] [PubMed]
41. Bai, M.; Cui, Y.; Sang, Z.; Gao, S.; Zhao, H.; Mei, X. Zinc Ions Regulate Mitochondrial Quality Control in Neurons under Oxidative Stress and Reduce PANoptosis in Spinal Cord Injury Models via the Lgals3-Bax Pathway. *Free Radic. Biol. Med.* **2024**, *221*, 169–180. [CrossRef] [PubMed]
42. Dabravolski, S.A.; Sadykhov, N.K.; Kartuesov, A.G.; Borisov, E.E.; Sukhorukov, V.N.; Orekhov, A.N. Interplay between Zn^{2+} Homeostasis and Mitochondrial Functions in Cardiovascular Diseases and Heart Ageing. *Int. J. Mol. Sci.* **2022**, *23*, 6890. [CrossRef]
43. Deng, H.; Qiao, X.; Xie, T.; Fu, W.; Li, H.; Zhao, Y.; Guo, M.; Feng, Y.; Chen, L.; Zhao, Y.; et al. SLC-30A9 Is Required for Zn^{2+} Homeostasis, Zn^{2+} Mobilization, and Mitochondrial Health. *Proc. Natl. Acad. Sci. USA* **2021**, *118*, e2023909118. [CrossRef]
44. Jiang, D.; Sullivan, P.G.; Sensi, S.L.; Steward, O.; Weiss, J.H. Zn^{2+} Induces Permeability Transition Pore Opening and Release of Pro-Apoptotic Peptides from Neuronal Mitochondria. *J. Biol. Chem.* **2001**, *276*, 47524–47529. [CrossRef]
45. Mendoza, A.D.; Dietrich, N.; Tan, C.-H.; Herrera, D.; Kasiah, J.; Payne, Z.; Cubillas, C.; Schneider, D.L.; Kornfeld, K. Lysosome-Related Organelles Contain an Expansion Compartment That Mediates Delivery of Zinc Transporters to Promote Homeostasis. *Proc. Natl. Acad. Sci. USA* **2024**, *121*, e2307143121. [CrossRef] [PubMed]
46. Lichtlen, P.; Schaffner, W. Putting Its Fingers on Stressful Situations: The Heavy Metal-regulatory Transcription Factor MTF-1. *BioEssays* **2001**, *23*, 1010–1017. [CrossRef] [PubMed]
47. Hotamisligil, G.S. Foundations of Immunometabolism and Implications for Metabolic Health and Disease. *Immunity* **2017**, *47*, 406–420. [CrossRef]
48. Clemente-Suárez, V.J.; Beltrán-Velasco, A.I.; Redondo-Flórez, L.; Martín-Rodríguez, A.; Tornero-Aguilera, J.F. Global Impacts of Western Diet and Its Effects on Metabolism and Health: A Narrative Review. *Nutrients* **2023**, *15*, 2749. [CrossRef]
49. Diet Review: Anti-Inflammatory Diet. The Nutrition Source. 2021. Available online: https://nutritionsource.hsph.harvard.edu/healthy-weight/diet-reviews/anti-inflammatory-diet/ (accessed on 18 January 2025).
50. Milanski, M.; Degasperi, G.; Coope, A.; Morari, J.; Denis, R.; Cintra, D.E.; Tsukumo, D.M.L.; Anhe, G.; Amaral, M.E.; Takahashi, H.K.; et al. Saturated Fatty Acids Produce an Inflammatory Response Predominantly through the Activation of TLR4 Signaling in Hypothalamus: Implications for the Pathogenesis of Obesity. *J. Neurosci.* **2009**, *29*, 359–370. [CrossRef]
51. Finucane, O.M.; Lyons, C.L.; Murphy, A.M.; Reynolds, C.M.; Klinger, R.; Healy, N.P.; Cooke, A.A.; Coll, R.C.; McAllan, L.; Nilaweera, K.N.; et al. Monounsaturated Fatty Acid–Enriched High-Fat Diets Impede Adipose NLRP3 Inflammasome–Mediated IL-1β Secretion and Insulin Resistance Despite Obesity. *Diabetes* **2015**, *64*, 2116–2128. [CrossRef]
52. Li, X.; Ren, Y.; Chang, K.; Wu, W.; Griffiths, H.R.; Lu, S.; Gao, D. Adipose Tissue Macrophages as Potential Targets for Obesity and Metabolic Diseases. *Front. Immunol.* **2023**, *14*, 1153915. [CrossRef] [PubMed]
53. Barman, S.; Srinivasan, K. Zinc Supplementation Ameliorates Diabetic Cataract Through Modulation of Crystallin Proteins and Polyol Pathway in Experimental Rats. *Biol. Trace Elem. Res.* **2019**, *187*, 212–223. [CrossRef]
54. Izgilov, R.; Kislev, N.; Omari, E.; Benayahu, D. Advanced Glycation End-Products Accelerate Amyloid Deposits in Adipocyte's Lipid Droplets. *Cell Death Dis.* **2024**, *15*, 846. [CrossRef]
55. Asadipooya, K.; Lankarani, K.B.; Raj, R.; Kalantarhormozi, M. RAGE Is a Potential Cause of Onset and Progression of Nonalcoholic Fatty Liver Disease. *Int. J. Endocrinol.* **2019**, *2019*, 2151302. [CrossRef]
56. Barman, S.; Srinivasan, K. Attenuation of Oxidative Stress and Cardioprotective Effects of Zinc Supplementation in Experimental Diabetic Rats. *Br. J. Nutr.* **2017**, *117*, 335–350. [CrossRef]
57. Barman, S.; Pradeep, S.R.; Srinivasan, K. Zinc Supplementation Alleviates the Progression of Diabetic Nephropathy by Inhibiting the Overexpression of Oxidative-Stress-Mediated Molecular Markers in Streptozotocin-Induced Experimental Rats. *J. Nutr. Biochem.* **2018**, *54*, 113–129. [CrossRef]

https://doi.org/10.3390/targets4010005

58. Hotamisligil, G.S. Inflammation, Metaflammation and Immunometabolic Disorders. *Nature* **2017**, *542*, 177–185. [CrossRef] [PubMed]
59. Marreiro, D.d.N.; Cruz, K.J.C.; Morais, J.B.S.; Beserra, J.B.; Severo, J.S.; de Oliveira, A.R.S. Zinc and Oxidative Stress: Current Mechanisms. *Antioxidants* **2017**, *6*, 24. [CrossRef]
60. Wessels, I.; Maywald, M.; Rink, L. Zinc as a Gatekeeper of Immune Function. *Nutrients* **2017**, *9*, 1286. [CrossRef]
61. Wessels, I.; Haase, H.; Engelhardt, G.; Rink, L.; Uciechowski, P. Zinc Deficiency Induces Production of the Proinflammatory Cytokines IL-1β and TNFα in Promyeloid Cells via Epigenetic and Redox-Dependent Mechanisms. *J. Nutr. Biochem.* **2013**, *24*, 289–297. [CrossRef] [PubMed]
62. Dierichs, L.; Kloubert, V.; Rink, L. Cellular Zinc Homeostasis Modulates Polarization of THP-1-Derived Macrophages. *Eur. J. Nutr.* **2018**, *57*, 2161–2169. [CrossRef] [PubMed]
63. Kulik, L.; Maywald, M.; Kloubert, V.; Wessels, I.; Rink, L. Zinc Deficiency Drives Th17 Polarization and Promotes Loss of Treg Cell Function. *J. Nutr. Biochem.* **2019**, *63*, 11–18. [CrossRef]
64. Liuzzi, J.P.; Lichten, L.A.; Rivera, S.; Blanchard, R.K.; Aydemir, T.B.; Knutson, M.D.; Ganz, T.; Cousins, R.J. Interleukin-6 Regulates the Zinc Transporter Zip14 in Liver and Contributes to the Hypozincemia of the Acute-Phase Response. *Proc. Natl. Acad. Sci. USA* **2005**, *102*, 6843–6848. [CrossRef]
65. Merriman, C.; Fu, D. Down-Regulation of the Islet-Specific Zinc Transporter-8 (ZnT8) Protects Human Insulinoma Cells against Inflammatory Stress. *J. Biol. Chem.* **2019**, *294*, 16992–17006. [CrossRef]
66. Liu, M.-J.; Bao, S.; Gálvez-Peralta, M.; Pyle, C.J.; Rudawsky, A.C.; Pavlovicz, R.E.; Killilea, D.W.; Li, C.; Nebert, D.W.; Wewers, M.D.; et al. ZIP8 Regulates Host Defense through Zinc-Mediated Inhibition of NF-κB. *Cell Rep.* **2013**, *3*, 386–400. [CrossRef]
67. Summersgill, H.; England, H.; Lopez-Castejon, G.; Lawrence, C.B.; Luheshi, N.M.; Pahle, J.; Mendes, P.; Brough, D. Zinc Depletion Regulates the Processing and Secretion of IL-1β. *Cell Death Dis.* **2014**, *5*, e1040. [CrossRef]
68. Jimenez-Rondan, F.R.; Ruggiero, C.H.; Cousins, R.J. Long Noncoding RNA, MicroRNA, Zn Transporter Zip14 (Slc39a14) and Inflammation in Mice. *Nutrients* **2022**, *14*, 5114. [CrossRef]
69. Taslim, N.A.; Graciela, A.M.; Harbuwono, D.S.; Syauki, A.Y.; Anthony, A.N.; Ashari, N.; Aman, A.M.; Tjandrawinata, R.R.; Hardinsyah, H.; Bukhari, A.; et al. Zinc as a Modulator of miRNA Signaling in Obesity. *Nutrients* **2025**, *17*, 3375. [CrossRef] [PubMed]
70. Ye, Z.; Cheng, M.; Fan, L.; Ma, J.; Zhang, Y.; Gu, P.; Xie, Y.; You, X.; Zhou, M.; Wang, B.; et al. Plasma microRNA Expression Profiles Associated with Zinc Exposure and Type 2 Diabetes Mellitus: Exploring Potential Role of miR-144-3p in Zinc-Induced Insulin Resistance. *Environ. Int.* **2023**, *172*, 107807. [CrossRef] [PubMed]
71. Barman, S.; Srinivasan, K. Zinc Supplementation Alleviates Hyperglycemia and Associated Metabolic Abnormalities in Streptozotocin-Induced Diabetic Rats. *Can. J. Physiol. Pharmacol.* **2016**, *94*, 1356–1365. [CrossRef]
72. Mitchell, S.B.; Aydemir, T.B. Roles of Zinc and Zinc Transporters in Development, Progression, and Treatment of Inflammatory Bowel Disease (IBD). *Front. Nutr.* **2025**, *12*, 1649658. [CrossRef] [PubMed]
73. Wan, Y.; Zhang, B. The Impact of Zinc and Zinc Homeostasis on the Intestinal Mucosal Barrier and Intestinal Diseases. *Biomolecules* **2022**, *12*, 900. [CrossRef]
74. Fan, Y.; Zhang, X.; Yang, L.; Wang, J.; Hu, Y.; Bian, A.; Liu, J.; Ma, J. Zinc Inhibits High Glucose-Induced NLRP3 Inflammasome Activation in Human Peritoneal Mesothelial Cells. *Mol. Med. Rep.* **2017**, *16*, 5195–5202. [CrossRef] [PubMed]
75. Aydemir, T.B.; Cousins, R.J. The Multiple Faces of the Metal Transporter ZIP14 (SLC39A14). *J. Nutr.* **2018**, *148*, 174–184. [CrossRef]
76. Winslow, J.W.W.; Limesand, K.H.; Zhao, N. The Functions of ZIP8, ZIP14, and ZnT10 in the Regulation of Systemic Manganese Homeostasis. *Int. J. Mol. Sci.* **2020**, *21*, 3304. [CrossRef]
77. Aydemir, T.B.; Chang, S.-M.; Guthrie, G.J.; Maki, A.B.; Ryu, M.-S.; Karabiyik, A.; Cousins, R.J. Zinc Transporter ZIP14 Functions in Hepatic Zinc, Iron and Glucose Homeostasis during the Innate Immune Response (Endotoxemia). *PLoS ONE* **2012**, *7*, e48679. [CrossRef]
78. Gartmann, L.; Wex, T.; Grüngreiff, K.; Reinhold, D.; Kalinski, T.; Malfertheiner, P.; Schütte, K. Expression of Zinc Transporters ZIP4, ZIP14 and ZnT9 in Hepatic Carcinogenesis—An Immunohistochemical Study. *J. Trace Elem. Med. Biol.* **2018**, *49*, 35–42. [CrossRef] [PubMed]
79. Su, W.; Feng, M.; Liu, Y.; Cao, R.; Liu, Y.; Tang, J.; Pan, K.; Lan, R.; Mao, Z. ZnT8 Deficiency Protects From APAP-Induced Acute Liver Injury by Reducing Oxidative Stress Through Upregulating Hepatic Zinc and Metallothioneins. *Front. Pharmacol.* **2021**, *12*, 721471. [CrossRef]
80. Germanos, M.; Gao, A.; Taper, M.; Yau, B.; Kebede, M.A. Inside the Insulin Secretory Granule. *Metabolites* **2021**, *11*, 515. [CrossRef]
81. Lemaire, K.; Ravier, M.A.; Schraenen, A.; Creemers, J.W.M.; Van de Plas, R.; Granvik, M.; Van Lommel, L.; Waelkens, E.; Chimienti, F.; Rutter, G.A.; et al. Insulin Crystallization Depends on Zinc Transporter ZnT8 Expression, but Is Not Required for Normal Glucose Homeostasis in Mice. *Proc. Natl. Acad. Sci. USA* **2009**, *106*, 14872–14877. [CrossRef]
82. da Silva Xavier, G.; Bellomo, E.A.; McGinty, J.A.; French, P.M.; Rutter, G.A. Animal Models of GWAS-Identified Type 2 Diabetes Genes. *J. Diabetes Res.* **2013**, *2013*, 906590. [CrossRef]

https://doi.org/10.3390/targets4010005

83. Sladek, R.; Rocheleau, G.; Rung, J.; Dina, C.; Shen, L.; Serre, D.; Boutin, P.; Vincent, D.; Belisle, A.; Hadjadj, S.; et al. A Genome-Wide Association Study Identifies Novel Risk Loci for Type 2 Diabetes. *Nature* **2007**, *445*, 881–885. [CrossRef]
84. Cai, Y.; Kirschke, C.P.; Huang, L. SLC30A Family Expression in the Pancreatic Islets of Humans and Mice: Cellular Localization in the β-Cells. *J. Mol. Histol.* **2018**, *49*, 133–145. [CrossRef]
85. Liu, Y.; Batchuluun, B.; Ho, L.; Zhu, D.; Prentice, K.J.; Bhattacharjee, A.; Zhang, M.; Pourasgari, F.; Hardy, A.B.; Taylor, K.M.; et al. Characterization of Zinc Influx Transporters (ZIPs) in Pancreatic β Cells. *J. Biol. Chem.* **2015**, *290*, 18757–18769. [CrossRef] [PubMed]
86. Egefjord, L.; Jensen, J.L.; Bang-Berthelsen, C.H.; Petersen, A.B.; Smidt, K.; Schmitz, O.; Karlsen, A.E.; Pociot, F.; Chimienti, F.; Rungby, J.; et al. Zinc Transporter Gene Expression Is Regulated by Pro-Inflammatory Cytokines: A Potential Role for Zinc Transporters in Beta-Cell Apoptosis? *BMC Endocr. Disord.* **2009**, *9*, 7. [CrossRef]
87. Troche, C.; Beker Aydemir, T.; Cousins, R.J. Zinc Transporter Slc39a14 Regulates Inflammatory Signaling Associated with Hypertrophic Adiposity. *Am. J. Physiol.-Endocrinol. Metab.* **2016**, *310*, E258–E268. [CrossRef] [PubMed]
88. Aydemir, T.B.; Kim, M.-H.; Kim, J.; Colon-Perez, L.M.; Banan, G.; Mareci, T.H.; Febo, M.; Cousins, R.J. Metal Transporter Zip14 (Slc39a14) Deletion in Mice Increases Manganese Deposition and Produces Neurotoxic Signatures and Diminished Motor Activity. *J. Neurosci.* **2017**, *37*, 5996–6006. [CrossRef] [PubMed]
89. Mitchell, S.B.; Thorn, T.L.; Lee, M.-T.; Kim, Y.; Comrie, J.M.C.; Bai, Z.S.; Johnson, E.L.; Aydemir, T.B. Metal Transporter SLC39A14/ZIP14 Modulates Regulation between the Gut Microbiome and Host Metabolism. *Am. J. Physiol.-Gastrointest. Liver Physiol.* **2023**, *325*, G593–G607. [CrossRef]
90. McCabe, S.; Limesand, K.; Zhao, N. Recent Progress toward Understanding the Role of ZIP14 in Regulating Systemic Manganese Homeostasis. *Comput. Struct. Biotechnol. J.* **2023**, *21*, 2332–2338. [CrossRef]
91. Tepaamorndech, S.; Kirschke, C.P.; Pedersen, T.L.; Keyes, W.R.; Newman, J.W.; Huang, L. Zinc Transporter 7 Deficiency Affects Lipid Synthesis in Adipocytes by Inhibiting Insulin-Dependent Akt Activation and Glucose Uptake. *FEBS J.* **2016**, *283*, 378–394. [CrossRef]
92. Fukunaka, A.; Fukada, T.; Bhin, J.; Suzuki, L.; Tsuzuki, T.; Takamine, Y.; Bin, B.-H.; Yoshihara, T.; Ichinoseki-Sekine, N.; Naito, H.; et al. Zinc Transporter ZIP13 Suppresses Beige Adipocyte Biogenesis and Energy Expenditure by Regulating C/EBP-β Expression. *PLoS Genet.* **2017**, *13*, e1006950. [CrossRef]
93. Barman, S.; Srinivasan, K. Enhanced Intestinal Absorption of Micronutrients in Streptozotocin-Induced Diabetic Rats Maintained on Zinc Supplementation. *J. Trace Elem. Med. Biol.* **2018**, *50*, 182–187. [CrossRef]
94. Andrews, G.K. Regulation and Function of Zip4, the Acrodermatitis Enteropathica Gene. *Biochem. Soc. Trans.* **2008**, *36*, 1242–1246. [CrossRef]
95. Kuliyev, E.; Zhang, C.; Sui, D.; Hu, J. Zinc Transporter Mutations Linked to Acrodermatitis Enteropathica Disrupt Function and Cause Mistrafficking. *J. Biol. Chem.* **2021**, *296*, 100269. [CrossRef]
96. Hu, X.; Xiao, W.; Lei, Y.; Green, A.; Lee, X.; Maradana, M.R.; Gao, Y.; Xie, X.; Wang, R.; Chennell, G.; et al. Aryl Hydrocarbon Receptor Utilises Cellular Zinc Signals to Maintain the Gut Epithelial Barrier. *Nat. Commun.* **2023**, *14*, 5431. [CrossRef] [PubMed]
97. Jimenez-Rondan, F.R.; Ruggiero, C.H.; Riva, A.; Yu, F.; Stafford, L.S.; Cross, T.R.; Larkin, J.; Cousins, R.J. Deletion of Metal Transporter Zip14 Reduces Major Histocompatibility Complex II Expression in Murine Small Intestinal Epithelial Cells. *Proc. Natl. Acad. Sci. USA* **2025**, *122*, e2422321121. [CrossRef]
98. Podany, A.B.; Wright, J.; Lamendella, R.; Soybel, D.I.; Kelleher, S.L. ZnT2-Mediated Zinc Import into Paneth Cell Granules Is Necessary for Coordinated Secretion and Paneth Cell Function in Mice. *Cell. Mol. Gastroenterol. Hepatol.* **2016**, *2*, 369–383. [CrossRef]
99. Shao, Y.-X.; Lei, Z.; Wolf, P.G.; Gao, Y.; Guo, Y.-M.; Zhang, B.-K. Zinc Supplementation, via GPR39, Upregulates PKCζ to Protect Intestinal Barrier Integrity in Caco-2 Cells Challenged by Salmonella Enterica Serovar Typhimurium. *J. Nutr.* **2017**, *147*, 1282–1289. [CrossRef] [PubMed]
100. Franco, C.; Canzoniero, L.M.T. Zinc Homeostasis and Redox Alterations in Obesity. *Front. Endocrinol.* **2024**, *14*, 1273177. [CrossRef] [PubMed]
101. Lichten, L.A.; Ryu, M.-S.; Guo, L.; Embury, J.; Cousins, R.J. MTF-1-Mediated Repression of the Zinc Transporter Zip10 Is Alleviated by Zinc Restriction. *PLoS ONE* **2011**, *6*, e21526. [CrossRef]
102. Jou, M.-Y.; Hall, A.G.; Philipps, A.F.; Kelleher, S.L.; Lönnerdal, B. Tissue-Specific Alterations in Zinc Transporter Expression in Intestine and Liver Reflect a Threshold for Homeostatic Compensation during Dietary Zinc Deficiency in Weanling Rats. *J. Nutr.* **2009**, *139*, 835–841. [CrossRef]
103. Ziatdinova, M.M.; Valova, Y.V.; Mukhammadiyeva, G.F.; Fazlieva, A.S.; Karimov, D.D.; Kudoyarov, E.R. Analysis of MT1 and ZIP1 Gene Expression in the Liver of Rats with Chronic Poisoning with Cadmium Chloride. *Hyg. Sanit.* **2021**, *100*, 1298–1302. [CrossRef]

https://doi.org/10.3390/targets4010005

104. Dufner-Beattie, J.; Huang, Z.L.; Geiser, J.; Xu, W.; Andrews, G.K. Mouse ZIP1 and ZIP3 Genes Together Are Essential for Adaptation to Dietary Zinc Deficiency during Pregnancy. *Genesis* **2006**, *44*, 239–251. [CrossRef] [PubMed]
105. Fukada, T.; Civic, N.; Furuichi, T.; Shimoda, S.; Mishima, K.; Higashiyama, H.; Idaira, Y.; Asada, Y.; Kitamura, H.; Yamasaki, S.; et al. The Zinc Transporter SLC39A13/ZIP13 Is Required for Connective Tissue Development; Its Involvement in BMP/TGF-β Signaling Pathways. *PLoS ONE* **2008**, *3*, e3642. [CrossRef]
106. Bahadoran, Z.; Ghafouri-Taleghani, F.; Todorčević, M. Zinc and Adipose Organ Dysfunction: Molecular Insights into Obesity and Metabolic Disorders. *Curr. Nutr. Rep.* **2025**, *14*, 117. [CrossRef]
107. Samuelson, D.R.; Haq, S.; Knoell, D.L. Divalent Metal Uptake and the Role of ZIP8 in Host Defense Against Pathogens. *Front. Cell Dev. Biol.* **2022**, *10*, 924820. [CrossRef]
108. McMahon, R.J.; Cousins, R.J. Regulation of the Zinc Transporter ZnT-1 by Dietary Zinc. *Proc. Natl. Acad. Sci. USA* **1998**, *95*, 4841–4846. [CrossRef] [PubMed]
109. Cai, Z.; Wu, X.; Wang, T.; Song, Z.; Ni, P.; Zhong, M.; Su, Y.; Xie, E.; Sun, S.; Lin, Y.; et al. SLC39A8-Mediated Zinc Dyshomeostasis Potentiates Kidney Disease. *Proc. Natl. Acad. Sci. USA* **2025**, *122*, e2426352122. [CrossRef]
110. Chen, W.; Lu, H.; Ying, Y.; Li, H.; Shen, H.; Cai, J. Zinc and Chronic Kidney Disease: A Review. *J. Nutr. Sci. Vitaminol.* **2024**, *70*, 98–105. [CrossRef]
111. Chi, Y.; Zhang, X.; Liang, D.; Wang, Y.; Cai, X.; Dong, J.; Li, L.; Chi, Z. ZnT8 Exerts Anti-Apoptosis of Kidney Tubular Epithelial Cell in Diabetic Kidney Disease Through TNFAIP3-NF-κB Signal Pathways. *Biol. Trace Elem. Res.* **2023**, *201*, 2442–2457. [CrossRef]
112. Liuzzi, J.P.; Blanchard, R.K.; Cousins, R.J. Differential Regulation of Zinc Transporter 1, 2, and 4 mRNA Expression by Dietary Zinc in Rats. *J. Nutr.* **2001**, *131*, 46–52. [CrossRef] [PubMed]
113. Grubman, A.; Lidgerwood, G.E.; Duncan, C.; Bica, L.; Tan, J.-L.; Parker, S.J.; Caragounis, A.; Meyerowitz, J.; Volitakis, I.; Moujalled, D.; et al. Deregulation of Subcellular Biometal Homeostasis through Loss of the Metal Transporter, Zip7, in a Childhood Neurodegenerative Disorder. *Acta Neuropathol. Commun.* **2014**, *2*, 25. [CrossRef]
114. Sabouri, S.; Rostamirad, M.; Dempski, R.E. Unlocking the Brain's Zinc Code: Implications for Cognitive Function and Disease. *Front. Biophys.* **2024**, *2*, 1406868. [CrossRef]
115. Mao, Z.; Lin, H.; Su, W.; Li, J.; Zhou, M.; Li, Z.; Zhou, B.; Yang, Q.; Zhou, M.; Pan, K.; et al. Deficiency of ZnT8 Promotes Adiposity and Metabolic Dysfunction by Increasing Peripheral Serotonin Production. *Diabetes* **2019**, *68*, 1197–1209. [CrossRef]
116. Alshawaf, A.J.; Alnassar, S.A.; Al-Mohanna, F.A. The Interplay of Intracellular Calcium and Zinc Ions in Response to Electric Field Stimulation in Primary Rat Cortical Neurons In Vitro. *Front. Cell. Neurosci.* **2023**, *17*, 1118335. [CrossRef]
117. Jarosz, M.; Olbert, M.; Wyszogrodzka, G.; Młyniec, K.; Librowski, T. Antioxidant and Anti-Inflammatory Effects of Zinc. Zinc-Dependent NF-κB Signaling. *Inflammopharmacology* **2017**, *25*, 11–24. [CrossRef]
118. Anson, K.J.; Corbet, G.A.; Palmer, A.E. Zn^{2+} Influx Activates ERK and Akt Signaling Pathways. *Proc. Natl. Acad. Sci. USA* **2021**, *118*, e2015786118. [CrossRef] [PubMed]
119. Zhu, B.; Yang, C.; Liu, D.; Zhi, Q.; Hua, Z.-C. Zinc Depletion Induces JNK/P38 Phosphorylation and Suppresses Akt/mTOR Expression in Acute Promyelocytic NB4 Cells. *J. Trace Elem. Med. Biol.* **2023**, *79*, 127264. [CrossRef] [PubMed]
120. Maywald, M.; Wessels, I.; Rink, L. Zinc Signals and Immunity. *Int. J. Mol. Sci.* **2017**, *18*, 2222. [CrossRef]
121. Fessler, M.B.; Rudel, L.L.; Brown, M. Toll-like Receptor Signaling Links Dietary Fatty Acids to the Metabolic Syndrome. *Curr. Opin. Lipidol.* **2009**, *20*, 379–385. [CrossRef]
122. Kim, B.; Lee, W.-W. Regulatory Role of Zinc in Immune Cell Signaling. *Mol. Cells* **2021**, *44*, 335–341. [CrossRef]
123. Prasad, A.S.; Bao, B. Molecular Mechanisms of Zinc as a Pro-Antioxidant Mediator: Clinical Therapeutic Implications. *Antioxidants* **2019**, *8*, 164. [CrossRef]
124. Kim, Y.-H.; Eom, J.-W.; Koh, J.-Y. Mechanism of Zinc Excitotoxicity: A Focus on AMPK. *Front. Neurosci.* **2020**, *14*, 577958. [CrossRef]
125. Wong, V.V.T.; Nissom, P.M.; Sim, S.-L.; Yeo, J.H.M.; Chuah, S.-H.; Yap, M.G.S. Zinc as an Insulin Replacement in Hybridoma Cultures. *Biotechnol. Bioeng.* **2006**, *93*, 553–563. [CrossRef]
126. Huang, Q.; Du, J.; Merriman, C.; Gong, Z. Genetic, Functional, and Immunological Study of ZnT8 in Diabetes. *Int. J. Endocrinol.* **2019**, *2019*, 1524905. [CrossRef] [PubMed]
127. Chim, S.M.; Howell, K.; Dronzek, J.; Wu, W.; Hout, C.V.; Ferreira, M.A.R.; Ye, B.; Li, A.; Brydges, S.; Arunachalam, V.; et al. Genetic Inactivation of Zinc Transporter SLC39A5 Improves Liver Function and Hyperglycemia in Obesogenic Settings. *eLife* **2023**, *12*, RP90419. [CrossRef]
128. Fukunaka, A.; Fujitani, Y. Role of Zinc Homeostasis in the Pathogenesis of Diabetes and Obesity. *Int. J. Mol. Sci.* **2018**, *19*, 476. [CrossRef]

https://doi.org/10.3390/targets4010005

129. Horton, T.M.; Allegretti, P.A.; Lee, S.; Moeller, H.P.; Smith, M.; Annes, J.P. Zinc-Chelating Small Molecules Preferentially Accumulate and Function within Pancreatic β Cells. *Cell Chem. Biol.* **2019**, *26*, 213–222.e6. [CrossRef]
130. Ranasinghe, P.; Wathurapatha, W.S.; Galappatthy, P.; Katulanda, P.; Jayawardena, R.; Constantine, G.R. Zinc Supplementation in Prediabetes: A Randomized Double-Blind Placebo-Controlled Clinical Trial. *J. Diabetes* **2018**, *10*, 386–397. [CrossRef] [PubMed]
131. Ranasinghe, P.; Jayawardena, R.; Pigera, A.; Katulanda, P.; Constantine, G.R.; Galappaththy, P. Zinc Supplementation in Pre-Diabetes: Study Protocol for a Randomized Controlled Trial. *Trials* **2013**, *14*, 52. [CrossRef] [PubMed]
132. Chim, S.M.; Howell, K.; Dronzek, J.; Wu, W.; Hout, C.V.; Ferreira, M.A.R.; Ye, B.; Li, A.; Brydges, S.; Arunachalam, V.; et al. Genetic Inactivation of Zinc Transporter SLC39A5 Improves Liver Function and Hyperglycemia in Obesogenic Settings. *medRxiv* **2021**, 2021.12.08.21267440. [CrossRef]
133. Bin, B.-H.; Hojyo, S.; Ryong Lee, T.; Fukada, T. Spondylocheirodysplastic Ehlers-Danlos Syndrome (SCD-EDS) and the Mutant Zinc Transporter ZIP13. *Rare Dis.* **2014**, *2*, e974982. [CrossRef]
134. Hu, R.; Ma, Q.; Kong, Y.; Wang, Z.; Xu, M.; Chen, X.; Su, Y.; Xiao, T.; He, Q.; Wang, X.; et al. A Compound Screen Based on Isogenic hESC-Derived β Cell Reveals an Inhibitor Targeting ZnT8-Mediated Zinc Transportation to Protect Pancreatic β Cell from Stress-Induced Cell Death. *Adv. Sci.* **2025**, *12*, 2413161. [CrossRef]
135. Batta, I.; Sharma, G. Molecular Docking Simulation to Predict Inhibitors Against Zinc Transporters. *ChemRxiv* **2024**. [CrossRef]
136. Jangid, H.; Shah, N.H.; Dar, M.A.; Wani, A.K. Identification of Potent Phytochemical Inhibitors against Mobilized Colistin Resistance: Molecular Docking, MD Simulations, ADMET, and Toxicity Predictions. *Silico Res. Biomed.* **2025**, *1*, 100013. [CrossRef]
137. Kim, B.; Kim, H.Y.; Yoon, B.R.; Yeo, J.; In Jung, J.; Yu, K.-S.; Kim, H.C.; Yoo, S.-J.; Park, J.K.; Kang, S.W.; et al. Cytoplasmic Zinc Promotes IL-1β Production by Monocytes and Macrophages through mTORC1-Induced Glycolysis in Rheumatoid Arthritis. *Sci. Signal.* **2022**, *15*, eabi7400. [CrossRef]
138. Kang, J.-A.; Kwak, J.-S.; Park, S.-H.; Sim, K.-Y.; Kim, S.K.; Shin, Y.; Jung, I.J.; Yang, J.-I.; Chun, J.-S.; Park, S.-G. ZIP8 Exacerbates Collagen-Induced Arthritis by Increasing Pathogenic T Cell Responses. *Exp. Mol. Med.* **2021**, *53*, 560–571. [CrossRef]
139. Kim, J.-H.; Jeon, J.; Shin, M.; Won, Y.; Lee, M.; Kwak, J.-S.; Lee, G.; Rhee, J.; Ryu, J.-H.; Chun, C.-H.; et al. Regulation of the Catabolic Cascade in Osteoarthritis by the Zinc-ZIP8-MTF1 Axis. *Cell* **2014**, *156*, 730–743. [CrossRef] [PubMed]
140. Gálvez-Peralta, M.; Wang, Z.; Bao, S.; Knoell, D.L.; Nebert, D.W. Tissue-Specific Induction of Mouse ZIP8 and ZIP14 Divalent Cation/Bicarbonate Symporters by, and Cytokine Response to, Inflammatory Signals. *Int. J. Toxicol.* **2014**, *33*, 246–258. [CrossRef]
141. Briassoulis, G.; Briassoulis, P.; Ilia, S.; Miliaraki, M.; Briassouli, E. The Anti-Oxidative, Anti-Inflammatory, Anti-Apoptotic, and Anti-Necroptotic Role of Zinc in COVID-19 and Sepsis. *Antioxidants* **2023**, *12*, 1942. [CrossRef] [PubMed]
142. Barman, N.; Haque, M.A.; Ridwan, M.; Ghosh, D.; Islam, A.B.M.M.K. Loss-of-Function Variant of *SLC30A8* Rs13266634 (C > T) Protects against Type 2 Diabetes by Stabilizing ZnT8: Insights from Epidemiological and Computational Analyses. *J. Genet. Eng. Biotechnol.* **2025**, *23*, 100565. [CrossRef]
143. Thomsen, S.K.; Gloyn, A.L. Human Genetics as a Model for Target Validation: Finding New Therapies for Diabetes. *Diabetologia* **2017**, *60*, 960–970. [CrossRef] [PubMed]
144. Yang, J.C.; Zhao, M.; Chernikova, D.; Arias-Jayo, N.; Zhou, Y.; Situ, J.; Gutta, A.; Chang, C.; Liang, F.; Lagishetty, V.; et al. ZIP8 A391T Crohn's Disease-Linked Risk Variant Induces Colonic Metal Ion Dyshomeostasis, Microbiome Compositional Shifts, and Inflammation. *Dig. Dis. Sci.* **2024**, *69*, 3760–3772. [CrossRef]
145. Hall, S.C.; Smith, D.R.; Dyavar, S.R.; Wyatt, T.A.; Samuelson, D.R.; Bailey, K.L.; Knoell, D.L. Critical Role of Zinc Transporter (Zip8) in Myeloid Innate Immune Cell Function and the Host Response against Bacterial Pneumonia. *J. Immunol.* **2021**, *207*, 1357–1370. [CrossRef]
146. Miyai, T.; Hojyo, S.; Ikawa, T.; Kawamura, M.; Irié, T.; Ogura, H.; Hijikata, A.; Bin, B.-H.; Yasuda, T.; Kitamura, H.; et al. Zinc Transporter SLC39A10/ZIP10 Facilitates Antiapoptotic Signaling during Early B-Cell Development. *Proc. Natl. Acad. Sci. USA* **2014**, *111*, 11780–11785. [CrossRef]
147. Wang, J.; Cheng, X.; Zhao, H.; Yang, Q.; Xu, Z. Downregulation of the Zinc Transporter SLC39A13 (ZIP13) Is Responsible for the Activation of CaMKII at Reperfusion and Leads to Myocardial Ischemia/Reperfusion Injury in Mouse Hearts. *J. Mol. Cell. Cardiol.* **2021**, *152*, 69–79. [CrossRef] [PubMed]
148. Weaver, B.; Zhang, Y.; Hiscox, S.; Guo, G.; Apte, U.; Taylor, K.; Sheline, C.; Wang, L.; Andrews, G. Zip4 (Slc39a4) Expression Is Activated in Hepatocellular Carcinomas and Functions to Repress Apoptosis, Enhance Cell Cycle and Increase Migration. *PLoS ONE* **2010**, *5*, e13158. [CrossRef]
149. Sun, S.; Xie, E.; Xu, S.; Ji, S.; Wang, S.; Shen, J.; Wang, R.; Shen, X.; Su, Y.; Song, Z.; et al. The Intestinal Transporter SLC30A1 Plays a Critical Role in Regulating Systemic Zinc Homeostasis. *Adv. Sci.* **2024**, *11*, 2406421. [CrossRef]
150. Trial | NCT01062347. Available online: https://cdek.pharmacy.purdue.edu/trial/NCT01062347/ (accessed on 19 November 2025).
151. Hara, T.; Yoshigai, E.; Ohashi, T.; Fukada, T. Zinc Transporters as Potential Therapeutic Targets: An Updated Review. *J. Pharmacol. Sci.* **2022**, *148*, 221–228. [CrossRef] [PubMed]

https://doi.org/10.3390/targets4010005

152. Liu, L.; Yang, J.; Wang, C. Analysis of the Prognostic Significance of Solute Carrier (SLC) Family 39 Genes in Breast Cancer. *Biosci. Rep.* **2020**, *40*, BSR20200764. [CrossRef]
153. Liu, Y.; Wei, L.; Zhu, Z.; Ren, S.; Jiang, H.; Huang, Y.; Sun, X.; Sui, X.; Jin, L.; Sun, X. Zinc Transporters Serve as Prognostic Predictors and Their Expression Correlates with Immune Cell Infiltration in Specific Cancer: A Pan-Cancer Analysis. *J. Cancer* **2024**, *15*, 939–954. [CrossRef] [PubMed]

https://doi.org/10.3390/targets4010005

targets

MDPI

Review

The Indole Scaffold in Biochemistry and Therapeutics: A Privileged Structure with Diverse Chemical, Biological, and Clinical Significance

Cristina Manuela Drăgoi [1], Alina-Crenguţa Nicolae [1,*] and Ion-Bogdan Dumitrescu [2]

1 Department of Biochemistry, Faculty of Pharmacy, Carol Davila University of Medicine and Pharmacy, RO-020956 Bucharest, Romania; cristina.dragoi@umfcd.ro

2 Department of Physics and Informatics, Faculty of Pharmacy, Carol Davila University of Medicine and Pharmacy, RO-020956 Bucharest, Romania; ion.dumitrescu@umfcd.ro

* Correspondence: alina.nicolae@umfcd.ro

Abstract

The indole scaffold represents a privileged structural motif in medicinal chemistry, celebrated for its remarkable chemical versatility, biological ubiquity, and clinical relevance. This review provides a comprehensive analysis of the recent research on the indole nucleus, emphasizing its physicochemical properties, reactivity patterns, and capacity to interact with a wide array of biological targets. Found in key endogenous compounds such as serotonin and melatonin, indole serves as a cornerstone in neurochemical signaling, circadian regulation, and chrono-metabolic homeostasis. Beyond its physiological roles, synthetic indole derivatives have shown extensive therapeutic potential across diverse domains, including oncology, infectious diseases, neurodegenerative disorders, immunomodulation, and metabolic syndromes. The review explores structure–activity relationships (SAR), pharmacokinetics, and the molecular mechanisms by which indole-based compounds exert their tremendous effects, that are ranging from enzyme inhibition to receptor modulation. Special focus is given to current clinical applications and emerging strategies for enhancing drug specificity, bioavailability, and safety through indolic frameworks. Additionally, we highlight the translational potential of indole-containing molecules in personalized medicine, underscoring opportunities for future drug discovery. By integrating insights from medicinal chemistry, biochemistry, pharmacology, and clinical science, this review affirms the indole ring's enduring value as a central scaffold in therapeutic innovation.

Keywords: indole; melatonin; tryptophan; serotonin; drug discovery; therapeutic innovation

1. Introduction

Few molecular frameworks in medicinal chemistry are as ubiquitous and versatile as indole. This bicyclic heterocycle—formally a benzene ring fused to a pyrrole ring—was first synthesized by Adolf Baeyer in 1866 during studies on the indigo dye. Since then, the indole nucleus has come to be recognized as a privileged scaffold in drug discovery, a term denoting structural motifs that can bind to multiple biological targets with high affinity [1]. Indole's planar aromatic system and its ability to engage in π–π stacking and hydrogen bonding, through the indole N–H, enable it to mimic diverse protein motifs and bind reversibly to enzymes and receptors. This remarkable capacity for multi-target interactions underlies the frequent appearance of indoles in natural products, biochemical pathways, and pharmaceuticals [1–13].

Given indole's prominence, several reviews in the past decade have examined indole-containing compounds and their biological activities [1,2,12,14]. However, many of these prior works adopted either an overly broad survey or focused on specific subtopics, which can obscure the unifying principles that make indole such a privileged structure. The present review aims to provide a comprehensive yet focused analysis that clearly articulates those unifying principles through a target-driven lens.

To ensure up-to-date and comprehensive coverage, we performed a systematic literature search of the PubMed database spanning the last ten years (2015–2025), retrieving relevant publications on indole derivatives, their molecular targets, and mechanisms of action. Our review is organized around specific biological targets and pathways—from neurotransmitter receptors to enzymes and beyond—to highlight how the indole scaffold can be tailored to engage a wide array of proteins. By structuring the discussion in this target-centric manner, we seek to emphasize our central thesis: the indole nucleus is a uniquely adaptable pharmacophore leveraged by nature and chemists alike to yield diverse therapeutic agents. This organization not only underscores the versatility of indole in modulating different biological systems but also delivers a clearer narrative and added value against the backdrop of existing literature. In doing so, we hope to provide readers with a transparent and insightful understanding of why indole has achieved such a prominent place in medicinal chemistry, and how this knowledge can guide future drug discovery efforts.

1.1. Natural Occurrence of Indoles

Indole moieties occur widely across living systems, underscoring their fundamental biological roles. The essential amino acid tryptophan contains an indole side chain and serves as the biosynthetic precursor to numerous biomolecules. For example, tryptophan is hydroxylated and decarboxylated to yield serotonin (5-hydroxytryptamine), a neurotransmitter that regulates mood, appetite, and digestion; serotonin is further enzymatically N-acetylated and O-methylated to produce melatonin, a neurohormone governing circadian rhythms [3,4]. In the plant kingdom, the indole scaffold appears in the growth hormone indole-3-acetic acid (IAA, or auxin), which controls cell division and elongation. These examples illustrate how evolution has incorporated the indole nucleus into core biochemical pathways in both animals and plants [6]. Even free indole itself functions as a biologically active messenger. At high concentrations, indole is a bacterial metabolite that contributes to the malodor of feces, whereas at trace levels it has a flowery aroma and finds use in perfumery. Many bacteria actively secrete indole as a cell-to-cell signaling molecule that influences processes such as spore formation, plasmid stability, biofilm development, and virulence. This ubiquitous presence of indole—from an amino acid building block and neurotransmitters to plant hormones and microbial signals—highlights its role as a foundational chemical communicator in nature [5,6].

The indole scaffold pervades the realm of secondary metabolites across diverse life forms. Indole alkaloids are widespread in the plant kingdom—for example, the Madagascar periwinkle (*Catharanthus roseus*) produces the vinca alkaloids vincristine and vinblastine (potent anti-cancer agents), while *Rauwolfia serpentina* yields reserpine, an indole alkaloid used historically as an antihypertensive and antipsychotic. Fungi provide another notable source: ergot alkaloids such as ergotamine and bromocriptine contain indole moieties and have been used to treat migraines and Parkinson's disease. Marine organisms also contribute indole derivatives; certain sponges (e.g., *Topsentin* species) produce indole-based metabolites with antibacterial, cytotoxic, and even anti-HIV activities. The prevalence of indoles in such a wide array of natural products underscores the privileged status of

https://doi.org/10.3390/targets4010004

this scaffold—in essence, evolution appears to have "pre-selected" the indole ring for biologically relevant chemistry across different kingdoms of life [7–10,15,16].

1.2. Indoles in Pharmaceuticals

The indole scaffold's prevalence in nature is paralleled by its success in modern medicine. Indole-based pharmaceuticals span virtually every major therapeutic area, illustrating the scaffold's adaptability. Notably, surveys of drug databases indicate that indole and its derivatives rank among the most common scaffolds found in FDA-approved small-molecule drugs [7,8,11–13]. Prominent examples include: vincristine and vinblastine—indole-containing vinca alkaloids from periwinkle, used as anticancer agents (introduced in the 1960s), indomethacin—a non-steroidal anti-inflammatory drug (NSAID) featuring an indole acetic acid moiety, pindolol—a β-adrenergic blocker for hypertension that contains an indole core, sumatriptan—a migraine medication acting as a serotonin (5-HT) receptor agonist, structurally derived from an indole tryptamine, melatonin—an indole-based neurohormone used therapeutically as a sedative and sleep-regulating agent, delavirdine—an HIV-1 reverse transcriptase inhibitor featuring an indole scaffold, and many other summarized in Table 1.

These and many other approved drugs contain an indole nucleus, effectively making it a nearly "universal" pharmacophore present in treatments for central nervous, cardiovascular, endocrine, and immune system disorders. Medicinal chemists are drawn to indole not only for its broad spectrum of bioactivity but also for its favorable drug-like properties (such as moderate molecular weight and appropriate lipophilicity) and the myriad possibilities for functionalization that allow fine-tuning of potency and selectivity. Consequently, numerous synthetic indole derivatives have advanced into clinical trials or preclinical development in recent years, underscoring the scaffold's continued prominence in drug discovery [1,2,12,14].

Table 1. FDA/EMA-Approved Drugs Containing an Indole Nucleus.

Drug	Year (Agency)	Chemical Structure	Therapeutic Indications	Mechanism of Action	Notes
Dihydroergotamine (ergot alkaloid)	1946 (FDA)	*Structure:* Ergot indole derivative (indole fused to lysergic scaffold).	Acute treatment of migraine	**$5\text{-HT}_{1B/1D}$ receptor agonist** on cranial blood vessels (causing vasoconstriction); also antagonizes trigeminal neurotransmission to relieve migraine.	–
Tegaserod	2002 (FDA)	*Structure:* Indole carbinol derivative	Irritable bowel syndrome (IBS) with constipation	**5-HT_4 receptor partial agonist**, stimulating GI peristalsis and secretion to relieve constipation.	Withdrawn 2007 (CV risks); reapproved 2019 (restricted use)
Ondansetron	1991 (FDA)	*Structure:* Indole-3-carboxamide	Prevention of chemotherapy- and surgery-induced nausea and vomiting	**5-HT_3 receptor antagonist** on vagal afferents and in the CNS chemoreceptor trigger zone, blocking serotonin-mediated emetic signals.	–
Dolasetron	1997 (FDA)	*Structure:* Indole-3-carboxylate	Prevention of chemotherapy- and post-operative nausea/vomiting	**5-HT_3 receptor antagonist** (prodrug converted to hydrodolasetron) blocking serotonin in GI tract and CNS (antiemetic).	–

https://doi.org/10.3390/targets4010004

Table 1. *Cont.*

Drug	Year (Agency)	Chemical Structure	Therapeutic Indications	Mechanism of Action	Notes
Tropisetron (Navoban)	1992 (EU)	*Structure:* Indole-3-carboxamide	Prevention of chemotherapy-induced nausea/vomiting (used in EU)	**5-HT$_3$ receptor antagonist**—blocks peripheral and central serotonin receptors to prevent emesis.	(Not FDA-approved; launched in EU 1992)
Vilazodone	2011 (FDA)	*Structure:* Indole-piperazine	Major depressive disorder (antidepressant)	**Serotonin reuptake inhibitor** and **5-HT$_{1A}$ receptor partial agonist**, enhancing serotonergic neurotransmission.	–
Triptans (e.g., **Sumatriptan**, Naratriptan, Zolmitriptan, Rizatriptan, Almotriptan, Eletriptan, Frovatriptan)	1992 (FDA)	*Structure:* (Class example Sumatriptan	Acute treatment of migraine attacks (all triptans)	**5-HT$_{1B/1D}$ receptor agonists** on cranial blood vessels and trigeminal nerve terminals, causing vasoconstriction and reduced neuropeptide release to alleviate migraine.	Sumatriptan (first triptan) FDA-approved 1992; others followed through 2002.
Alosetron	2000 (FDA)	*Structure:* Indole-3-carboxylic acid amidine	Diarrhea-predominant IBS in women	**5-HT$_3$ receptor antagonist** on enteric neurons, modulating visceral pain and GI transit to relieve IBS-D symptoms.	Withdrawn 2000 (ischemic colitis); reintroduced 2002 with restrictions.
Methysergide (Sansert)	1962 (FDA)	*Structure:* Ergot alkaloid (indole nucleus fused in lysergic structure)	Migraine prophylaxis (vascular headaches)	**5-HT$_2$ receptor antagonist** (and partial agonist); modulates serotonin-mediated vasodilation and inflammation in migraines.	Withdrawn (2002, fibrosis complications)
Dihydroergocornine (Ergoloid)	1953 (FDA)	*Structure:* Ergoloid mesylate component (indole fused)	Symptomatic treatment of age-related cognitive decline (as part of **ergoloid mesylates**/Hydergine)	**Adrenergic and serotonin receptor modulator** (ergot alkaloid); enhances cerebral blood flow and metabolism (mechanism not fully specific).	Used in combination (ergoloid mix)
Indomethacin	1965 (FDA)	*Structure:* Indole-3-acetic acid NSAID	Pain and inflammation in rheumatoid arthritis, osteoarthritis, acute gout, etc.	**Non-selective COX-1/2 inhibitor**—blocks prostaglandin synthesis, providing anti-inflammatory, analgesic, and antipyretic effects.	–
Acemetacin	1980 (UK)	*Structure:* Indole acetic acid prodrug	Pain and inflammation in rheumatoid arthritis, osteoarthritis (prodrug of indomethacin)	**Non-selective COX inhibitor** (metabolized to indomethacin)—same mechanism as indomethacin.	Approved in Europe (not FDA); less GI toxicity than indomethacin.
Etodolac	1991 (FDA)	*Structure:* Indole acetic acid NSAID (indole)	Osteoarthritis, rheumatoid arthritis pain; acute pain	**COX-2 preferential inhibitor**—reduces prostaglandins to alleviate inflammation and pain.	–
Carprofen (Rimadyl)	1988 (FDA)	*Structure:* Carbazole-propionic acid (fused indole ring)	Relief of pain and inflammation in osteoarthritis (veterinary use; formerly in humans)	**Non-selective COX inhibitor**—NSAID action to reduce inflammation and pain.	Human use 1988–1993 (withdrawn for safety); now veterinary NSAID.
Cabergoline	1996 (FDA)	*Structure:* Ergoline (indole fused in tetracyclic ergoline core)	Hyperprolactinemia (pituitary tumors, prolactin disorders)	**Dopamine D$_2$ agonist** on pituitary lactotroph cells, suppressing prolactin release.	–

https://doi.org/10.3390/targets4010004

Table 1. *Cont.*

Drug	Year (Agency)	Chemical Structure	Therapeutic Indications	Mechanism of Action	Notes
Lisuride	1985 (EU)	*Structure:* Ergoline (fused indole)	Parkinson's disease; migraine prophylaxis (in EU)	**Dopamine receptor agonist** (D_2 family) similar to bromocriptine; also 5-HT receptor effects.	Not FDA-approved (used in EU since ~1983).
Bromocriptine	1978 (FDA)	*Structure:* Ergoline (fused indole polycyclic)	Hyperprolactinemia (e.g., prolactinomas), acromegaly; Parkinson's (adjunct)	**Dopamine D_2 agonist**—stimulates dopamine receptors, inhibiting prolactin secretion; in Parkinson's, substitutes for dopamine.	–
Pergolide (Permax)	1988 (FDA)	*Structure:* Ergoline (indole fused in polycyclic ergot structure)	Parkinson's disease (adjunct to levodopa)	**Dopamine D_2 agonist**—stimulates striatal dopamine receptors to improve motor function in Parkinson's.	Withdrawn 2007 (valvular heart disease risk).
Methylergometrine (Methylergonovine)	1945 (FDA)	*Structure:* Ergot alkaloid (indole fused)	Postpartum hemorrhage (uterotonic to prevent/treat bleeding)	**Oxytocic/uterotonic**—stimulates uterine smooth muscle via serotonin and adrenergic receptors, causing contraction to reduce hemorrhage.	In use for decades; only oral uterotonic in US
Sertindole (Serdolect)	1996 (EMA)	*Structure:* Phenylindole antipsychotic (indole ring)	Schizophrenia (atypical antipsychotic)	**5-HT_{2A} and D_2 receptor antagonist**—balances serotonin/dopamine neurotransmission in cortex and limbic system.	Not FDA-approved; EMA-approved 1996, withdrawn 1998 (QT prolongation), reapproved 2005.
Pindolol	1982 (FDA)	*Structure:* Indole-2-propanol β-blocker (indole structure)	Hypertension; angina; arrhythmias	**Non-selective $\beta_1\beta_2$ adrenergic receptor blocker** (with intrinsic sympathomimetic activity)—reduces heart rate and contractility, lowering blood pressure.	–
Carvedilol	1995 (FDA)	*Structure:* Carbazole-propanolamine (fused indole)	Heart failure; hypertension; post-MI left ventricular dysfunction	**Non-selective β-adrenergic blocker** (β_1, β_2) **and α_1-blocker**—decreases heart rate and myocardial oxygen demand, and causes vasodilation.	Racemic drug (both enantiomers active: β-blockade and α-blockade).
Nicergoline	1978 (IT, EU)	*Structure:* Ergoline derivative (indole fused)	Senile dementia and vascular cognitive impairment (vasodilator)	**α_1-Adrenergic antagonist** in cerebral vessels—improves arterial blood flow and metabolism in the brain.	Not FDA-approved (used in EU/Asia)
Dihydroergocristine (Ergoloid)	1953 (FDA)	*Structure:* Ergoloid mesylate component (indole fused)	Senile cognitive impairment (as part of ergoloid combination)	**Adrenergic/serotonergic modulator** (see DHE and ergoloids above)—enhances cerebral perfusion and metabolism.	Part of **Hydergine** (with dihydroergocornine, etc.).
Ergometrine (Ergonovine)	1940 (FDA)	*Structure:* Simple lysergamide (indole fused)	Postpartum uterine atony and hemorrhage	**Partial agonist at α-adrenergic, 5-HT, and dopaminergic receptors** in uterine muscle—causes sustained uterine contraction to prevent bleeding.	Historic use in obstetrics
Yohimbine	1980 (FDA)	*Structure:* **Indole alkaloid** (indole fused in **quinolizidine** ring)	Erectile dysfunction (historical use); also used in neurogenic orthostatic hypotension	**α_2-Adrenergic receptor antagonist**—increases sympathetic outflow and blood flow (erectile effect).	FDA-approved for impotence, use supplanted by PDE5 inhibitors; now found in supplements.

https://doi.org/10.3390/targets4010004

Table 1. *Cont.*

Drug	Year (Agency)	Chemical Structure	Therapeutic Indications	Mechanism of Action	Notes
Vinblastine (Vinca alkaloid)	1965 (FDA)	*Structure:* **Bis-indole** (vindoline + catharanthine dimer; indole moieties fused)	Hodgkin's lymphoma, lymphomas, breast, testicular cancers	**Antimitotic agent (microtubule inhibitor)**—binds tubulin and prevents spindle formation, causing mitotic arrest in dividing cells.	Naturally derived from *Catharanthus roseus.*
Vincristine (Vinca alkaloid)	1963 (FDA)	*Structure:* **Bis-indole** (dimeric indole–indoline; indole)	Acute leukemias, lymphomas, childhood cancers	**Antimitotic (vinca alkaloid)**—similar mechanism to vinblastine, inhibiting microtubule polymerization.	Neurotoxic side effects dose-limiting.
Vinorelbine (Navelbine)	1994 (FDA)	*Structure:* Semi-synthetic vinca (bis-indole, indole nucleus)	Non-small cell lung cancer; metastatic breast cancer	**Antimitotic (vinca alkaloid)**—binds tubulin, causing mitotic arrest (like vincristine).	Semisynthetic vincristine derivative (less neurotoxic).
Vindesine (Eldisine)	1979 (EU)	*Structure:* Semi-synthetic vinca (bis-indole)	Acute lymphoblastic leukemia (ALL); other cancers (investigational)	**Antimitotic**—vinca alkaloid mechanism (tubulin binding, mitosis inhibition).	Not FDA-approved (used in Canada/EU).
Vinflunine (Javlor)	2009 (EMA)	*Structure:* Fluorinated vinca (bis-indole)	Second-line therapy for metastatic urothelial (bladder) cancer	**Antimitotic (vinca)**—microtubule inhibitor (similar to vinblastine) for cancer cell cycle arrest.	EMA-approved; not approved in US.
Tadalafil (Cialis)	2003 (FDA)	*Structure:* **Beta-carboline** (indole fused to pyridone)	Erectile dysfunction; BPH; pulmonary arterial hypertension	**PDE-5 inhibitor**—selectively inhibits phosphodiesterase-5, raising cGMP in smooth muscle and causing vasodilation (e.g., penile erection, pulmonary vasodilation).	–
Melatonin	1992 (FDA)	*Structure:* Endogenous **indoleamine** (N-acetyl-5-methoxytryptamine, indole core)	Jet lag, insomnia (circadian rhythm sleep disorders)	**$MT_{1/2}$ melatonin receptor agonist** in suprachiasmatic nucleus—synchronizes circadian sleep–wake cycle.	Available OTC as dietary supplement in US (FDA-approved prolonged-release form in EU, 2007).
Alectinib (Alecensa)	2015 (FDA)	*Structure:* Indole-carbazole derivative (indole fused in multi-ring)	ALK-positive metastatic non-small cell lung cancer (NSCLC)	**ALK tyrosine kinase inhibitor**—targets ALK fusion oncoproteins in cancer cells, inhibiting proliferation.	Active against crizotinib-resistant ALK mutations.
Tezacaftor (Symdeko combo)	2018 (FDA)	*Structure:* Indole-amide CFTR corrector (indole nucleus)	Cystic fibrosis (F508del mutation, with ivacaftor)	**CFTR folding corrector**—improves processing and trafficking of defective F508del-CFTR channels to cell surface.	–
Panobinostat (Farydak)	2015 (FDA)	*Structure:* Hydroxyamic **indole** (indole ring)	Multiple myeloma (refractory, in combination therapy)	**Pan-HDAC inhibitor**—inhibits histone deacetylases, altering gene expression and inducing cancer cell apoptosis.	Accelerated approval 2015; indication withdrawn 2021 (due to newer therapies).
Bazedoxifene (Duavee with conjugated estrogens)	2013 (FDA)	*Structure:* Indole-based **benzoxazine** (fused indole)	Postmenopausal osteoporosis prevention; menopausal vasomotor symptoms (with conjugated estrogens)	**Selective estrogen receptor modulator (SERM)**—estrogen agonist on bone (reduces resorption) and antagonist on uterine/breast tissue.	Approved as combination (with estrogen) in 2013.

https://doi.org/10.3390/targets4010004

Table 1. *Cont.*

Drug	Year (Agency)	Chemical Structure	Therapeutic Indications	Mechanism of Action	Notes
Osimertinib (Tagrisso)	2015 (FDA)	*Structure:* **Aza-indole** EGFR inhibitor (indole ring)	EGFR T790M mutation-positive NSCLC (3rd-gen EGFR TKI)	**Irreversible EGFR tyrosine kinase inhibitor**—covalently binds mutant EGFR kinase, blocking signaling in tumor cells.	Orally active; penetrates CNS (useful for brain metastases).
Rucaparib (Rubraca)	2016 (FDA)	*Structure:* Tricyclic indole (indole-phenyl ring,)	BRCA-mutated ovarian cancer; prostate cancer (PARP inhibitor)	**PARP inhibitor**—traps PARP on DNA single-strand breaks, preventing DNA repair in BRCA-deficient cancer cells (leading to cell death).	FDA accelerated approval 2016; EMA approval 2018.
Fluvastatin (Lescol)	1993 (FDA)	*Structure:* **Indole**-based statin (indole ring)	Hyperlipidemia—lowers LDL cholesterol; prevention of cardiovascular events	**HMG-CoA reductase inhibitor**—blocks cholesterol synthesis in liver, upregulating LDL receptors to clear plasma LDL.	First fully synthetic statin (indole class), FDA-approved 1993.
Delavirdine (Rescriptor)	1997 (FDA)	*Structure:* **Bis-indole** (indole-2-carboxamide with indole sulfonamide)	HIV-1 infection (antiretroviral NNRTI)	**Non-nucleoside reverse transcriptase inhibitor (NNRTI)**—binds HIV-1 RT allosterically to block RNA-dependent DNA polymerase activity, preventing viral DNA synthesis.	Often used in combination HAART (though less potent than newer NNRTIs).
Midostaurin (Rydapt)	2017 (FDA)	*Structure:* **Indolocarbazole** (two indoles fused)	FLT3-mutated acute myeloid leukemia; systemic mastocytosis	**Multi-kinase inhibitor** (FLT3, KIT, etc.)—inhibits signaling in FLT3-ITD mutant leukemic cells, inducing apoptosis.	Derived from staurosporine (natural bis-indole); first targeted FLT3 inhibitor.
Zafirlukast (Accolate)	1996 (FDA)	*Structure:* Indole-3-acyl sulfonamide (indole)	Asthma prophylaxis and chronic treatment	**$CysLT_1$ leukotriene receptor antagonist** in airway—blocks LTC_4/LTD_4 mediated bronchoconstriction and inflammation.	First leukotriene antagonist (FDA 1996); taken orally for mild-moderate asthma.
Reserpine	1955 (FDA)	*Structure:* **Indole alkaloid** (indole fused in yohimban nucleus)	Hypertension (obsolete); antipsychotic (historic)	**VMAT inhibitor**—irreversibly blocks vesicular monoamine transporter in adrenergic neurons, depleting NE, dopamine, serotonin stores and thus lowering blood pressure and causing sedation.	*Rauwolfia serpentina* alkaloid; early anti-hypertensive (now rarely used).
Metergoline	1960s (FDA)	*Structure:* Ergoline (indole fused)	Migraine prophylaxis; investigational in hyperprolactinemia	**Serotonin antagonist** (5-HT_2 and 5-HT_2)—reduces serotonin-mediated vasodilation (in migraines) and blocks prolactin release (via dopamine agonism).	Largely superseded by newer agents; sometimes used in seasonal affective disorder.
Lurbinectedin (Zepzelca)	2020 (FDA)	*Structure:* **Tetrahydroisoquinoline** (related to indole alkaloid)	Second-line treatment of metastatic **small-cell lung cancer**	**DNA minor-groove binder/alkylator**—covalently binds DNA and inhibits oncogenic transcription, leading to cancer cell death	FDA accelerated approval Jun 2020.

Important observation: Many drugs above feature fused indole systems. For example, ergot alkaloids (e.g., bromocriptine, ergometrine) contain an indole nucleus fused to a polycyclic lysergamide structure, and vinca alkaloids (vinblastine, etc.) are dimers of indole units (bis-indoles). Despite their structural complexity, the indole pharmacophore is crucial to their biological activity.

https://doi.org/10.3390/targets4010004

2. Chemical Properties and Interactions with Endogenous and Exogenous Molecules

The indole ring system imparts a unique set of chemical properties [1,17,18]. As a planar aromatic heterocycle with a conjugated 10 π-electron system, indole's nitrogen lone pair is delocalized, making the N–H group weakly basic and the molecule neutral at physiological pH (Figure 1). The indole NH can donate hydrogen bonds, while the electron-rich five-membered ring acts as a π-donor to electrophiles. These features support π–π stacking with aromatic residues (e.g., phenylalanine, tyrosine, tryptophan), commonly observed in protein binding sites through T-shaped interactions. Indole's planarity also promotes electron delocalization and facilitates intercalation into DNA, stabilizing the double helix and modulating transcriptional activity through both π-stacking and hydrogen bonding interactions [19,20]. Notably, endogenous indole derivatives such as melatonin, serotonin and tryptophan have demonstrated the ability to interact with DNA through non-covalent mechanisms, contributing to genomic stability [21]. Melatonin, in particular, exhibits high affinity for DNA via both electrostatic and hydrophobic interactions, forming protective complexes that shield the molecule from oxidative damage. Tryptophan, through its aromatic side chain, can also engage in stabilizing interactions with DNA, particularly under oxidative or inflammatory conditions, thus contributing to the preservation of genomic integrity. These molecular features underscore the unique duality of indole compounds as both signaling molecules and structural protectors at the nucleic acid interface [3,19,20,22].

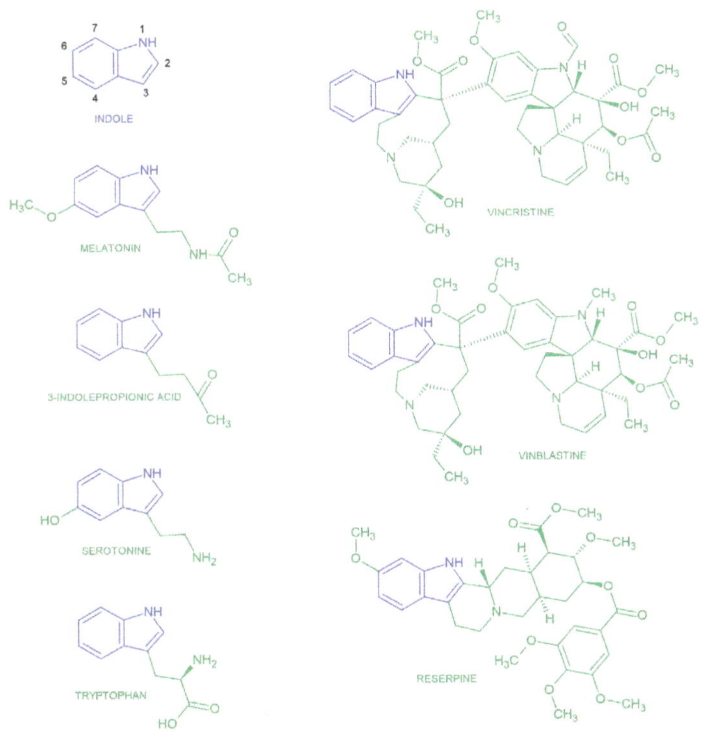

Figure 1. Indole scaffold (blue color) and some biologically active derivatives (green color with the indole highlighted in blue) of endogenous and exogenous origin, encompassing tryptophan, serotonin, melatonin, indole-3-propionic acid, and clinically relevant indole alkaloids such as reserpine, vincristine, and vinblastine.

https://doi.org/10.3390/targets4010004

3. Reactivity and Synthetic Strategies

Indole's electron-rich pyrrole ring dictates much of its chemistry. The indole scaffold is chemically versatile at multiple positions without damaging the core, allowing medicinal chemistry to explore broad structure–activity relationships. A multitude of synthetic routes to indoles have been developed, reflecting the ring's importance. Traditional methods date back to the 19th and early 20th centuries and include named reactions that form indole from various precursors. The Fischer indole synthesis is the quintessential example, converting aryl hydrazones to indoles under acid catalysis. Many Fischer-based syntheses are used industrially to produce tryptamine derivatives and indole-3-acetic acid. Other classical routes include the Bischler–Möhlau (from α-halo-ketones and anilines), Baeyer–Emmerling (from ortho-nitrocinnamic acids), Madelung (intramolecular cyclization of N-phenylamides), Reissert (cyclization of α-aryl-β-aminoketones), Bartoli (addition of vinyl Grignard to nitroanilines), Larock (Pd-catalyzed annulation of anilines with alkynes), and Leimgruber–Batcho (synthesis of indoles via nitrotoluene intermediates) reactions. Each offers distinct entry points depending on available starting materials [23,24].

Contemporary research continues to expand indole synthetic methods. One trend has been one-pot and multi-component reactions to assemble indoles efficiently. Another major advance is transition-metal catalysis: C–H activation and cyclization strategies allow indole construction under mild conditions with diverse substituents [25,26].

In summary, chemically, the indole ring is stable, readily synthesized or sourced from nature, and amenable to substitution. These factors, combined with its favorable size and aromatic character, make indole an attractive core for medicinal chemistry.

4. Endogenous Indole-Based Compounds

Tryptophan and its metabolites: L-Tryptophan is the sole amino acid containing an indole nucleus, highlighting nature's adoption of indole in core biochemistry. As an essential amino acid, tryptophan must be obtained from the diet and incorporated into proteins, where its indole side chain often contributes to structural stability and enzymatic function [27]. Beyond its role in protein synthesis, tryptophan serves as a precursor to key indole-containing metabolites via two main pathways. Approximately 95% is metabolized through the hepatic kynurenine pathway, where enzymes like IDO and TDO convert it to N-formylkynurenine and ultimately to NAD+ precursors such as quinolinic acid—though these metabolites lack the indole ring, they exert important biological effects. The remaining ~5% enters the serotonin/melatonin pathway, beginning with hydroxylation of the indole ring to form 5-hydroxytryptophan, which is then decarboxylated to serotonin (5-HT). As an indoleamine neurotransmitter, serotonin regulates mood, appetite, and gastrointestinal motility, acting through a broad family of receptors that specifically recognize its indole ethylamine structure. Serotonin is inactivated primarily by oxidative deamination (monoamine oxidase) to yield 5-hydroxyindoleacetic acid (5-HIAA), an indole carboxylic acid excreted in urine [28,29].

From serotonin, one additional enzymatic step in the pineal gland produces melatonin (N-acetyl-5-methoxytryptamine), another endogenous indole. Melatonin features an indole core with a methoxy substitution at C5 and an N-acetyl group on the side chain. This small modification endows melatonin with distinct properties: it is highly lipophilic, can easily cross cell membranes and the blood–brain barrier, and primarily acts as a hormone regulating circadian rhythms [21,30,31]. Melatonin is synthesized mainly at night by the pineal gland, via the enzymes arylalkylamine N-acetyltransferase and hydroxyindole-O-methyltransferase acting on serotonin. It binds to G-protein-coupled melatonin receptors in the brain and periphery to signal darkness and induce physiological changes associated with sleep onset [32,33]. Besides circadian regulation, melatonin has antioxidant

https://doi.org/10.3390/targets4010004

properties—its indole ring can scavenge free radicals, and its metabolites (e.g., cyclic 3-hydroxymelatonin) also act as radical scavengers [34,35]. Melatonin is metabolized in the liver to 6-hydroxymelatonin via cytochrome P450 and conjugated with sulfate; its half-life is short (~30–60 min). Both serotonin and melatonin illustrate how nature exploits the indole ring for signaling: the former as a neurotransmitter with fast synaptic actions, the latter as a hormone with systemic chronobiological effects [36–38].

Humans are also exposed to indole derivatives produced by gut microbiota, which convert dietary tryptophan into indole via bacterial tryptophanase. These microbial indoles enter circulation and, while high levels (e.g., indoxyl sulfate) are associated with renal toxicity in chronic kidney disease, lower concentrations exert beneficial effects. For example, indole-3-propionic acid (IPA), produced by Clostridium sporogenes, acts as a neuroprotective antioxidant and activates intestinal pregnane X receptors (PXR), enhancing barrier integrity and dampening inflammation. Similarly, indole-3-aldehyde (IAld) engages the aryl hydrocarbon receptor (AhR) in immune cells, promoting IL-22 secretion and maintaining mucosal homeostasis. Thus, even exogenous indoles from our microbiome participate in human physiology, mediating the cross-talk between diet, microbes, and the host's immune-metabolic balance [39,40].

In summary, the indole scaffold is embedded in multiple layers of human biology: as part of essential nutrients and hormones, as neurotransmitters and signal mediators, and as a chemical currency between microbes and host. Tryptophan, serotonin, and melatonin exemplify the range from metabolic precursor to neurotransmitter to hormone, all unified by the indole core.

5. Receptor Pharmacology and Molecular Targets

One reason indoles are "privileged" in pharmacology is their ability to engage a wide spectrum of receptors and enzymes. The indole ring's size, planarity, and electronics allow it to fit into many binding sites, from G protein-coupled receptors (GPCRs) to nuclear receptors to enzymes. A classic example is the serotonin receptor family: serotonin (5-HT) itself is an indoleamine, and many drugs targeting 5-HT receptors are indole or tryptamine analogs. The 5-HT1 subclass (which includes the 5-HT1A receptor involved in anxiety/depression) is activated by indolealkylamines like buspirone's active metabolite and the antidepressant vilazodone (which contains an indole core) [41,42]. The 5-HT1B/1D receptors, which mediate cranial vasoconstriction, are the targets of triptans (e.g., sumatriptan, rizatriptan)—these anti-migraine drugs are indole derivatives that mimic serotonin's indoleethylamine structure to selectively agonize 5-HT receptors in cranial blood vessels. The 5-HT2 family, notably 5-HT2A, is potently stimulated by indole hallucinogens like lysergic acid diethylamide (LSD) and psilocybin (a 4-phosphorylated tryptamine), demonstrating how indole scaffolds can achieve high affinity at CNS receptors. Even the lone ligand-gated ionotropic serotonin receptor, 5-HT3, has potent antagonists with indole cores (e.g., ondansetron is a carbazole, structurally related to indole) used as antiemetics. Thus, indoles are intimately tied to serotonergic signaling as both native ligands and drugs [43–45].

Another set of GPCRs recognizing indoles is the melatonin receptors (MT1 and MT2). These receptors naturally bind melatonin, and synthetic analogs have been designed on the indole scaffold to treat insomnia or depression. For example, ramelteon is a high-affinity MT1/MT2 agonist used for insomnia that, while not an indole by structure, was developed by structurally mimicking melatonin's indole pharmacophore. Agomelatine, an atypical antidepressant, was modeled after melatonin as well. These cases highlight that even when the final drug is not an indole per se, indole analogs often serve as starting points or inspiration due to their excellent fit in the binding site [46–49].

https://doi.org/10.3390/targets4010004

Indole scaffolds also modulate adrenergic receptors, key GPCRs for norepinephrine and epinephrine. Yohimbine, an indole alkaloid from *Pausinystalia johimbe*, acts as an $\alpha 2$-adrenergic antagonist, historically used for erectile dysfunction and neuropharmacological research. Pindolol, a β-blocker with an indole core, inhibits $\beta 1/\beta 2$ receptors to lower blood pressure and also exhibits partial agonism at 5-HT1A receptors, illustrating multi-receptor activity. Carvedilol, used in heart failure, includes a carbazole moiety (a dibenzopyrrole structure related to indole) and blocks both beta and $\alpha 1$ receptors. These examples highlight the adaptability of indole frameworks in targeting various aminergic GPCRs, including adrenergic and potentially dopaminergic receptors, despite dopamine's distinct chemical structure [50–52].

Beyond GPCRs, indoles can target ligand-gated ion channels and transporters. For example, the plant alkaloid strychnine (from *Strychnos nux-vomica*) is a well-known indole alkaloid that acts as an antagonist at glycine receptors in the spinal cord, causing powerful convulsions. Similarly, ibogaine (an indole alkaloid from *Tabernanthe iboga*) binds to multiple sites—it blocks NMDA-type glutamate receptors (ion channels) and interacts with serotonin and opioid receptors—it has been investigated for treating addiction. The diversity of these targets underscores the indole's flexibility: in strychnine, it presents a rigid polyvalent scaffold to a chloride channel receptor, whereas in ibogaine, its indole is part of a larger polycyclic system that can lodge into several protein pockets [28,53–55].

Indoles also engage nuclear receptors and other intracellular targets. The aryl hydrocarbon receptor (AhR) is a cytosolic receptor/transcription-factor that senses aromatic ligands; while classically activated by environmental toxins like dioxin, AhR is also activated by natural indoles. Endogenously, indole metabolites such as indoxyl sulfate, indole-3-aldehyde (IAld), and kynurenine bind AhR and modulate gene expression involved in xenobiotic metabolism and immune regulation. For instance, in the gut, bacterial IAld acting on AhR induces IL-22 production, which helps maintain mucosal immunity [17,56]. Indole-3-carbinol (I3C), a compound from cruciferous vegetables, and its dimer 3,3′-diindolylmethane (DIM) are well-known dietary indoles that activate AhR; through AhR and downstream pathways, they can alter estrogen metabolism and exert anti-proliferative effects. AhR activation by indoles exemplifies a ligand–receptor pair likely honed by nature to sense dietary or microbial metabolites [57,58].

Peroxisome proliferator-activated receptors (PPARs) are nuclear receptors involved in metabolic regulation. Some indole derivatives have been developed as PPARγ agonists, such as thiazolidinedione–indole hybrids reported by Corigliano et al. (2018) for potential anticancer use. The indole scaffold facilitated proper alignment within the PPARγ ligand-binding pocket, and SAR analysis revealed that modifications like a 5-methoxy group significantly improved activity. Although traditional PPARγ drugs (e.g., rosiglitazone) lack an indole core, these findings support the potential of indoles as effective scaffolds for nuclear receptor targeting [59].

Enzymes are yet another class of targets: indoles can inhibit enzymes either by mimicking a substrate or by binding allosterically. A historical example is physostigmine (eserine), an indole alkaloid from the Calabar bean, which reversibly inhibits acetylcholinesterase (AChE) and was used to treat glaucoma and myasthenia gravis. Its indole core is thought to bind in the AChE gorge similarly to tryptophan residues and stabilize the enzyme–inhibitor complex [60,61]. Indomethacin, one of the earliest synthetic NSAIDs, is an indole-3-acetic acid derivative that potently inhibits cyclooxygenase (prostaglandin G/H synthase) to exert anti-inflammatory effects. Interestingly, indomethacin's indole ring (with a 5-methoxy and 4-chlorobenzoyl substitution) binds deep in the COX active site, and the drug's orientation resembles that of an arachidonic acid segment—again highlighting indole's protein-mimicking ability [62–64].

https://doi.org/10.3390/targets4010004

Another enzyme target is monoamine oxidase (MAO): tryptamine analogs (indole ethylamines) can act as MAO inhibitors; for example, some β-carbolines (which contain an indole fused to a pyridine) inhibit MAO-A in the brain, prolonging monoamine neurotransmitter action—this is relevant to antidepressant and psychoactive effects (the hallucinogen harmine from *Banisteriopsis caapi* is a β-carboline MAO-A inhibitor) [65,66]. Indole-containing compounds are also prominent as kinase inhibitors in oncology. The multi-kinase inhibitor sunitinib (for renal cancer and GIST) is built on an indole (indolinone) core; indirubin, a bis-indole isomer of indigo, is a known cyclin-dependent kinase inhibitor used in chronic myelogenous leukemia in traditional Chinese medicine. Medicinal chemists have created numerous indole-based kinase inhibitors, given that the flat heterocycle can snugly occupy the adenine-binding pocket of kinases. In fact, a survey of recent clinical candidates shows indole/indazole cores in inhibitors of Aurora kinases, JAK kinases, etc., attesting that indoles remain relevant scaffolds for enzyme active sites in drug discovery [67–71].

These examples demonstrate that the indole scaffold exhibits remarkable target promiscuity, effectively engaging diverse classes of pharmacological targets, including G protein–coupled receptors (e.g., serotonin, melatonin, adrenergic receptors), ligand-gated and voltage-gated ion channels (e.g., glycine, NMDA receptors), membrane transporters and metabolic enzymes (e.g., acetylcholinesterase, monoamine oxidase, cyclooxygenases), as well as intracellular transcriptional regulators such as the aryl hydrocarbon receptor (AhR) and various nuclear hormone receptors.

6. Therapeutic Applications

6.1. Oncology

Indole-based compounds have a storied and ongoing role in cancer therapy. Several natural indole alkaloids were among the earliest effective chemotherapeutics and remain in use. The vinca alkaloids, isolated from *Catharanthus roseus* in the 1950s, include vinblastine and vincristine—complex bis-indole compounds (dimeric indole–indoline structures) that bind tubulin and inhibit microtubule polymerization, causing mitotic arrest of cancer cells. Vincristine and vinblastine were FDA-approved in the early 1960s and dramatically improved cure rates for leukemias, lymphomas, and other malignancies. Semisynthetic vincristine analogs like vinorelbine and vindesine were later developed to broaden activity and reduce neurotoxicity [16,72]. These drugs exemplify how an indole scaffold can interfere with a fundamental process (spindle formation) in rapidly dividing cells. Another renowned natural product is camptothecin, a quinoline-indole alkaloid from *Camptotheca acuminata*. Camptothecin's planar indole-containing structure intercalates into DNA at topoisomerase I cleavage sites, stabilizing the DNA–enzyme complex and leading to lethal DNA breaks. Its derivatives irinotecan and topotecan are important agents in colorectal and ovarian cancer, respectively [73,74].

Indole-3-carbinol (I3C), found in cruciferous vegetables, represents a nutraceutical approach to oncology. I3C itself is not the active agent in vivo—in the acidic stomach, it dimerizes into 3,3′-diindolylmethane (DIM) and other oligomeric products. These indoles have been shown to induce phase I/II drug-metabolizing enzymes, modulate estrogen metabolism (favoring less estrogenic catechol estrogens), and cause cell-cycle arrest and apoptosis in cancer cells. For example, I3C/DIM can downregulate cyclin-dependent kinases and upregulate tumor suppressor p21, particularly in hormone-responsive tumors. Though I3C/DIM are available as supplements and have shown anti-cancer effects in preclinical models of breast, prostate, and colon cancer, clinical evidence is still limited, and high doses are required. Nonetheless, they highlight the potential of indole diet-derived agents as cancer chemopreventives [75–77].

https://doi.org/10.3390/targets4010004

Another prominent class of indole-based anticancer agents is kinase inhibitors. The oxindole (indolin-2-one) scaffold, exemplified by sunitinib, serves as a versatile ATP-competitive motif targeting multiple receptor tyrosine kinases (e.g., VEGFR, PDGFR, c-Kit), thereby inhibiting tumor angiogenesis and cell growth. Sunitinib is approved for renal cell carcinoma and GIST. Other clinically relevant kinase inhibitors, such as abemaciclib (a CDK4/6 inhibitor), and candidates targeting DYRK and Aurora kinases, frequently incorporate indole frameworks for optimal fit within the ATP-binding pocket. SAR in these series shows the indole N–H or N–H substituents can form key H-bonds with kinase hinge residues, and substitutions at C-5 or C-6 can reach into specificity pockets [67,78,79].

Several indole derivatives target hormone-related pathways in cancer. Tamoxifen (for breast cancer) is not an indole, but researchers have explored indoles as selective estrogen receptor modulators (SERMs) and as aromatase inhibitors. For instance, some indole-3-carboxyethyl piperazines showed SERM activity, and indole-derivatives like aminoglutethimide (though not a true indole) pioneered aromatase inhibition. Indole-structured antiandrogens have also been investigated for prostate cancer [80,81]. Additionally, as noted earlier, indoleamine 2,3-dioxygenase (IDO1) is an immunosuppressive enzyme often overexpressed in tumors; inhibitors such as indoximod (1-methyl-tryptophan) aim to counteract the "tryptophan starvation" strategy of tumors and restore T-cell activity. Indoximod and analogs entered clinical trials in melanoma and other cancers. While the first major trial of an IDO1 inhibitor (epacadostat, a hydroxyamidine that is not indole-based) in combination immunotherapy did not meet its endpoint, interest remains in modulating the Trp–kynurenine–AhR axis in cancer. Endogenously, accumulations of kynurenine (from Trp via IDO) activate AhR in tumor cells and regulatory T-cells, aiding immune evasion. Indole-based drugs that block IDO or AhR are being explored to tilt the immune balance against tumors [28,82,83].

In oncology-related indoles, clear structure–activity relationships (SAR) have emerged. Minor substitutions on the indole ring can significantly impact potency and selectivity [59,84–86]. For example, N-substitutions and catharanthine modifications in vinblastine/vincristine derivatives modulate efficacy and toxicity. Electron-withdrawing groups and halogenation (e.g., 6-bromoindoles) enhance antiproliferative and antimicrobial activity by altering electronics and lipophilicity. A 5-methoxy group on thiazolidinedione–indole hybrids improved activity against breast cancer cells, likely via favorable electronic and steric effects. The indole scaffold allows extensive derivatization at N-1 and C-2 to C-7, with medicinal chemists also employing heterocyclic bioisosteres (e.g., indazole, benzimidazole) to refine drug-like properties while maintaining target affinity [59,84,85]. In conclusion, the indole scaffold's contribution to oncology spans cytotoxic natural products, targeted kinase inhibitors, hormonal pathway modulators, and emerging immunotherapeutics. Its privileged structure has enabled binding to diverse cancer targets (tubulin, kinases, receptors, enzymes), and ongoing SAR refinements continue to produce novel anticancer leads.

6.2. Infectious Diseases: Antibacterial, Antimalarial, Antiviral

Indole-based compounds also exhibit a broad spectrum of anti-infective activities. In bacteriology, the indole ring is both a messenger and a weapon. As a signaling molecule, indole, endogenously produced by commensal *E. coli* in the gut microbiome, can induce biofilm formation or antibiotic tolerance in bacterial communities. But as a scaffold for antibiotics, indole has yielded candidates against some of the toughest pathogens. A prominent example is the search for new tuberculosis (TB) drugs: researchers have extensively explored indole derivatives as anti-tubercular agents [8]. The rationale is that indole's flat aromatic nature might intercalate or bind to mycobacterial enzymes and cell wall components. Indeed, a novel indole-2-carboxamide (azaindole) was found to inhibit the

https://doi.org/10.3390/targets4010004

essential TB enzyme DprE1 (decaprenylphosphoryl-β-D-ribose epimerase) involved in cell wall arabinan synthesis—this compound (known as GSK-2556286 or 1,4-azaindole) showed potent activity and has entered clinical trials for tuberculosis [87,88]. Similarly, an indazole sulfonamide DG167 has shown potent anti-TB activity in preclinical studies. These compounds are non-covalent inhibitors that likely occupy enzyme active sites, with the indole/indazole ring providing key π-stacking with NAD or aromatic residues. Beyond these, numerous indole derivatives have been synthesized and screened against *M. tuberculosis*, yielding hits that disrupt various pathways: cell wall assembly, DNA gyrase, protein synthesis, etc. For example, some indole-3-carboxamides inhibit the TB gyrase enzyme, while indole-containing bis-cations have been reported to physically disrupt the mycobacterial cell membrane. SAR in these series indicates that cationic or amphiphilic decorations on indole can enhance penetration and binding—e.g., adding basic amine chains to indole (making it an indole-based amphiphile) increased membrane depolarizing ability against TB mycobacteria [8,88,89].

Indole scaffolds have also shown promise as antibacterial agents. Although linezolid is not an indole, indole-based analogs of oxazolidinones have demonstrated enhanced potency. The natural product indolmycin, a selective inhibitor of bacterial tryptophanyl-tRNA synthetase, has modest activity against species like Mycobacterium and Helicobacter, but its spectrum can be expanded through indole ring modifications. Halogenated and hydrophobic indole derivatives, such as halogenated indole hydrazides, have shown superior activity against *Staphylococcus* and *E. coli*, in some cases outperforming conventional antibiotics [90,91]. Indole-derivatives have also shown synergy with known antibiotics, helping to overcome resistance, although none are yet in clinical use. One interesting anti-biofilm agent is iodoindole (5-iodoindole), which can inhibit biofilm formation by *Pseudomonas* by interfering with bacterial signaling.

In the fight against malaria and parasitic infections, indole-derived structures have shown promise as well. Some antimalarial agents have been inspired by natural indole alkaloids. For example, the bis-indole alkaloid cryptolepine (from *Cryptolepis sanguinolenta*) has antiplasmodial activity; analogs modifying its indole share improved selectivity and reduced toxicity. Synthetic efforts led to indole-chalcone hybrids, indole-quinoline hybrids, and other dual pharmacophore compounds targeting *Plasmodium falciparum*. A noteworthy case: incorporating an indole moiety into 4-aminoquinoline antimalarials (chloroquine analogs) produced active hybrids that tackle chloroquine-resistant strains [92,93]. Moreover, because melatonin modulates the cell cycle of malaria parasites, researchers synthesized melatonin–indole derivatives to disrupt parasite rhythms. One study found an indole-quinoline hybrid that showed low-nanomolar antimalarial potency and proposed that the indole part was engaging a parasite receptor, possibly the parasite's melatonin-binding protein [94–96].

For viral infections, indole scaffolds feature in several antiviral agents, both approved and experimental. Arbidol (umifenovir) is a broad-spectrum antiviral used in Russia and China for influenza; it is an indole derivative with a bulky hydrophobic substituent. Arbidol is thought to inhibit viral entry by stabilizing the viral glycoprotein–host membrane interaction. The indole ring in arbidol is essential for its conformation and binding, and medicinal chemistry optimization around that core led to improved analogs against hepatitis C virus as well [97,98]. In recent years, indole-based compounds have been investigated against emerging viruses: for instance, some indole-containing protease inhibitors for dengue and SARS-CoV-2 have been reported [99,100]. An illustrative study by Zhang et al. (2015) reviewed indole-containing antivirals and noted that many such molecules target viral enzymes or attachment proteins [14]. Indole's synthetic flexibility enables its integration with other pharmacophores to create dual-action agents—for ex-

https://doi.org/10.3390/targets4010004

ample, indole-carboxamides that inhibit both HIV reverse transcriptase and integrase. In influenza research, a spiro-indole compound was shown to block viral entry by inhibiting hemagglutinin-mediated membrane fusion. Structural studies revealed the indole ring π-stacked with a tryptophan residue in the HA2 subunit, preventing the conformational shift essential for fusion. These examples highlight indole's ability to mimic tryptophan and interfere with key protein–protein interactions [101].

In summary, the indole scaffold is emerging as an increasingly valuable framework in anti-infective drug discovery. It has yielded promising candidates for tuberculosis, serves as a structural template for novel antibacterial agents, and features prominently in the design of antiviral compounds. Ongoing medicinal chemistry efforts, often inspired by naturally occurring indole metabolites with evolutionary antimicrobial functions, continue to expand their therapeutic potential against a broad spectrum of pathogens.

6.3. Central Nervous System (CNS) Disorders

A variety of CNS-active drugs incorporate the indole framework, exploiting its ability to interact with serotonin, melatonin, and other neural pathways. As mentioned, serotonergic systems are a major focus. The early antidepressant indalpine (a selective 5-HT reuptake inhibitor introduced in France in the 1980s) featured an indole core [102]. Though indalpine was withdrawn due to its toxicity, it demonstrated that an indole-ethylamine structure could be tuned to selectively inhibit the serotonin transporter. The multimodal antidepressant vilazodone is an indole: it combines SSRI activity with 5-HT_{1A} partial agonism. Structurally, vilazodone is essentially a tryptamine analog with an indole nucleus linked to a piperazine—this highlights that adding a bulky amine to an indole yields high affinity at serotonin receptors [103,104]. Benzodiazepines dominated anxiolysis for years, but buspirone, an azaspirodecanedione, became a notable non-sedating anxiolytic acting as a 5-HT_{1A} partial agonist; buspirone's metabolite 1-(2-pyrimidinyl)piperazine is the active 5-HT_{1A} agent, but buspirone's structure was initially developed by considering indole analogs of reserpine [105].

Indole scaffolds are less common in antipsychotics than in antidepressants, but have notable examples. Reserpine, an early indole alkaloid, irreversibly inhibits vesicular monoamine transporters (VMAT), depleting monoamines like dopamine and serotonin—its depressive side effects historically supported the monoamine hypothesis. Atypical antipsychotics such as ziprasidone incorporate an indole-like moiety and act as multi-target antagonists (e.g., 5-HT2A, D2). Melatonin, an endogenous indole, and its synthetic agonist ramelteon are also used for sleep-related disorders, reflecting indole's broader neuromodulatory potential [46,106–109].

In neurodegenerative diseases, indoles have been explored as neuroprotectants. Tryptophan metabolites like kynurenic acid (though not an indole) and indole-3-propionic acid (IPA) have neuroprotective roles—IPA, in particular, is a potent scavenger of hydroxyl radicals and has shown protective effects in models of Alzheimer's disease and Huntington's disease [28,110–113]. While IPA itself is not a drug yet, its presence in human brain and blood (derived from gut microbes) suggests a natural neuroprotective indole. Melatonin, beyond sleep regulation, is a neuroprotectant in models of stroke, Parkinson's, and AD, due to its antioxidative actions and mitigation of excitotoxicity. Clinically, melatonin is used off-label for certain headache disorders (e.g., cluster headache) and jet lag, reflecting its CNS effects [114–116].

Analgesic pathways also intersect with indoles. The opioid receptor family doesn't have endogenous indole ligands, but the plant alkaloid mitragynine (from *Mitragyna speciosa*, "kratom") is an indole-based molecule that agonizes μ-opioid receptors [117]. It produces analgesia with a different side-effect profile than classic opioids, and semi-

https://doi.org/10.3390/targets4010004

synthetic analogs are being studied as novel analgesics with potentially less respiratory depression. Its SAR indicates the indole with a corynanthe skeleton is required for opioid activity.

Overall, the indole scaffold has delivered multiple CNS drugs from antidepressants and anxiolytics to antipsychotics and migraine agents. It continues to be relevant, as seen in the resurgence of psychedelic-assisted therapy and the investigation of microbial indoles for brain health. In particular, its ability to modulate monoaminergic systems makes it invaluable in neuropsychopharmacology.

6.4. Metabolic and Endocrine Disorders

Indole derivatives have found use in metabolic and endocrine conditions, though perhaps less prominently than in oncology or CNS. One area is antidiabetic therapy. While most current antidiabetics (metformin, sulfonylureas, DPP-4 inhibitors, GLP-1 analogs) do not feature indole cores, research has produced indole-based insulin sensitizers and insulin secretagogues [118,119]. As noted earlier, some PPARγ agonists have been built on indole scaffolds [120]. For example, a series of 2,4-thiazolidinedione-indoles was made to activate PPARγ; the lead compound showed low micromolar efficacy in upregulating PPARγ-dependent genes and improving glucose uptake in cells [59]. Although thiazolidinediones (e.g., rosiglitazone) themselves are not indoles, this work suggests that indoles could provide alternative scaffolds with different safety profiles.

In type 2 diabetes and obesity, increasing attention has been given to the gut microbiome and its metabolites. Among these, indole-3-propionic acid (IPA) and indole-3-acetic acid (IAA), produced by commensal bacteria, show promising metabolic benefits. Elevated IPA levels correlate with reduced risk of developing type 2 diabetes and improved insulin secretion. Mechanistically, IPA enhances mitochondrial function and fatty acid oxidation in liver and muscle, helping to reduce insulin resistance. It also activates the pregnane X receptor (PXR) to strengthen gut barrier integrity and lower systemic inflammation associated with metabolic syndrome. In animal models, IPA supplementation prevented weight gain and liver fat accumulation by promoting an anti-inflammatory immune profile in adipose tissue [111,121–126]. Although IPA itself is not a drug, these findings have stimulated interest in indole derivatives as postbiotics or leads for metabolic disease treatment. One could envision derivatives of IPA or IAA that have improved pharmacokinetics but retain receptor activities (AhR, PXR activation) to harness these effects.

Another metabolic indole is melatonin, which influences energy balance [127,128]. Beyond sleep, melatonin receptors are expressed in pancreatic islets and adipose tissue. Melatonin has been shown to affect insulin secretion (generally inhibitory at night) and to influence brown fat activation (melatonin can promote thermogenesis in animal studies). Some studies linked melatonin supplementation to prevention of beta-cell loss in diabetes models, likely due to its antioxidant properties. While melatonin is not an antidiabetic drug per se, circadian misalignment (low melatonin output) correlates with higher diabetes risk, and there is interest in melatonin analogues for metabolic syndrome, especially to improve nocturnal metabolic profiles [121,129–136].

Indoles have shown hepatoprotective effects in various contexts. Indole-3-carbinol/3,3′-diindolylmethane (DIM) again appears in non-alcoholic fatty liver disease (NAFLD) models, as I3C/DIM supplementation reduced liver fat and fibrosis, partly by modulating estrogen receptors and inducing detoxifying enzymes that reduce lipid peroxidation. Additionally, microbial indoles like indole-3-acetate can activate AhR in hepatic immune cells (Kupffer cells), shifting them to an anti-inflammatory phenotype and mitigating liver injury [137–139].

Indoles have also shown potential in anti-obesity research. Traditional compounds like ajmalicine, an indole alkaloid from *Rauwolfia serpentina*, have demonstrated blood glucose-

https://doi.org/10.3390/targets4010004

lowering effects and were historically used in Eastern medicine to manage diabetes. More recently, indole itself, when administered chronically to obese mice, was found to reduce adipose tissue inflammation and improve glucose tolerance. These effects were linked to its role in modulating the gut microbiota, suggesting that indole, a microbial metabolite, may act as a natural regulator of gut hormones and intestinal permeability—factors that influence metabolic homeostasis and body weight regulation [15,140].

In metabolic disorders, indole-based agents target a range of proteins—including enzymes, nuclear receptors, and GPCRs like melatonin receptors—making SAR highly target-specific. For chronic use, these compounds often incorporate metabolically stabilizing groups (e.g., halogens at the 5-position) to prevent CYP-mediated degradation and extend half-life. Prodrug strategies also enhance efficacy; for instance, melatonin's rapid metabolism has led to controlled-release formulations. While not yet standard in metabolic therapy, emerging evidence supports the role of dietary and microbiota-derived indoles in improving metabolic health, fueling interest in indole analogs and microbiome-targeted strategies for personalized intervention.

6.5. Inflammation and Immune Modulation

The indole scaffold has been central to several anti-inflammatory and immunomodulatory agents. A landmark is indomethacin, an indole-3-acetic acid derivative that remains one of the most potent cyclooxygenase (COX) inhibitors known. Indomethacin's indole ring with a 4-chlorobenzoyl substitution and a methyl at N-1 fits into the COX enzyme's arachidonate binding site, blocking prostaglandin synthesis. It was introduced in the 1960s and is still used for gout flares and pericarditis due to its strong anti-inflammatory effect. SAR around indomethacin found that the indole N-benzoyl is essential; replacing indole with indene led to sulindac, a prodrug NSAID that is activated in vivo. Other indole NSAIDs include tolmetin (an indole acetic acid analog) and etodolac (an indole-like pyranocarboxylic acid), demonstrating indole's privileged status in anti-inflammatory design [141–146].

Indoles also appear in coxib-like molecules and dual COX/LOX inhibitors studied for inflammation. A series of indole-5,6-fused heterocycles was reported as 5-LOX inhibitors for asthma and allergic inflammation. The rationale is that the indole ring can mimic the arachidonic acid backbone in the 5-LOX active site, similar to COX [147].

An important immune enzyme target in which indoles play a role is indoleamine 2,3-dioxygenase (IDO1). IDO1 is upregulated in chronic infections, cancer, and during inflammation as a negative feedback mechanism—by catabolizing tryptophan to kynurenine, it has the potential to restrict metabolic resources for T cells and shift their differentiation in favor of Treg. Inhibiting IDO1 has been a strategy in cancer immunotherapy, but it could also boost antimicrobial immunity, since many pathogens like *Toxoplasma* are controlled by IDO-mediated tryptophan depletion. Several IDO1 inhibitors are indole derivatives (e.g., 1-methyl-D-tryptophan, also known as indoximod) or related heterocycles, leveraging the fact that IDO's active site binds indoleamine substrates. These inhibitors aim to preserve tryptophan levels and prevent the generation of immunosuppressive kynurenines. Though a phase III trial failed as mentioned, there are ongoing efforts to refine patient selection for IDO inhibitor therapy in cancer [28,112,148].

Indole compounds have also been shown to modulate NF-κB and other inflammatory signaling pathways. For example, 3,3′-diindolylmethane (DIM) in macrophages can inhibit NF-κB activation and thereby reduce expression of proinflammatory cytokines like TNF-α and IL-6. It also activates Nrf2, an anti-oxidant response factor, providing cytoprotection. In adipose tissue, indole (from gut bacteria) was found to reduce NF-κB and chemokine

https://doi.org/10.3390/targets4010004

levels, correlating with fewer infiltrating inflammatory macrophages in adipose of obese mice [149–152].

The connection between gut microbiota, indoles, and inflammation is a hot area. Conditions like inflammatory bowel disease (IBD) have been linked to reduced populations of indole-producing gut flora. Restoring indole metabolites, either via probiotics that produce indole-3-propionic acid/indole-3-aldehyde or via synthetic indole analogs, has shown benefit in animal IBD models by strengthening the mucosal barrier and inducing IL-22, reconstructing the epithelium [123]. These findings suggest a therapeutic strategy: indole-based drugs or diet supplements to induce immune homeostasis in the gut.

Indole-based agents contribute to inflammation control at multiple levels: direct enzyme inhibition (COX, IDO), receptor modulation (AhR, PXR), and altering immune cell differentiation. The gut-derived indoles act as natural immunosuppressants or immunostimulants depending on context, and harnessing these could lead to more physiological ways to treat inflammatory diseases. The privileged structure of indole enables these diverse modes of action, reinforcing its value in designing anti-inflammatory and immunotherapeutic compounds.

To better illustrate the pharmacological versatility of the indole scaffold, a comprehensive schematic representation is provided in Figure 2. It maps the major therapeutic classes of indole-based drugs, their molecular targets, and representative compounds.

In recent years, computational methodologies have become indispensable tools for rational drug design, enabling early prediction of biological activity, selectivity, and physicochemical property profiles before extensive experimental validation. In this context, QSAR and statistical modeling approaches have been successfully applied to support target-driven ligand optimization, as illustrated by recent studies demonstrating how descriptor-based models can guide structural refinement toward improved potency and selectivity profiles [153]. The increasing integration of machine learning techniques further strengthens this paradigm by enabling more robust activity prediction and rational structuring of chemical space, particularly within translationally oriented frameworks that aim to bridge in silico design with biological relevance [154]. Complementary to these approaches, structure-based modeling strategies combining molecular docking with predictive modeling have been shown to provide mechanistic insights into ligand–target interactions while simultaneously informing optimization pathways [155]. Moreover, practical in silico workflows incorporating docking, binding mode analysis, and virtual screening have proven valuable for rationalizing observed biological activity and guiding targeted structural modifications of bioactive scaffolds, including nitrogen-containing heterocycles [156]. Given the ubiquity of indoles in pharmacology, extensive SAR knowledge has accumulated on how substitutions of the indole ring affect biological activity. Medicinal chemists often view indole as a pharmacophore nucleus that presents multiple interaction points: the pyrrole NH (hydrogen bond donor and modest polar handle), the π-system (for aromatic stacking and van der Waals contacts), and substitutable positions (C-2 through C-7 and N-1) that can be optimized for target binding and pharmacokinetics [157–159]. SAR studies affirm that the indole nucleus tolerates a broad range of modifications while maintaining bioactivity. This "forgiveness" is part of why indole is privileged—it provides a robust scaffold that can be decorated to interact with many targets. Substitution at different ring positions allows fine-tuning of properties: N-1 and C-3 for direct binding interactions and pharmacokinetic tweaks, C-5/C-6/C-7 for electronic and lipophilic modulation, and bridging or extending from indole for hybrid multi-functional drugs. By understanding these SAR principles, chemists can rationally modify indole leads to optimize efficacy, selectivity, and drug-likeness for a given therapeutic goal.

https://doi.org/10.3390/targets4010004

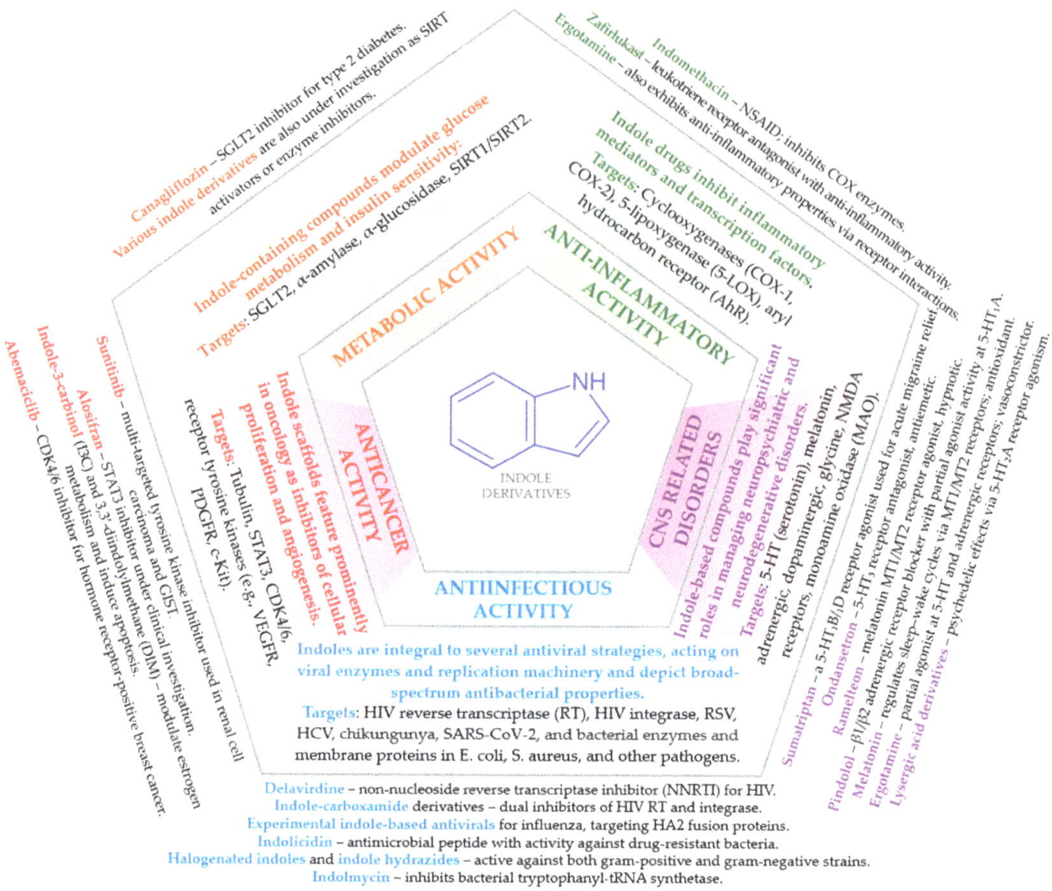

Figure 2. Overview of the main therapeutic domains of indole-containing compounds. The figure highlights key pharmacological classes (CNS, oncology, anti-infective, metabolic, and anti-inflammatory), representative drugs, and their principal molecular targets such as GPCRs, nuclear receptors, enzymes, and ion channels. This schematic representation underscores the structural adaptability of the indole scaffold and its privileged status in drug discovery across diverse disease areas. Figure adapted from [12].

7. Translational and Personalized Medicine Perspectives

Looking ahead, the indole scaffold is assured to remain a cornerstone in drug design, especially as we enter an era of personalized medicine. One reason is the inherent versatility of indoles—medicinal chemists can relatively easily generate indole libraries with diverse substitutions, leveraging the many synthetic routes and the tolerance of the scaffold to modification. This allows rapid prototyping of drug candidates tailored to emerging targets, including those identified through genomic and proteomic studies of diseases [160–162].

In personalized medicine, therapies are tailored to patient-specific factors (e.g., genetic markers, biomarker profiles). Indole-based drugs can be excellent candidates for such tailoring. For instance, if a subgroup of cancer patients is identified by high expression of an indole-interacting target, one can design indole analogs to selectively bind that target. The indole scaffold's privileged binding means it is often a good starting point for orphan or novel protein targets—screening libraries tend to have many indoles for this reason [1,12].

https://doi.org/10.3390/targets4010004

We also see personalized approaches in the context of the microbiome and diet. As described, individuals have different capacities to produce indole metabolites (like IPA, IAA) based on their gut flora [140,163–165]. These metabolites can serve as biomarkers of gut health and disease risk (e.g., low serum IPA predicting diabetes). In the future, one could envision measuring a patient's indole metabolite profile and then "prescribing" an indole-based intervention accordingly—such as a probiotic that boosts beneficial indole production or a synthetic indole analog that compensates for a deficiency.

In pharmacogenomics, polymorphisms in drug-metabolizing enzymes can affect indole drug efficacy and safety. For example, CYP2C9 metabolizes indomethacin; poor metabolizer patients may have higher indomethacin levels, increasing side effect risk. If a patient's genome indicates slow CYP2C9, a physician might choose a different NSAID or a lower indomethacin dose—a personalized dosing facilitated by understanding the indole's metabolic pathway. Another example: CYP2D6 polymorphisms could affect the metabolism of some indole alkaloids or neurodrugs (though many indole drugs like SSRIs are metabolized by CYP2D6). As genetic testing becomes routine, adjusting indole-drug choice and dosing based on genotype could improve outcomes. Indole-based antidepressants, for instance, might be preferentially used or avoided depending on a patient's CYP2D6/2C19 status (these CYPs handle many antidepressants) [166–171].

Targeted drug delivery and formulation is another translational angle. Some indole drugs suffer from solubility or stability issues: indoles can be prone to light oxidation. Modern formulation techniques—nanocarriers, prodrugs, etc.—can enhance their clinical performance.

Indoles are also being integrated into theranostic agents—compounds that have both therapeutic and imaging capabilities. Tryptophan and serotonin analogs labeled with radionuclides are used in PET imaging to track IDO activity in tumors or serotonin synthesis in the brain. For instance, 5-HTP labeled with C-11 can image neuroendocrine tumors' serotonin production. These imaging agents are indole-based and could guide therapy decisions (e.g., whether to use an IDO inhibitor in immunotherapy, based on a PET scan showing high tryptophan metabolism in the tumor) [172]. Alpha-[^{11}C]-methyl-L-tryptophan PET is another example, used to localize epileptic foci by detecting regions of high tryptophan metabolism. If a patient shows high uptake of this indole tracer in a brain region, it may guide surgical resection of that area to control epilepsy—a personalized surgical plan informed by an indole radiotracer [173].

The concept of using "indole templates" in drug design reflects the frequent strategy of building new compounds around established indole scaffolds. Medicinal chemists often develop second-generation drugs by modifying existing indole cores to enhance potency or selectivity, aided by well-established synthetic methods. Indole's predictable behavior also supports computational approaches, where docking and pharmacophore models commonly feature indole motifs. For example, virtual screens for anti-inflammatory agents often prioritize indole-containing structures, as seen with drugs like indomethacin or etodolac, improving the likelihood of identifying active hits [174–179].

As our understanding of disease biomarkers deepens, indole-based therapies can be tailored to specific patient profiles. In oncology, elevated kynurenine—a marker of IDO activity and an AhR ligand linked to immunosuppression—may indicate the need for IDO1 inhibitors or AhR antagonists, some of which are indole derivatives. In inflammatory bowel disease, low IL-22 levels and dysbiosis marked by reduced fecal indoles could justify probiotic interventions that restore indole-3-aldehyde levels and promote mucosal immunity via AhR activation.

Personalized approaches also enhance safety. For instance, patients genetically predisposed to NSAID-induced gastropathy may benefit from safer indole-based alternatives like

https://doi.org/10.3390/targets4010004

etodolac or require gastroprotective co-therapy. Pharmacogenomic data can guide the use of drugs like perindopril, an indole-containing ACE inhibitor, allowing dose adjustments based on ACE gene variants to optimize efficacy and minimize side effects.

Looking forward, indole scaffolds are increasingly integrated into emerging therapeutic platforms such as PROTACs, where indole-based ligands are used to recruit target proteins or E3 ligases for selective degradation—underscoring the scaffold's ongoing relevance in next-generation drug design [180].

In summary, the indole scaffold stands as a bridge between classical medicinal chemistry and modern personalized approaches. Its chemical pliability and biological relevance make it an enduring template. As we stratify patients by molecular markers and harness advanced delivery systems, indole-based compounds can be tailored in structure and formulation to meet individual needs. The flexibility of the scaffold—both in synthetic derivatization and in multi-target binding—aligns well with the future of medicine, which demands adaptable and precise therapeutics.

8. Limitations and Future Directions

While the indole scaffold is celebrated for its versatility, it is not without challenges. Recognizing these limitations is important for guiding future research and drug development. Each limitation discussed below is paired with optimization strategies and concrete design approaches to mitigate it.

8.1. Specificity vs. Promiscuity

Indole is a privileged scaffold known for binding many targets, but this same quality can lead to off-target effects. Indole-based drugs sometimes hit multiple receptors or enzymes, which can cause side effects. For example, an indole compound developed as a kinase inhibitor might unintentionally bind to serotonin receptors or hERG cardiac ion channels, leading to undesirable effects. Achieving high specificity with an indole can be difficult when closely related proteins have binding pockets that readily accommodate indoles [181].

To overcome off-target promiscuity, medicinal chemists employ structure-based design and careful substituent optimization. Structural data can reveal unique pocket features near the binding site of the intended target that are absent in off-target proteins. By adding an appropriate substituent to the indole scaffold to fill that extra pocket, the compound can bind the desired target more tightly while avoiding the off-targets. This concrete design approach exploits subtle structural differences between proteins to enhance specificity. Modern AI-driven modeling and virtual screening further aid in navigating the fine line between broad activity and selectivity, suggesting modifications that maximize target affinity and minimize off-target interactions [182].

Another optimization strategy is to control where the indole drug acts in the body through prodrug design or targeted delivery. For instance, in inflammatory bowel disease (IBD), an indole-based drug can be administered as a colon-targeted prodrug that only releases the active indole in the gut, thereby limiting systemic exposure and reducing off-target effects elsewhere. By localizing the drug's action to specific tissues or organs, one can harness the indole's potent activity where needed while minimizing interactions with unintended biological targets [183].

8.2. Metabolic Liabilities

Indole rings can undergo extensive metabolism in the body, which may lead to rapid clearance or toxic metabolites. A common pathway for indoles is phase I oxidation: for example, the benzene ring of the indole can be oxidized to form reactive epoxides or

https://doi.org/10.3390/targets4010004

dihydrodiols, and the pyrrole nitrogen can undergo N-oxidation or N-glucuronidation. Such metabolism can create reactive species—for instance, liver enzymes might convert an indole into a reactive arene oxide that can bind to proteins and cause toxicity. Indole itself is converted in the liver to indoxyl, then to indoxyl sulfate, a uremic toxin implicated in kidney damage [184,185]. For drug candidates, these metabolic transformations pose safety risks and can also reduce efficacy if the drug is quickly broken down.

Future indole drugs may require structural modifications to avoid problematic metabolism. One concrete strategy is to block metabolic "hot spots" on the indole scaffold. For example, if a particular ring position is prone to epoxidation by CYP enzymes, adding a small substituent like a fluorine or chlorine at that position can prevent the formation of a reactive epoxide (halogens often block or slow oxidative metabolism at that site). Similarly, incorporating electron-withdrawing groups on the indole ring can make it less electron-rich and thus less susceptible to phase I oxidation, thereby improving metabolic stability. Another approach is to design indole derivatives that preferentially undergo safer phase II metabolism (direct conjugation such as glucuronidation or sulfation) rather than forming reactive intermediates. By favoring benign metabolic pathways, one can achieve a cleaner metabolic profile with reduced toxic byproducts [12].

Early drug-design testing and computational metabolite prediction are also key design approaches. Medicinal chemists now often perform metabolite profiling in vitro (e.g., with liver microsomes or hepatocytes) for new indole compounds to identify any reactive metabolites early [186]. If a harmful metabolite is detected, the indole's structure can be tweaked by blocking the vulnerable site or removing the metabolically labile fragment, before the compound progresses further. Indeed, several indole-based drug candidates have failed in development due to metabolic liabilities such as rapid clearance or toxic metabolite formation, underscoring the importance of addressing metabolism through smart molecular design [187].

8.3. Solubility and Formulation

Planar heterocycles like indole are often quite lipophilic and poorly soluble in water. This poor solubility can limit a drug's oral bioavailability or necessitate high doses to achieve therapeutic plasma levels. Many indole-based drugs have required special formulation techniques or salt forms to be administered effectively. Without adequate solubility, even a potent indole drug may fail to reach its target in the body at sufficient concentrations [188,189].

An ideal solution is to build solubility into the molecule through structural modifications. One approach is to introduce polar or ionizable substituents onto the indole scaffold in positions that do not disrupt target binding. For example, adding a hydrophilic side chain or functional groups like amines, carboxylic acids, or hydroxyls can increase water solubility by enabling hydrogen bonding or ionization, thus improving dissolution in biological fluids. Such modifications must be balanced so as not to interfere with the indole's biological activity, but with careful medicinal chemistry this balance can often be achieved [3].

Another common strategy is the use of prodrugs specifically designed for solubility. Attaching a solubilizing moiety that will later be cleaved in the body can dramatically improve formulation characteristics. For instance, if an indole compound has a free –OH or –NH group, it can be derivatized into a phosphate ester or hemisuccinate prodrug. These derivatives are much more water-soluble, allowing the drug to be delivered in aqueous solution; once administered, enzymes cleave the prodrug, releasing the active indole in vivo. This tactic has been used in multiple drugs to enable intravenous formulations or to enhance oral absorption of otherwise insoluble molecules.

https://doi.org/10.3390/targets4010004

In addition to molecular modifications, formulation technologies offer complementary solutions [190]. Nanotechnology-based delivery systems, for example, can encapsulate indole drugs in nano-carriers like liposomes, polymeric nanoparticles, or micelles. Encapsulation can improve apparent solubility and protect the drug as it travels through the bloodstream, releasing the indole at the target site. While such formulation approaches are valuable, the preferred approach in drug development is often to modify the chemical structure early so that the compound is reasonably soluble on its own [191]. In summary, by either modifying the indole scaffold to include solubilizing features or employing innovative formulation strategies, researchers can ensure that indole-based drug candidates have acceptable solubility and bioavailability profiles [192,193].

8.4. Toxicology and Safety Considerations

Some indole-containing drugs and natural products have demonstrated organ-specific toxicities or serious side effects, highlighting safety as a critical limitation. For instance, indomethacin is effective for pain and inflammation, but chronic use is limited by gastrointestinal ulceration and bleeding, partly due to COX-1 enzyme inhibition in the gut and the direct irritant effect of the indole acetic acid on the stomach lining [194,195]. Another example is the indole alkaloid reserpine, an antihypertensive agent that fell out of favor because it caused severe depression by depleting neurotransmitters in the brain [107]. Similarly, certain indole alkaloids like ajmaline can paradoxically cause arrhythmias or other cardiac issues at higher doses [196–198]. These examples illustrate that indole scaffolds can sometimes interact with biological systems in unintended ways, leading to toxicity that limits their therapeutic use.

When developing new indole-based therapeutics, a proactive approach to safety is essential. One key strategy is early and broad off-target profiling. Rather than waiting for adverse effects to emerge later, modern drug discovery includes screening indole candidates against panels of critical receptors and enzymes (for example, a broad CNS receptor panel to catch any indole binding to neuronal receptors that might indicate neuropsychiatric side effects, or hERG channel assays to flag potential cardiac arrhythmia risk) [199–202]. If an indole compound shows activity on one of these off-targets, medicinal chemists can modify the molecule's structure to reduce that unwanted activity—for instance, by reducing lipophilicity or steric complementarity so that it no longer fits the off-target binding site as well. In some cases, reducing a drug's ability to cross the blood–brain barrier by increasing polarity is a deliberate design strategy to avoid CNS side effects, especially if the therapeutic target is outside the CNS [203,204].

Another modern design approach to improve safety is utilizing human cell-derived models early in development. Techniques using induced pluripotent stem cell (iPSC)-derived organoids (miniaturized tissues grown in vitro) are becoming invaluable for toxicity testing. For example, liver organoids or hepatocyte cultures can be exposed to a new indole compound to monitor for signs of liver toxicity (such as enzyme induction or cell injury) and to identify toxic metabolites. Cardiac organoids or human stem-cell-derived cardiomyocytes can be used to detect any propensity of an indole drug to cause arrhythmias or cardiac cell damage [205–208]. Similarly, gut organoid models can reveal if a compound is likely to irritate or harm the gastrointestinal lining, important for indole analogs similar to indomethacin. By integrating these predictive toxicology tools, researchers can gather concrete safety data and make informed design adjustments or terminate a problematic candidate early before reaching clinical trials [209,210].

Furthermore, pharmacogenomic insights can guide safer indole drug design and use. If certain structural features of an indole lead to toxicity only in individuals with specific genetic profiles, this information can be used to refine the molecule or to develop companion

https://doi.org/10.3390/targets4010004

diagnostics. In practice, understanding genetic variability in drug-metabolizing enzymes might prompt chemists to design an indole that bypasses a particular metabolic route associated with idiosyncratic toxicity. At the very least, it can inform how a drug is prescribed (personalized dosing or excluding at-risk populations), but ideally these considerations feed back into structural design to create inherently safer molecules [211–213].

8.5. Conclusion and Future Outlook

The indole scaffold's journey in medicinal chemistry is far from over. Its limitations—off-target activities, metabolic quirks, poor solubility, and potential toxicities—are well-recognized, and each is an active area of research aimed at turning these weaknesses into strengths. As discussed above, researchers are now explicitly linking each challenge with targeted solutions: improved computational modeling, including AI-driven drug design, enables the prediction and rational optimization of indole–target interactions to enhance binding specificity; smart chemistry modifications and prodrug approaches help us sidestep metabolic and solubility issues; and advanced biological models together with pharmacogenomic data enable us to flag and mitigate toxicity risks early.

Encouragingly, the enduring presence of indoles in many successful drugs attests that this scaffold will continue to be a fertile ground for pharmaceutical innovation. By addressing current challenges with the optimization strategies outlined and harnessing new technologies, the field can ensure that indole-based therapeutics remain at the forefront of treating human disease for years to come. The versatility of indole, guided by rational design and modern techniques, promises an exciting future where we retain its benefits while minimizing its drawbacks, ultimately delivering safer and more effective indole-derived medicines.

Author Contributions: Conceptualization, C.M.D. and A.-C.N.; methodology, C.M.D. and A.-C.N.; software, I.-B.D.; validation, C.M.D. and A.-C.N. and I.-B.D.; formal analysis, C.M.D.; investigation, C.M.D. and I.-B.D.; resources, C.M.D.; data curation, A.-C.N.; writing—original draft preparation, C.M.D.; writing—review and editing, C.M.D., A.-C.N. and I.-B.D.; visualization, I.-B.D.; supervision, I.-B.D.; project administration, I.-B.D.; funding acquisition, C.M.D. and A.-C.N. All authors have read and agreed to the published version of the manuscript.

Funding: This research received no external funding.

Institutional Review Board Statement: Not applicable.

Informed Consent Statement: Not applicable.

Data Availability Statement: The original contributions presented in this study are included in the article. Further inquiries can be directed to the corresponding author.

Conflicts of Interest: The authors declare no conflicts of interest.

References

1. Naim, M.J.; Alam, O.; Alam, J.; Bano, F.; Alam, P.; Shrivastava, N. Recent Review on Indole: A Privileged Structure Scaffold. *Int. J. Pharm. Sci. Res.* **2016**, *7*, 51–62.
2. Taber, D.F.; Tirunahari, P.K. Indole synthesis: A review and proposed classification. *Tetrahedron* **2011**, *67*, 7195–7210. [CrossRef]
3. Drăgoi, C.M.; Zanfirescu, A.; Dumitrescu, I.B.; Ungurianu, A.; Margină, D.M.; Nicolae, A.C. An Experimental Dynamic Investigation of the Influence of Melatonin, Serotonin and Tryptophan on the Stability of the DNA Structure. *Chemistry* **2024**, *6*, 922–940. [CrossRef]
4. Bravo, R.; Matito, S.; Cubero, J.; Paredes, S.D.; Franco, L.; Rivero, M.; Rodríguez, A.B.; Barriga, C. Tryptophan-enriched cereal intake improves nocturnal sleep, melatonin, serotonin, and total antioxidant capacity levels and mood in elderly humans. *Age* **2013**, *35*, 1277–1285. [CrossRef]
5. Wyatt, M.; Greathouse, K.L. Targeting Dietary and Microbial Tryptophan-Indole Metabolism as Therapeutic Approaches to Colon Cancer. *Nutrients* **2021**, *13*, 1189. [CrossRef] [PubMed]

https://doi.org/10.3390/targets4010004

6. Hayashi, K.I. Chemical Biology in Auxin Research. *Cold Spring Harb. Perspect. Biol.* **2021**, *13*, a040105. [CrossRef]
7. Rosales, P.F.; Bordin, G.S.; Gower, A.E.; Moura, S. Indole alkaloids: 2012 until now, highlighting the new chemical structures and biological activities. *Fitoterapia* **2020**, *143*, 104558. [CrossRef] [PubMed]
8. Bajad, N.G.; Singh, S.K.; Singh, S.K.; Singh, T.D.; Singh, M. Indole: A promising scaffold for the discovery and development of potential anti-tubercular agents. *Curr. Res. Pharmacol. Drug Discov.* **2022**, *3*, 100119. [CrossRef] [PubMed]
9. Toiu, A.; Vlase, L.; Drăgoi, C.M.; Vodnar, D.; Oniga, I. Phytochemical analysis, antioxidant and antibacterial activities of *Hypericum humifusum* L. (Hypericaceae). *Farmacia* **2016**, *64*, 663–667.
10. Dumitrescu, I.B.; Drăgoi, C.M.; Nicolae, A.C. From Soil to Brain: Olive Oil Attributes, Consumer Choices, Intermittent Fasting, and Their Impact on Health. *Nutrients* **2025**, *17*, 1905. [CrossRef]
11. Nicolae, A.C.; Arsene, A.L.; Vuță, V.; Popa, D.E.; Sîrbu, C.A.; Burcea-Dragomiroiu, G.T.A.; Dumitrescu, I.B.; Velescu, B.S.; Gofiță, E.; Drăgoi, C.M.; et al. In vitro P-gp expression after administration of CNS active drugs. *Farmacia* **2016**, *64*, 844–850.
12. Mo, X.; Rao, D.P.; Kaur, K.; Hassan, R.; Abdel-Samea, A.S.; Farhan, S.M.; Bräse, S.; Hashem, H. Indole Derivatives: A Versatile Scaffold in Modern Drug Discovery—An Updated Review on Their Multifaceted Therapeutic Applications (2020–2024). *Molecules* **2024**, *29*, 4770. [CrossRef]
13. Rivera-Fonseca, J.L.; González-Rivas, N.; Unnamatla, M.V.B.; García-Eleno, M.A.; Reyes, H.; López-Valdez, L.G.; Corona-Becerril, D.; Cuevas-Yañez, E. Synthesis and Development of Indole Based 5-HT3 Receptor Antagonists as Anti-Emetic Drugs in Oncology: An Update. *Curr. Med. Chem.* **2021**, *28*, 8733–8754. [CrossRef]
14. Zhang, M.Z.; Chen, Q.; Yang, G.F. A review on recent developments of indole-containing antiviral agents. *Eur. J. Med. Chem.* **2015**, *89*, 421–441. [CrossRef]
15. Azmi, M.B.; Sultana, S.; Naeem, S.; Qureshi, S.A. In silico investigation on alkaloids of Rauwolfia serpentina as potential inhibitors of 3-hydroxy-3-methyl-glutaryl-CoA reductase. *Saudi J. Biol. Sci.* **2021**, *28*, 731–737. [CrossRef] [PubMed]
16. Pilla Reddy, V.; Fretland, A.J.; Zhou, D.; Sharma, S.; Chen, B.; Vishwanathan, K.; McGinnity, D.F.; Xu, Y.; Ware, J.A. Mechanistic physiology-based pharmacokinetic modeling to elucidate vincristine-induced peripheral neuropathy following treatment with novel kinase inhibitors. *Cancer Chemother. Pharmacol.* **2021**, *88*, 451–464. [CrossRef]
17. Umer, S.M.; Solangi, M.; Khan, K.M.; Saleem, R.S.Z. Indole-Containing Natural Products 2019–2022: Isolations, Reappraisals, Syntheses, and Biological Activities. *Molecules* **2022**, *27*, 7586. [CrossRef]
18. Arsene, A.L.; Uivaroși, V.; Mitrea, N.; Drăgoi, C.M.; Nicolae, A.C. The binding properties of some novel ruthenium (III) complexes with human serum transferrin. *Biopolym. Cell* **2011**, *27*, 141–146. [CrossRef]
19. Heravi, M.M.; Amiri, Z.; Kafshdarzadeh, K.; Zadsirjan, V. Synthesis of indole derivatives as prevalent moieties present in selected alkaloids. *RSC Adv.* **2021**, *11*, 33540–33612. [CrossRef] [PubMed]
20. Alves, J.E.F.; de Oliveira, J.F.; de Lima Souza, T.R.C.; de Moura, R.O.; de Carvalho Júnior, L.B.; Alves de Lima Mdo, C. Novel indole-thiazole and indole-thiazolidinone derivatives as DNA groove binders. *Int. J. Biol. Macromol.* **2021**, *170*, 622–635. [CrossRef] [PubMed]
21. Drăgoi, C.M.; Mitrea, N.; Arsene, A.L.; Ilie, M.; Nicolae, A.C. Jurkat E6.1 cell line studies regarding the effects of some bio-indols on the membrane fluidity. *Farmacia* **2012**, *60*, 13–20.
22. Drăgoi, C.M.; Nicolae, A.C.; Dumitrescu, I.B.; Popa, D.E.; Ritivoiu, M.; Arsene, A.L. DNA targeting as a molecular mechanism underlying endogenous indoles biological effects. *Farmacia* **2019**, *67*, 367–377. [CrossRef]
23. Fan, L.; Zhu, X.; Liu, X.; He, F.; Yang, G.; Xu, C.; Yang, X. Recent Advances in the Synthesis of 3,n-Fused Tricyclic Indole Skeletons via Palladium-Catalyzed Domino Reactions. *Molecules* **2023**, *28*, 1647. [CrossRef]
24. Bowman, C.; Denis, M.; Canesi, S. Recent strategy for the synthesis of indole and indoline skeletons in natural products. *Chem. Commun.* **2025**, *61*, 5563–5576. [CrossRef]
25. Petrone, D.A.; Kondo, M.; Zeidan, N.; Lautens, M. Pd(0)-Catalyzed Dearomative Diarylation of Indoles. *Chem.–A Eur. J.* **2016**, *22*, 5684–5691. [CrossRef]
26. Gršič, M.; Meden, A.; Knez, D.; Jukič, M.; Svete, J.; Gobec, S. Synthesis and Cholinesterase Inhibitory Activity of Selected Indole-Based Compounds. *Acta Chim. Slov.* **2023**, *70*, 545–559. [CrossRef] [PubMed]
27. Garrido, M.; Espino, J.; Toribio-Delgado, A.F.; Cubero, J.; Maynar-Mariño, J.I.; Barriga, C.; Paredes, S.D.; Rodríguez, A.B. A Jerte Valley Cherry-Based Product as a Supply of Tryptophan. *Int. J. Tryptophan Res.* **2012**, *5*, 9–14. [CrossRef]
28. Lu, Z.; Zhang, C.; Zhang, J.; Su, W.; Wang, G.; Wang, Z. The Kynurenine Pathway and Indole Pathway in Tryptophan Metabolism Influence Tumor Progression. *Cancer Med.* **2025**, *14*, e70703. [CrossRef]
29. Kong, Y.; Wang, Q.; Wang, J.; Qiu, X.; Yang, Y.; Liu, J.; Yang, F.; Qi, R. Indole-3-propionic acid enhances glycolytic myofiber formation in piglets through PI3K-mTOR activation and gut microbiota-driven tryptophan metabolic alteration. *Anim. Nutr.* **2025**, *22*, 363–374. [CrossRef] [PubMed]
30. Reiter, R.J.; Tan, D. Melatonin: An antioxidant in edible plants. *Ann. N. Y. Acad. Sci.* **2002**, *957*, 341–344. [CrossRef] [PubMed]
31. Verma, A.K.; Singh, S.; Rizvi, S.I. Therapeutic potential of melatonin and its derivatives in aging and neurodegenerative diseases. *Biogerontology* **2023**, *24*, 183–206. [CrossRef]

https://doi.org/10.3390/targets4010004

32. Nicolae, A.C.; Mitrea, N.; Arsene, A.L.; Constantinescu, M.Z.; Vuță, V.; Drăgoi, C.M. In vitro P-glycoprotein inhibition assay on N2a murine cell line. *Farmacia* **2013**, *61*, 481–491.
33. Lv, Q.; Tao, K.; Yao, X.; Pang, M.; Cao, B.; Liu, C.; Wang, F. Melatonin MT1 receptors regulate the Sirt1/Nrf2/Ho-1/Gpx4 pathway to prevent α-synuclein-induced ferroptosis in Parkinson's disease. *J. Pineal Res.* **2024**, *76*, e12948. [CrossRef] [PubMed]
34. Drăgoi, C.M.; Nicolae, A.C.; Ungurianu, A.; Margină, D.M.; Grădinaru, D.; Dumitrescu, I.B. Circadian Rhythms, Chrononutrition, Physical Training, and Redox Homeostasis—Molecular Mechanisms in Human Health. *Cells* **2024**, *13*, 138. [CrossRef] [PubMed]
35. Voiculescu, S.E.; Le Duc, D.; Roșca, A.E.; Zeca, V.; Chițimuș, D.M.; Arsene, A.L.; Drăgoi, C.M.; Nicolae, A.C.; Zăgrean, L.; Schöneberg, T.; et al. Behavioral and molecular effects of prenatal continuous light exposure in the adult rat. *Brain Res.* **2016**, *1650*, 51–59. [CrossRef]
36. Van Reeth, O.; Weibel, L.; Olivares, E.; Maccari, S.; Mocaer, E.; Turek, F.W. Melatonin or a melatonin agonist corrects age-related changes in circadian response to environmental stimulus. *Am. J. Physiol.-Regul. Integr. Comp. Physiol.* **2001**, *280*, R1582–R1591. [CrossRef]
37. Keskin-Aktan, A.; Akbulut, K.G.; Abdi, S.; Akbulut, H. SIRT2 and FOXO3a expressions in the cerebral cortex and hippocampus of young and aged male rats: Antioxidant and anti-apoptotic effects of melatonin. *Biol. Futur.* **2022**, *73*, 71–85. [CrossRef]
38. Drăgoi, C.M.; Moroșan, E.; Dumitrescu, I.B.; Nicolae, A.C.; Arsene, A.L.; Drăgănescu, D.; Lupuliasa, D.; Ioniță, A.C.; Stoian, A.P.; Nicolae, C.; et al. Insights into chrononutrition: The innermost interplay amongst nutrition, metabolism and the circadian clock, in the context of epigenetic reprogramming. *Farmacia* **2019**, *67*, 557–571. [CrossRef]
39. Wang, X.; Wang, Z.; Cao, J.; Dong, Y.; Chen, Y. Gut microbiota-derived metabolites mediate the neuroprotective effect of melatonin in cognitive impairment induced by sleep deprivation. *Microbiome* **2023**, *11*, 17. [CrossRef]
40. Kumar, P.; Lee, J.; Lee, J. Diverse roles of microbial indole compounds in eukaryotic systems. *Biol. Rev.* **2021**, *96*, 2522–2545. [CrossRef]
41. Hrabak, M.; Moon, C.; Bolaños-Guzmán, C.A.; Steiner, H. Vilazodone, a Selective Serotonin Reuptake Inhibitor with Diminished Impact on Methylphenidate-Induced Gene Regulation in the Striatum: Role of 5-HT1A Receptor. *Mol. Neurobiol.* **2024**, *61*, 1907–1919. [CrossRef]
42. Dawson, L.A.; Watson, J.M. Vilazodone: A 5-HT 1A Receptor Agonist/Serotonin Transporter Inhibitor for the Treatment of Affective Disorders. *CNS Neurosci. Ther.* **2009**, *15*, 107–117. [CrossRef]
43. Svane, N.; Bällgren, F.; Ginosyan, A.; Kristensen, M.; Brodin, B.; Loryan, I. Regional distribution of unbound eletriptan and sumatriptan in the CNS and PNS in rats: Implications for a potential central action. *J. Headache Pain* **2024**, *25*, 187. [CrossRef]
44. Deen, M.; Hougaard, A.; Hansen, H.D.; Schain, M.; Dyssegaard, A.; Knudsen, G.M.; Ashina, M. Association Between Sumatriptan Treatment During a Migraine Attack and Central 5-HT 1B Receptor Binding. *JAMA Neurol.* **2019**, *76*, 834. [CrossRef]
45. Gergs, U.; Jacob, H.; Braekow, P.; Hofmann, B.; Pockes, S.; Humphrys, L.J.; Kirchhefer, U.; Fehse, C.; Neumann, J. Lysergic acid diethylamide stimulates cardiac human H2 histamine and cardiac human 5-HT4-serotonin receptors. *Naunyn-Schmiedeberg's Arch. Pharmacol.* **2024**, *397*, 221–236. [CrossRef]
46. Stroethoff, M.; Goetze, L.; Torregroza, C.; Bunte, S.; Raupach, A.; Heinen, A.; Mathes, A.; Hollmann, M.W.; Huhn, R. The Melatonin Receptor Agonist Ramelteon Induces Cardioprotection that Requires MT2 Receptor Activation and Release of Reactive Oxygen Species. *Cardiovasc. Drugs Ther.* **2020**, *34*, 303–310. [CrossRef] [PubMed]
47. Wang, Q.; Lu, Q.; Guo, Q.; Teng, M.; Gong, Q.; Li, X.; Du, Y.; Liu, Z.; Tao, Y. Structural basis of the ligand binding and signaling mechanism of melatonin receptors. *Nat. Commun.* **2022**, *13*, 454. [CrossRef] [PubMed]
48. Farzin, K.; Kheiltash, A.; Tafakhori, A.; Nakhjiri, N.E.; Sabet, M.S.; Nayeri, N.D. The effectiveness of agomelatine on headache severity and frequency in episodic migraine without aura; a parallel randomized controlled trial study. *BMC Neurol.* **2024**, *24*, 2. [CrossRef]
49. Kim, H.K.; Yang, K.I. Melatonin and melatonergic drugs in sleep disorders. *Transl. Clin. Pharmacol.* **2022**, *30*, 163. [CrossRef] [PubMed]
50. Hwang, S.; Lee, S.; Yoon, J.; Chung, J.Y. Population Pharmacokinetic−Pharmacodynamic Modeling of Carvedilol to Evaluate the Effect of Cytochrome P450 2D6 Genotype on the Heart Rate Reduction. *J. Korean Med. Sci.* **2023**, *38*, e173. [CrossRef]
51. Misselwitz, F.; Henderson, D.; Menakuru, S.R.; Morten, E.; Roe, C.; Whitaker, G.; Wohlfeil, S.; McDermott, J. Pharmacokinetics, Pharmacodynamics and Bioavailability of ACM-001.1 (*S*-Pindolol Benzoate) in Healthy Volunteers. *J. Cachexia Sarcopenia Muscle* **2025**, *16*, e13651. [CrossRef]
52. Münster, A.; Huster, J.; Sommer, S.; Traxler, C.; Votteler, A.; Hauber, W. Enhanced Risky Choice in Male Rats Elicited by the Acute Pharmacological Stressor Yohimbine Involves Prefrontal Dopamine D1 Receptor Activation. *Int. J. Neuropsychopharmacol.* **2024**, *27*, pyae006. [CrossRef]
53. Hussain, K.; Jahangir Alam, M.; Hussain, A.; Ahmad Siddique, N. Therapeutic Significance of Indole Scaffold in Medicinal Chemistry. *Int. J. Pharm. Sci. Res.* **2020**, *11*, 5441. [CrossRef]
54. Omar, F.; Tareq, A.M.; Alqahtani, A.M.; Dhama, K.; Sayeed, M.A.; Emran, T.B.; Simal-Gandara, J. Plant-Based Indole Alkaloids: A Comprehensive Overview from a Pharmacological Perspective. *Molecules* **2021**, *26*, 2297. [CrossRef] [PubMed]

https://doi.org/10.3390/targets4010004

55. Drăgoi, C.M.; Olaru, O.T.; Dinu, M.; Popescu, C.; Arsene, A.L.; Dune, A.; Nicolae, A.; Ancuceanu, R.; Zbârcea, C.; Negreș, S.; et al. Characterisation, pharmacotoxicological and biochemical studies on Morus alba L. extract and its fractions. *Farmacia* **2018**, *66*, 120–128.
56. Wang, P.; Tao, W.; Li, Q.; Ma, W.; Jia, W.; Kang, Y. Indole-3-Aldehyde alleviates lung inflammation in COPD through activating Aryl Hydrocarbon Receptor to inhibit HDACs/NF-κB/NLRP3 signaling pathways. *J. Mol. Med.* **2025**, *103*, 157–174. [CrossRef]
57. Amalia, R.; Ohama, T.; Parhar, I.; Airin, C.; Sato, H.; Budiyanto, A.; Astuti, P. Potential of Indole-3-Carbinol compounds from broccoli (Brassica oleracea var. italica) as natural aromatase blockers: In silico prediction and in vivo studies. *Open Vet. J.* **2025**, *15*, 1663–1672. [CrossRef] [PubMed]
58. Ho, E.; Wong, C.P.; Bouranis, J.A.; Shannon, J.; Zhang, Z. Cruciferous Vegetables, Bioactive Metabolites, and Microbiome for Breast Cancer Prevention. *Annu. Rev. Nutr.* **2025**, *45*, 171–195. [CrossRef] [PubMed]
59. Corigliano, D.M.; Syed, R.; Messineo, S.; Lupia, A.; Patel, R.; Reddy, C.V.R.; Dubey, P.K.; Colica, C.; Amato, R.; De Sarro, G.; et al. Indole and 2,4-Thiazolidinedione conjugates as potential anticancer modulators. *PeerJ* **2018**, *6*, e5386. [CrossRef]
60. Hussein, B.A.; Karimi, I.; Yousofvand, N. Chemo- and bio-informatics insight into anti-cholinesterase potentials of berries and leaves of Myrtus communis L., Myrtaceae: An in vitro/in silico study. *BMC Complement. Med. Ther.* **2023**, *23*, 421. [CrossRef]
61. Blackstone, N.G.; Olson, A.; Ainapurapu, B. Physostigmine in Anticholinergic Poisoning: An Old Antidote With Resurgence. *Cureus* **2020**, *12*, e11739. [CrossRef] [PubMed]
62. Siddique, S.; Ahmad, K.R.; Nawaz, S.K.; Raza, A.R.; Ahmad, S.N.; Ali, R.; Inayat, I.; Suleman, S.; Kanwal, M.A.; Usman, M. Evaluation of the anti-inflammatory, analgesic, anti-pyretic and anti-ulcerogenic potentials of synthetic indole derivatives. *Sci. Rep.* **2023**, *13*, 8639. [CrossRef]
63. Bokhtia, R.M.; Panda, S.S.; Girgis, A.S.; Samir, N.; Said, M.F.; Abdelnaser, A.; Nasr, S.; Bekheit, M.S.; Dawood, A.S.; Sharma, H.; et al. New NSAID Conjugates as Potent and Selective COX-2 Inhibitors: Synthesis, Molecular Modeling and Biological Investigation. *Molecules* **2023**, *28*, 1945. [CrossRef]
64. Burcea Dragomiroiu, G.T.A.; Popa, D.E.; Velescu, B.Ș.; Andrieș, A.; Ordeanu, V.; Nicolae, A.C.; Drăgoi, C.M.; Bârcă, M.; Ginghină, O. Synthesis, characterization and microbiological activity evaluation of novel hard gelatine capsules with cefaclor and piroxicam. *Farmacia* **2016**, *64*, 887–895.
65. Tarpley, M.; Oladapo, H.O.; Strepay, D.; Caligan, T.B.; Chdid, L.; Shehata, H.; Roques, J.R.; Thomas, R.; Laudeman, C.P.; Onyenwoke, R.U.; et al. Identification of harmine and β-carboline analogs from a high-throughput screen of an approved drug collection; profiling as differential inhibitors of DYRK1A and monoamine oxidase A and for in vitro and in vivo anti-cancer studies. *Eur. J. Pharm. Sci.* **2021**, *162*, 105821. [CrossRef]
66. Chaurasiya, N.; Leon, F.; Muhammad, I.; Tekwani, B. Natural Products Inhibitors of Monoamine Oxidases—Potential New Drug Leads for Neuroprotection, Neurological Disorders, and Neuroblastoma. *Molecules* **2022**, *27*, 4297. [CrossRef] [PubMed]
67. Semenyuta, I.; Los, O.; Sinenko, V.; Zhirnov, V.; Potikha, L.; Kobzar, O.; Brovarets, V. Design, Synthesis, and Antitumor Potential of New Thiazolecontained 5-Fluoro-2-Oxindole Derivatives as Sunitinib Analogues. *Curr. Med. Chem.* **2025**, *32*, 5279–5291. [CrossRef]
68. Miao, H.; Chen, D.; Ropa, J.; Purohit, T.; Kim, E.; Sulis, M.L.; Ferrando, A.; Cierpicki, T.; Grembecka, J. Combination of menin and kinase inhibitors as an effective treatment for leukemia with NUP98 translocations. *Leukemia* **2024**, *38*, 1674–1687. [CrossRef] [PubMed]
69. Wan, Y.; Li, Y.; Yan, C.; Yan, M.; Tang, Z. Indole: A privileged scaffold for the design of anti-cancer agents. *Eur. J. Med. Chem.* **2019**, *183*, 111691. [CrossRef]
70. Niculae, D.; Dusman, R.; Leonte, R.A.; Chilug, L.E.; Dragoi, C.M.; Nicolae, A.; Serban, R.M.; Niculae, D.A.; Dumitrescu, I.B.; Draganescu, D. Biological Pathways as Substantiation of the Use of Copper Radioisotopes in Cancer Theranostics. *Front. Phys.* **2021**, *8*, 568296. [CrossRef]
71. Qin, R.; You, F.M.; Zhao, Q.; Xie, X.; Peng, C.; Zhan, G.; Han, B. Naturally derived indole alkaloids targeting regulated cell death (RCD) for cancer therapy: From molecular mechanisms to potential therapeutic targets. *J. Hematol. Oncol.* **2022**, *15*, 133. [CrossRef]
72. Iacopetta, D.; Catalano, A.; Ceramella, J.; Barbarossa, A.; Carocci, A.; Fazio, A.; La Torre, C.; Caruso, A.; Ponassi, M.; Rosano, C.; et al. Synthesis, anticancer and antioxidant properties of new indole and pyranoindole derivatives. *Bioorg. Chem.* **2020**, *105*, 104440. [CrossRef]
73. Kornienko, T.E.; Chepanova, A.A.; Zakharenko, A.L.; Filimonov, A.S.; Luzina, O.A.; Dyrkheeva, N.S.; Nikolin, V.P.; Popova, N.A.; Salakhutdinov, N.F.; Lavrik, O.I. Enhancement of the Antitumor and Antimetastatic Effect of Topotecan and Normalization of Blood Counts in Mice with Lewis Carcinoma by Tdp1 Inhibitors—New Usnic Acid Derivatives. *Int. J. Mol. Sci.* **2024**, *25*, 1210. [CrossRef]
74. Doan, T.N.; Le, T.D.; Ho, N.A.; Ho, T.T.; Do, T.T.; Hoang, H.; Nguyen, M.H.; Bui, T.M.; Chu, H.H. Isolation, anticancer potency, and camptothecin—Producing ability of endophytic fungi isolated from *Ixora chinensis*. *Sci. Prog.* **2024**, *107*, 368504241253675. [CrossRef]

https://doi.org/10.3390/targets4010004

75. Kattel, S.; Antonious, G.F. Glucosinolates in Cruciferous Vegetables: Genetic and Environmental Regulation, Metabolic Pathways, and Cancer-Preventive Mechanisms. *Int. J. Plant Biol.* **2025**, *16*, 58. [CrossRef]
76. Reyes-Hernández, O.D.; Figueroa-González, G.; Quintas-Granados, L.I.; Gutiérrez-Ruíz, S.C.; Hernández-Parra, H.; Romero-Montero, A.; Del Prado-Audelo, M.L.; Bernal-Chavez, S.A.; Cortés, H.; Peña-Corona, S.I.; et al. 3,3′-Diindolylmethane and indole-3-carbinol: Potential therapeutic molecules for cancer chemoprevention and treatment via regulating cellular signaling pathways. *Cancer Cell Int.* **2023**, *23*, 180. [CrossRef] [PubMed]
77. Williams, D.E. Indoles Derived From Glucobrassicin: Cancer Chemoprevention by Indole-3-Carbinol and 3,3′-Diindolylmethane. *Front. Nutr.* **2021**, *8*, 734334. [CrossRef] [PubMed]
78. Nozawa, K.; Terada, M.; Onishi, M.; Ozaki, Y.; Takano, T.; Fakhouri, W.; Novick, D.; Haro, J.M.; Faris, L.H.; Kawaguchi, T.; et al. Real-world treatment patterns and outcomes of abemaciclib for the treatment of HR+, HER2- metastatic breast cancer patients in Japan. *Breast Cancer* **2023**, *30*, 657–665. [CrossRef] [PubMed]
79. Chiorean, E.G.; Picozzi, V.; Li, C.; Peeters, M.; Maurel, J.; Singh, J.; Golan, T.; Blanc, J.; Chapman, S.C.; Hussain, A.M.; et al. Efficacy and safety of abemaciclib alone and with PI3K/mTOR inhibitor LY3023414 or galunisertib versus chemotherapy in previously treated metastatic pancreatic adenocarcinoma: A randomized controlled trial. *Cancer Med.* **2023**, *12*, 20353–20364. [CrossRef]
80. Zgarbová, E.; Vrzal, R. The Impact of Indoles Activating the Aryl Hydrocarbon Receptor on Androgen Receptor Activity in the 22Rv1 Prostate Cancer Cell Line. *Int. J. Mol. Sci.* **2022**, *24*, 502. [CrossRef] [PubMed]
81. Wróbel, T.M.; Grudzińska, A.; Yakubu, J.; du Toit, T.; Sharma, K.; Harrington, J.C.; Björkling, F.; Jørgensen, F.S.; Pandey, A.V. Pyridine indole hybrids as novel potent CYP17A1 inhibitors. *J. Enzym. Inhib. Med. Chem.* **2025**, *40*, 2463014. [CrossRef]
82. Li, W.; Guo, L.; Xing, Z.; Fang, X.; Liang, H.; Zhang, S.; Shi, L.; Kuang, C.; Shi, L.; Zheng, Y.; et al. Forty-three key gene expressions involved in the effect of indoleamine 2,3-dioxygenase 1 expression on cancer prognosis may be a potential indoleamine 2,3-dioxygenase 1 inhibitor biomarker. *Clin. Transl. Med.* **2021**, *11*, e330. [CrossRef] [PubMed]
83. Azimnasab-sorkhabi, P.; Soltani-asl, M.; Yoshinaga, T.T.; Zaidan Dagli, M.L.; Massoco Cde, O.; Kfoury Junior, J.R. Indoleamine-2,3 dioxygenase: A fate-changer of the tumor microenvironment. *Mol. Biol. Rep.* **2023**, *50*, 6133–6145. [CrossRef] [PubMed]
84. Butler-Fernández, K.M.; Ramos, Z.; Francis-Malavé, A.M.; Bloom, J.; Dharmawardhane, S.; Hernández, E. Synthesis, Anti-Cancer and Anti-Migratory Evaluation of 3,6-Dibromocarbazole and 5-Bromoindole Derivatives. *Molecules* **2019**, *24*, 2686. [CrossRef] [PubMed]
85. Novikov, R.A.; Platonov, D.N.; Belyy, A.Y.; Potapov, K.V.; Novikov, M.A.; Tomilov, Y.V.; Kechko, O.I.; Seregina, T.A.; Zemskaya, A.S.; Solyev, P.N.; et al. 6-Bromoindole- and 6-Bromoindazole-Based Inhibitors of Bacterial Cystathionine γ-Lyase Containing 3-Aminothiophene-2-Carboxylate Moiety. *Molecules* **2025**, *30*, 388. [CrossRef]
86. Li, S.A.; Cadelis, M.M.; Sue, K.; Blanchet, M.; Vidal, N.; Brunel, J.M.; Bourguet-Kondracki, M.-L.; Copp, B.R. 6-Bromoindolglyoxylamido derivatives as antimicrobial agents and antibiotic enhancers. *Bioorg. Med. Chem.* **2019**, *27*, 2090–2099. [CrossRef]
87. Zhang, H.; He, F.; Gao, G.; Lu, S.; Wei, Q.; Hu, H.; Wu, Z.; Fang, M.; Wang, X. Approved Small-Molecule ATP-Competitive Kinases Drugs Containing Indole/Azaindole/Oxindole Scaffolds: R&D and Binding Patterns Profiling. *Molecules* **2023**, *28*, 943.
88. Nuermberger, E.L.; Martínez-Martínez, M.S.; Sanz, O.; Urones, B.; Esquivias, J.; Soni, H.; Tasneen, R.; Tyagi, S.; Li, S.-Y.; Converse, P.J.; et al. GSK2556286 Is a Novel Antitubercular Drug Candidate Effective In Vivo with the Potential To Shorten Tuberculosis Treatment. *Antimicrob. Agents Chemother.* **2022**, *66*, e0013222. [CrossRef]
89. Singampalli, A.; Kumar, P.; Bandela, R.; Bellapukonda, S.M.; Nanduri, S.; Yaddanapudi, V.M. Indazole–an emerging privileged scaffold: Synthesis and its biological significance. *RSC Med. Chem.* **2025**, *16*, 5196–5221. [CrossRef]
90. Maansson, M.; Vynne, N.G.; Klitgaard, A.; Nybo, J.L.; Melchiorsen, J.; Nguyen, D.D.; Sanchez, L.M.; Ziemert, N.; Dorrestein, P.C.; Andersen, M.R.; et al. An Integrated Metabolomic and Genomic Mining Workflow To Uncover the Biosynthetic Potential of Bacteria. *mSystems* **2016**, *1*, e00028-15. [CrossRef]
91. Williams, T.L.; Yin, Y.W.; Carter, C.W. Selective Inhibition of Bacterial Tryptophanyl-tRNA Synthetases by Indolmycin Is Mechanism-based. *J. Biol. Chem.* **2016**, *291*, 255–265. [CrossRef]
92. Tudu, C.K.; Bandyopadhyay, A.; Kumar, M.; Radha; Das, T.; Nandy, S.; Ghorai, M.; Gopalakrishnan, A.V.; Proćków, J.; Dey, A. Unravelling the pharmacological properties of cryptolepine and its derivatives: A mini-review insight. *Naunyn Schmiedebergs Arch. Pharmacol.* **2023**, *396*, 229–238. [CrossRef]
93. Forkuo, A.D.; Ansah, C.; Mensah, K.B.; Annan, K.; Gyan, B.; Theron, A.; Mancama, D.; Wright, C.W. In vitro anti-malarial interaction and gametocytocidal activity of cryptolepine. *Malar. J.* **2017**, *16*, 496. [CrossRef]
94. Schuck, D.C.; Jordão, A.K.; Nakabashi, M.; Cunha, A.C.; Ferreira, V.F.; Garcia, C.R.S. Synthetic indole and melatonin derivatives exhibit antimalarial activity on the cell cycle of the human malaria parasite Plasmodium falciparum. *Eur. J. Med. Chem.* **2014**, *78*, 375–382. [CrossRef]
95. Chavan, N.D.; Sarveswari, S.; Vijayakumar, V. Quinoline derivatives' biological interest for anti-malarial and anti-cancer activities: An overview. *RSC Adv.* **2025**, *15*, 30576–30604. [CrossRef]

https://doi.org/10.3390/targets4010004

96. Wang, N.; Wicht, K.J.; Imai, K.; Wang, M.Q.; Anh Ngoc, T.; Kiguchi, R.; Kaiser, M.; Egan, T.J.; Inokuchi, T. Synthesis, β-haematin inhibition, and in vitro antimalarial testing of isocryptolepine analogues: SAR study of indolo[3,2-c]quinolines with various substituents at C2, C6, and N11. *Bioorg. Med. Chem.* **2014**, *22*, 2629–2642. [CrossRef]
97. Kang, Y.; Shi, Y.; Xu, S. Arbidol: The current demand, strategies, and antiviral mechanisms. *Immun. Inflamm. Dis.* **2023**, *11*, e984. [CrossRef] [PubMed]
98. Leneva, I.; Kartashova, N.; Poromov, A.; Gracheva, A.; Korchevaya, E.; Glubokova, E.; Borisova, O.; Shtro, A.; Loginova, S.; Shchukina, V.; et al. Antiviral Activity of Umifenovir In Vitro against a Broad Spectrum of Coronaviruses, Including the Novel SARS-CoV-2 Virus. *Viruses* **2021**, *13*, 1665. [CrossRef]
99. Suceveanu, A.I.; Mazilu, L.; Suceveanu, A.P.; Parepa, I.; Dumitrescu, I.B.; Drăgoi, C.M.; Nicolae, A.C.; Botea, F.; Voinea, F.; Burcea-Dragomiroiu, G.T.A. Assertion for montelukast in the COVID-19 pandemics. *Farmacia* **2020**, *68*, 579–585. [CrossRef]
100. Arsene, A.L.; Dumitrescu, I.B.; Dragoi, C.M.; Udeanu, D.I.; Lupuliasa, D.; Jinga, V.; Drăgănescu, D.; Dinu-Pirvu, C.E.; Burcea Dragomiroiu, G.T.; Blejan, I.E.; et al. A new era for the therapeutic management of the ongoing COVID-19 pandemic. *Farmacia* **2020**, *68*, 185–196. [CrossRef]
101. Cihan-Üstündağ, G.; Zopun, M.; Vanderlinden, E.; Ozkirimli, E.; Persoons, L.; Çapan, G.; Naesens, L. Superior inhibition of influenza virus hemagglutinin-mediated fusion by indole-substituted spirothiazolidinones. *Bioorg. Med. Chem.* **2020**, *28*, 115130. [CrossRef] [PubMed]
102. St Hilaire, S.K.D.; Merica, H.; Gaillard, J.M. The effects of indalpine—A selective inhibitor of 5-HT uptake—On rat paradoxical sleep. *Eur. J. Pharmacol.* **1984**, *98*, 413–418. [CrossRef] [PubMed]
103. Shi, L.; Wang, J.; Xu, S.; Lu, Y. Efficacy and tolerability of vilazodone for major depressive disorder: Evidence from phase III/IV randomized controlled trials. *Drug Des. Dev. Ther.* **2016**, *10*, 3899–3907. [CrossRef]
104. Kadam, R.; Sontakke, S.; Tiple, P.; Motghare, V.; Bajait, C.; Kalikar, M. Comparative evaluation of efficacy and tolerability of vilazodone, escitalopram, and amitriptyline in patients of major depressive disorder: A randomized, parallel, open-label clinical study. *Indian J. Pharmacol.* **2020**, *52*, 79. [CrossRef] [PubMed]
105. Warren, A.L.; Lankri, D.; Cunningham, M.J.; Serrano, I.C.; Parise, L.F.; Kruegel, A.C.; Duggan, P.; Zilberg, G.; Capper, M.J.; Havel, V.; et al. Structural pharmacology and therapeutic potential of 5-methoxytryptamines. *Nature* **2024**, *630*, 237–246. [CrossRef]
106. Platanić Arizanović, L.; Gligorijević, N.; Cvijetić, I.; Mijatović, A.; Ristivojević, M.K.; Minić, S.; Kokić, A.N.; Miljević, Č.; Nikolić, M. Human Hemoglobin and Antipsychotics Clozapine, Ziprasidone and Sertindole: Friends or Foes? *Int. J. Mol. Sci.* **2023**, *24*, 8921. [CrossRef]
107. Strawbridge, R.; Javed, R.R.; Cave, J.; Jauhar, S.; Young, A.H. The effects of reserpine on depression: A systematic review. *J. Psychopharmacol.* **2023**, *37*, 248–260. [CrossRef]
108. Dai, L.-H.; Zhang, G.-R.; Ou, Y.-H.; Liu, X.-J.; Yao, H.-L.; Hu, W.-H.; Li, H.-J.; Lan, W.-J. Five New Indole Alkaloid Derivatives from Deep-Sea Fungus *Aspergillus fumigatus* AF1. *Mar. Drugs* **2025**, *23*, 4. [CrossRef]
109. Nicolae, A.C.; Mitrea, N.; Drăgoi, C.M.; Constantinescu, M.Z.; Ciofrângeanu, C.; Bărboi, G.; Arsene, A.L. Murine studies regarding the variation of oxidative status in serum, hepatic and brain samples, after administration of some CNS active drugs. *Farmacia* **2013**, *61*, 658–669.
110. Owe-Larsson, M.; Drobek, D.; Iwaniak, P.; Kloc, R.; Urbanska, E.M.; Chwil, M. Microbiota-Derived Tryptophan Metabolite Indole-3-Propionic Acid-Emerging Role in Neuroprotection. *Molecules* **2025**, *30*, 3628. [CrossRef]
111. Xue, H.; Chen, X.; Yu, C.; Deng, Y.; Zhang, Y.; Chen, S.; Chen, X.; Chen, K.; Yang, Y.; Ling, W. Gut Microbially Produced Indole-3-Propionic Acid Inhibits Atherosclerosis by Promoting Reverse Cholesterol Transport and Its Deficiency Is Causally Related to Atherosclerotic Cardiovascular Disease. *Circ. Res.* **2022**, *131*, 404–420. [CrossRef] [PubMed]
112. Fathi, M.; Vakili, K.; Yaghoobpoor, S.; Tavasol, A.; Jazi, K.; Hajibeygi, R.; Shool, S.; Sodeifian, F.; Klegeris, A.; McElhinney, A.; et al. Dynamic changes in metabolites of the kynurenine pathway in Alzheimer's disease, Parkinson's disease, and Huntington's disease: A systematic Review and meta-analysis. *Front. Immunol.* **2022**, *13*, 997240. [CrossRef]
113. Barresi, E.; Baglini, E.; Poggetti, V.; Castagnoli, J.; Giorgini, D.; Salerno, S.; Taliani, S.; Da Settimo, F. Indole-Based Compounds in the Development of Anti-Neurodegenerative Agents. *Molecules* **2024**, *29*, 2127. [CrossRef] [PubMed]
114. Zhang, Z.; Xue, P.; Bendlin, B.B.; Zetterberg, H.; De Felice, F.; Tan, X.; Benedict, C. Melatonin: A potential nighttime guardian against Alzheimer's. *Mol. Psychiatry.* **2025**, *30*, 237–250. [CrossRef]
115. Sumsuzzman, D.M.d.; Choi, J.; Jin, Y.; Hong, Y. Neurocognitive effects of melatonin treatment in healthy adults and individuals with Alzheimer's disease and insomnia: A systematic review and meta-analysis of randomized controlled trials. *Neurosci. Biobehav. Rev.* **2021**, *127*, 459–473. [CrossRef]
116. Terao, I.; Kodama, W. Comparative Efficacy, Tolerability, and Acceptability of Donanemab, Lecanemab, Aducanumab, Melatonin, and Aerobic Exercise for a Short Time on Cognitive Function in Mild Cognitive Impairment and Mild Alzheimer's Disease: A Systematic Review and Network Meta-Analysis. *J. Alzheimer's Dis.* **2024**, *98*, 825–835.

https://doi.org/10.3390/targets4010004

117. Hiranita, T.; Leon, F.; Felix, J.S.; Restrepo, L.F.; Reeves, M.E.; Pennington, A.E.; Obeng, S.; Avery, B.A.; McCurdy, C.R.; McMahon, L.R.; et al. The effects of mitragynine and morphine on schedule-controlled responding and antinociception in rats. *Psychopharmacology* **2019**, *236*, 2725–2734. [CrossRef]
118. Ghiţă, M. Research regarding the effect of leptin upon the ratio of certain lymphocyte populations in rat. *Farmacia* **2021**, *69*, 1089–1093. [CrossRef]
119. Mohamad, M.A.; Mitrea, N.; Nicolae, A.-C.; Constantinescu, M.-Z.; Dragoi, C.M.; Arsene, A.L.; Barb, C.G. The dynamics of adiponectin and leptin on metabolic syndrome patients and age matched healthy subjects. *Farmacia* **2014**, *62*, 532–545.
120. Eeda, V.; Wu, D.; Lim, H.Y.; Wang, W. Design, synthesis, and evaluation of potent novel peroxisome proliferator-activated receptor γ indole partial agonists. *Bioorg. Med. Chem. Lett.* **2019**, *29*, 126664. [CrossRef] [PubMed]
121. Tuomainen, M.; Lindström, J.; Lehtonen, M.; Auriola, S.; Pihlajamäki, J.; Peltonen, M.; Tuomilehto, J.; Uusitupa, M.; de Mello, V.D.; Hanhineva, K. Associations of serum indolepropionic acid, a gut microbiota metabolite, with type 2 diabetes and low-grade inflammation in high-risk individuals. *Nutr. Diabetes* **2018**, *8*, 35. [CrossRef]
122. Kovacikova, L.; Prnova, M.S.; Majekova, M.; Bohac, A.; Karasu, C.; Stefek, M. Development of Novel Indole-Based Bifunctional Aldose Reductase Inhibitors/Antioxidants as Promising Drugs for the Treatment of Diabetic Complications. *Molecules* **2021**, *26*, 2867. [CrossRef]
123. Chen, L.; Yang, Y.; Sun, S.; Xie, Y.; Pan, C.; Li, M.; Li, C.; Liu, Y.; Xu, Z.; Liu, W.; et al. Indolepropionic acid reduces obesity-induced metabolic dysfunction through colonic barrier restoration mediated via tuft cell-derived IL-25. *FEBS J.* **2022**, *289*, 5985–6004. [CrossRef]
124. Drăgoi, C.M.; Mitrea, N.; Arsene, A.L.; Nicolae, A.C.; Ilie, M. In vitro effects of some bio-indoles on the transmembrane potential of Jurkat E6.1 limphoblasts. *Farmacia* **2012**, *60*, 240–248.
125. Bartyzel, A.; Kaczor, A.A.; Mahmoudi, G.; Masoudiasl, A.; Wróbel, T.M.; Pitucha, M.; Matosiuk, D. Experimental and Computational Structural Studies of 2,3,5-Trisubstituted and 1,2,3,5-Tetrasubstituted Indoles as Non-Competitive Antagonists of GluK1/GluK2 Receptors. *Molecules* **2022**, *27*, 2479. [CrossRef] [PubMed]
126. Kumari, A.; Singh, R.K. Medicinal chemistry of indole derivatives: Current to future therapeutic prospectives. *Bioorganic Chem.* **2019**, *89*, 103021. [CrossRef] [PubMed]
127. Margină, D.M.; Drăgoi, C.M. Intermittent Fasting on Human Health and Disease. *Nutrients* **2023**, *15*, 4491. [CrossRef] [PubMed]
128. Ungurianu, A.; Margină, D.; Mihai, D.P.; Nicolae, A.C.; Drăgoi, C.M.; Grădinaru, D.; Zanfirescu, A. Caloric restriction mimetics: Pinostilbene versus resveratrol regarding SIRT1 and SIRT6 interaction. *Adv. Med. Sci.* **2025**, *70*, 44–50. [CrossRef]
129. Givler, D.; Givler, A.; Luther, P.M.; Wenger, D.M.; Ahmadzadeh, S.; Shekoohi, S.; Edinoff, A.N.; Dorius, B.K.; Baptiste, C.J.; Cornett, E.M.; et al. Chronic Administration of Melatonin: Physiological and Clinical Considerations. *Neurol. Int.* **2023**, *15*, 518–533. [CrossRef]
130. Cardinali, D.P.; Pandi-Perumal, S.R.; Brown, G.M. Melatonin as a Chronobiotic and Cytoprotector in Non-communicable Diseases: More than an Antioxidant. *Subcell. Biochem.* **2024**, *107*, 217–244.
131. Martín Giménez, V.M.; de las Heras, N.; Lahera, V.; Tresguerres, J.A.F.; Reiter, R.J.; Manucha, W. Melatonin as an Anti-Aging Therapy for Age-Related Cardiovascular and Neurodegenerative Diseases. *Front. Aging Neurosci.* **2022**, *14*, 888292. [CrossRef]
132. Kuthati, Y.; Rao, V.N.; Mende, L.K.; Wong, C.S. Therapeutic Potential of Melatonin in Management of Diabetic Mellitus and Diabetic Neuropathic Pain: Underlying Mechanisms, Challenges and Future Perspectives. *J. Formos. Med. Assoc.* **2025**. [CrossRef] [PubMed]
133. Vlăsceanu, A.M.; Petraru, C.; Baconi, D.; Ghica, M.; Arsene, A.; Popa, L.; Nicolae, A.; Dragoi, C.M.; Pavalache, G. Quantitative relationships of urinary cotinine levels in smoking diabetic patients. *Farmacia* **2015**, *63*, 349–356.
134. Grădinaru, D.; Mitrea, N.; Margină, D.; Arsene, A.; Gruia, V.; Drăgoi, C.; Nicolae, A.; Borşa, C.; Gherasim, P. Evaluation of serum osteocalcin in elderly patients with type-2 diabetes mellitus. *Farmacia* **2009**, *57*, 331–338.
135. Gruia, V.; Aramă, C.; Mitrea, N.; Arsene, A.L.; Grădinaru, D.; Drăgoi, C. The HPLC plasmatic profile of some fat-soluble antioxidant micronutrients (all-trans-retinol, α-tocopherol, coenzime Q10) in diabetic and dyslipidemic patients. *Farmacia* **2009**, *57*, 630–638.
136. Rasmussen, C.H.; O, C.K.; Chan, W.S.; Magkos, F.; Kong, A.P. Sleep habits in the pathogenesis and management of diabesity. *J. Diabetes Investig.* **2025**, *16*, 1202–1216. [CrossRef]
137. Shaheen, N.; Miao, J.; Xia, B.; Zhao, Y.; Zhao, J. Multifaceted Role of Microbiota-Derived Indole-3-Acetic Acid in Human Diseases and Its Potential Clinical Application. *FASEB J.* **2025**, *39*, e70574. [CrossRef]
138. Zhang, C.; Fu, Q.; Shao, K.; Liu, L.; Ma, X.; Zhang, F.; Zhang, X.; Meng, L.; Yan, C.; Zhao, X. Indole-3-acetic acid improves the hepatic mitochondrial respiration defects by PGC1a up-regulation. *Cell Signal.* **2022**, *99*, 110442. [CrossRef]
139. Sax, S.L.; Centomo, M.L.; Centofanti, F.; Rizzacasa, B.; Cox, S.; Cox, C.; Latini, A.; D'apice, M.R.; Mannucci, L.; Novelli, G.; et al. The Senolytic Effect of Indole-3-Carbinol (I3C) on Mouse Embryonic (MEF) and Human Fibroblast Cell Lines. *Int. J. Mol. Sci.* **2024**, *25*, 11652. [CrossRef]

https://doi.org/10.3390/targets4010004

140. Cani, P.D.; Possemiers, S.; Van de Wiele, T.; Guiot, Y.; Everard, A.; Rottier, O.; Geurts, L.; Naslain, D.; Neyrinck, A.; Lambert, D.M.; et al. Changes in gut microbiota control inflammation in obese mice through a mechanism involving GLP-2-driven improvement of gut permeability. *Gut* **2009**, *58*, 1091–1103. [CrossRef]
141. Sánchez-Luquez, K.; Reis Silveira, A.M.; Sánchez-Vinces, S.; Rosini Silva AAp Barreto, J.; Lemos de Brito, R.B.S.; Garcia, C.d.M.; Vieira, A.L.; Antonio, M.A.; Carvalho, P.d.O. Etodolac Single Dose Metabolic Profile Elucidation: Pharmacokinetics and Adverse Events in Healthy Volunteers. *Pharmaceuticals* **2025**, *18*, 82. [CrossRef]
142. Konovalova, I.S.; Kovalenko, S.M.; Kravchenko, D.V.; Chuev, V.P. Crystal structure of the non-steroidal anti-inflammatory drug (NSAID) tolmetin sodium. *Acta Crystallogr. E Crystallogr. Commun.* **2021**, *77*, 134–137. [CrossRef]
143. Galenko-Yaroshevsky, P.A.; Zelenskaya, A.V.; Suzdalev, K.F.; Popova, T.N.; Kvetkina, A.N.; Shamatova, M.M.; Chuyan, E.N.; Ravaeva, M.Y.; Murashko, R.A.; Glechyan, T.R.; et al. The TRPV1 Channel Modulator Imidazo[1,2-a]Indole Derivative Exhibits Pronounced and Versatile Anti-Inflammatory Activity In Vivo. *Biomedicines* **2026**, *14*, 60. [CrossRef]
144. Bhat, M.A.; Al-Omar, M.A.; Raish, M.; Ansari, M.A.; Abuelizz, H.A.; Bakheit, A.H.; Naglah, A.M. Indole Derivatives as Cyclooxygenase Inhibitors: Synthesis, Biological Evaluation and Docking Studies. *Molecules* **2018**, *23*, 1250. [CrossRef] [PubMed]
145. Zheng, X.; Zhang, L.; Zhai, J.; Chen, Y.; Luo, H.; Hu, X. The molecular basis for inhibition of sulindac and its metabolites towards human aldose reductase. *FEBS Lett.* **2012**, *586*, 55–59. [CrossRef] [PubMed]
146. Yin, T.; Wang, G.; Ye, T.; Wang, Y. Sulindac, a non-steroidal anti-inflammatory drug, mediates breast cancer inhibition as an immune modulator. *Sci. Rep.* **2016**, *6*, 19534. [CrossRef]
147. Zheng, M.; Zheng, M.; Ye, D.; Deng, Y.; Qiu, S.; Luo, X.; Chen, K.; Liu, H.; Jiang, H. Indole derivatives as potent inhibitors of 5-lipoxygenase: Design, synthesis, biological evaluation, and molecular modeling. *Bioorg. Med. Chem. Lett.* **2007**, *17*, 2414–2420. [CrossRef] [PubMed]
148. Brincks, E.L.; Adams, J.; Wang, L.; Turner, B.; Marcinowicz, A.; Ke, J.; Essmann, M.; Mautino, L.M.; Van Allen, C.; Kumar, S.; et al. Indoximod opposes the immunosuppressive effects mediated by IDO and TDO via modulation of AhR function and activation of mTORC1. *Oncotarget* **2020**, *11*, 2438–2461. [CrossRef]
149. Zhu, P.; Yu, H.; Zhou, K.; Bai, Y.; Qi, R.; Zhang, S. 3,3′-Diindolylmethane modulates aryl hydrocarbon receptor of esophageal squamous cell carcinoma to reverse epithelial-mesenchymal transition through repressing RhoA/ROCK1-mediated COX2/PGE2 pathway. *J. Exp. Clin. Cancer Res.* **2020**, *39*, 113. [CrossRef]
150. Liu, M.; Yasmeen, R.; Fukagawa, N.; Yu, L.; Kim, Y.; Wang, T. Dose-Dependent Responses of I3C and DIM on T-Cell Activation in the Human T Lymphocyte Jurkat Cell Line. *Int. J. Mol. Sci.* **2017**, *18*, 1409. [CrossRef]
151. Cao, J.; Bao, Q.; Hao, H. Indole-3-Carboxaldehyde Alleviates LPS-Induced Intestinal Inflammation by Inhibiting ROS Production and NLRP3 Inflammasome Activation. *Antioxidants* **2024**, *13*, 1107. [CrossRef]
152. Stanciu, A.; Zamfir-Chiru-Anton, A.; Stanciu, M.; Stoian, A.; Jinga, V.; Nitipir, C.; Bucur, A.; Pituru, T.S.; Arsene, A.L.; Dragoi, C.M.; et al. Clinical significance of serum melatonin in predicting the severity of oral squamous cell carcinoma. *Oncol Lett.* **2019**, *19*, 1537–1543. [CrossRef]
153. Mądra-Gackowska, K.; Baumgart, S.; Jędrzejewski, M.; Studzińska, R.; Szeleszczuk, Ł.; Gackowski, M. Computational QSAR study of novel 2-aminothiazol-4(5H)-one derivatives as 11β-HSD1 inhibitors. *J. Comput. Aided Mol. Des.* **2025**, *39*, 67. [CrossRef]
154. Gackowski, M.; Pluskota, R.; Koba, M. Predicting Antitumor Activity of Anthrapyrazole Derivatives using Advanced Machine Learning Techniques. *Curr. Comput. Aided Drug Des.* **2024**, *20*, 798–810. [CrossRef] [PubMed]
155. Gackowski, M.; Madriwala, B.; Koba, M. In silico design, docking simulation, and ANN-QSAR model for predicting the anticoagulant activity of thiourea isosteviol compounds as FXa inhibitors. *Chem. Pap.* **2023**, *77*, 7027–7044. [CrossRef]
156. Gackowski, M.; Madriwala, B.; Studzińska, R.; Koba, M. Novel Isosteviol-Based FXa Inhibitors: Molecular Modeling, In Silico Design and Docking Simulation. *Molecules* **2023**, *28*, 4977. [CrossRef]
157. Taliani, S.; Da Settimo, F.; Martini, C.; Laneri, S.; Novellino, E.; Greco, G. Exploiting the Indole Scaffold to Design Compounds Binding to Different Pharmacological Targets. *Molecules* **2020**, *25*, 2331. [CrossRef]
158. Meyer, M.E.; Doshi, A.; Polgar, W.E.; Zaveri, N.T. Discovery and structure-activity relationships (SAR) of a novel class of 2-substituted N-piperidinyl indole-based nociceptin opioid receptor ligands. *Bioorg. Med. Chem.* **2023**, *92*, 117421. [CrossRef] [PubMed]
159. Iqbal, T.; Khan, S.; Hussain, R.; Khan, Y.; Shoaib, K.; Rozeena; Saeed, M.; Darwish, H.W. Novel indole based fused triazole-thiadiazole derivatives as anti-diabetic agents: In vitro and in silico approaches. *Future Med. Chem.* **2024**, *16*, 2475–2486. [CrossRef]
160. Cornelissen, G.; Watanabe, Y.; Beaty, L.A.; Otsuka, K. Toward a personalized chronotherapy of blood pressure. *Biomed. J.* **2025**, *48*, 100849. [CrossRef]
161. Creangă, E.C.; Stan, R.; Nicolae, A.C.; Drăgoi, C.M.; Dumitrescu, I.B. Personalized Therapeutic Advances in Erythropoietin Signaling: From Anemia Management to Extensive Clinical Applications. *Pharmaceutics* **2025**, *17*, 1190. [CrossRef]
162. Drăgoi, C.M.; Nicolae, A.C.; Dumitrescu, I.B. Emerging Strategies in Drug Development and Clinical Care in the Era of Personalized and Precision Medicine. *Pharmaceutics* **2024**, *16*, 1107. [CrossRef]

https://doi.org/10.3390/targets4010004

163. Matenchuk, B.A.; Mandhane, P.J.; Kozyrskyj, A.L. Sleep, circadian rhythm, and gut microbiota. *Sleep Med. Rev.* **2020**, *53*, 101340. [CrossRef]
164. Bautista, J.; Ojeda-Mosquera, S.; Altamirano-Colina, A.; Hidalgo-Tinoco, C.; Di Capua Delgado, M.; López-Cortés, A. Bidirectional interactions between circadian rhythms and the gut microbiome. *Appl. Microbiol. Biotechnol.* **2025**, *109*, 218. [CrossRef]
165. Wang, Z.; Wang, Z.; Lu, T.; Chen, W.; Yan, W.; Yuan, K.; Shi, L.; Liu, X.; Zhou, X.; Shi, J.; et al. The microbiota-gut-brain axis in sleep disorders. *Sleep Med. Rev.* **2022**, *65*, 101691. [CrossRef]
166. Seripa, D.; Lozupone, M.; Miscio, G.; Stella, E.; La Montagna, M.; Gravina, C.; Urbano, M.; di Mauro, L.; Daniele, A.; Greco, A.; et al. CYP2D6 genotypes in revolving door patients with bipolar disorders. *Medicine* **2018**, *97*, e11998. [CrossRef] [PubMed]
167. Xin, Y.; Gao, L.; Li, S.; Wang, J.; Chen, C.; Tuo, Y.; Nie, G.; Li, R.; Sun, D.; Fu, Y.; et al. Effect of CYP2D6 and ABCB1 polymorphisms on pharmacokinetics and efficacy of aripiprazole in pediatric tic disorders. *BMC Pediatr.* **2025**, *25*, 495. [CrossRef]
168. Kehinde, O.; Vaughn, S.E.; Amaeze, O.; Toren, P.; Retke, B.; Oni-Orisan, A.; Ramsey, L.B. Cytochrome P450 2D6 *17 and *29 Allele Activity for Risperidone Metabolism: Advancing Precision Medicine Health Equity. *Clin. Pharmacol. Ther.* **2025**, *118*, 1152–1160. [CrossRef] [PubMed]
169. Theken, K.N.; Lee, C.R.; Gong, L.; Caudle, K.E.; Formea, C.M.; Gaedigk, A.; Klein, T.E.; Agúndez, J.A.; Grosser, T. Clinical Pharmacogenetics Implementation Consortium Guideline (CPIC) for CYP2C9 and Nonsteroidal Anti-Inflammatory Drugs. *Clin. Pharmacol. Ther.* **2020**, *108*, 191–200. [CrossRef]
170. Tarţa-Arsene, O.; Leanca, M.; Dică, A.; Bran, E.; Rad, F.; Timnea, O.; Păcurar, D.; Velescu, B.; Nicolae, A.; Dragoi, C.M. Dietary omega-3 fatty acids supplimentation for attention deficit with hyperactivity disorder in epileptic children. *Farmacia* **2017**, *65*, 550–556.
171. Ritivoiu, M.E.; Drăgoi, C.M.; Matei, D.; Stan, I.V.; Nicolae, A.C.; Craiu, M.; Dumitrescu, I.-B.; Ciolpan, A.A. Current and Future Therapeutic Approaches of Exocrine Pancreatic Insufficiency in Children with Cystic Fibrosis in the Era of Personalized Medicine. *Pharmaceutics* **2023**, *15*, 162. [CrossRef]
172. Jacobsen, J.P.R. Use of 5-Hydroxytryptophan Labeled With Carbon 11 in Social Anxiety Disorder. *JAMA Psychiatry* **2016**, *73*, 177. [CrossRef] [PubMed]
173. Juhász, C.; Chugani, D.C.; Muzik, O.; Wu, D.; Sloan, A.E.; Barger, G.; Watson, C.; Shah, A.K.; Sood, S.; Ergun, E.L.; et al. In Vivo Uptake and Metabolism of α-[11C]Methyl- l -Tryptophan in Human Brain Tumors. *J. Cereb. Blood Flow. Metab.* **2006**, *26*, 345–357. [CrossRef] [PubMed]
174. Bawazir, W.A.; Ain, Q. Indole–Imidazole Hybrids as Emerging Therapeutic Scaffolds: Synthetic Advances and Biomedical Applications. *Molecules* **2025**, *30*, 4164. [CrossRef]
175. Naaz, F.; Neha, K.; Haider, M.R.; Shafi, S. Indole Derivatives (2010–2020) As Versatile Tubulin Inhibitors: Synthesis and structure–activity Relationships. *Future Med. Chem.* **2021**, *13*, 1795–1828. [CrossRef]
176. Kaur, B.; Venugopal, S.; Verma, A.; Sahu, S.K.; Wadhwa, P.; Kumar, D. Recent Developments in the Synthesis and Anticancer Activity of Indole and Its Derivatives. *Curr. Org. Synth.* **2023**, *20*, 376–394. [CrossRef] [PubMed]
177. Baruah, B.; Pegu, C.D.; Deb, M.L. Indole as a Versatile Building Block in Cycloaddition Reactions: Synthesis of Diverse Heterocyclic Frameworks. *Top. Curr. Chem.* **2024**, *382*, 18. [CrossRef]
178. Mathada, B.S.; Somappa, S.B. An insight into the recent developments in anti-infective potential of indole and associated hybrids. *J. Mol. Struct.* **2022**, *1261*, 132808. [CrossRef]
179. Nicolae, A.C.; Drăgoi, C.M.; Ceaușu, I.; Poalelungi, C.; Iliescu, D.; Arsene, A.L. Clinical implications of the indolergic system and oxidative stress in physiological gestational homeostasis. *Farmacia* **2015**, *63*, 46–51.
180. Tran, N.L.; Leconte, G.A.; Ferguson, F.M. Targeted Protein Degradation: Design Considerations for PROTAC Development. *Curr. Protoc.* **2022**, *2*, e611. [CrossRef]
181. Muthukumar, V.; Vashishth, A.; Maniam, S.; Ghosh, B. Molecular Frameworks for ERK1/2 Inhibition: Lessons from Synthetic and SAR Explorations. *ACS Omega* **2025**, *10*, 62407–62426. [CrossRef]
182. Omage, F.B.; Salim, J.A.; Mazoni, I.; Yano, I.H.; Borro, L.; Gonzalez, J.E.H.; de Moraes, F.R.; Giachetto, P.F.; Tasic, L.; Arni, R.K.; et al. Protein allosteric site identification using machine learning and per amino acid residue reported internal protein nanoenvironment descriptors. *Comput. Struct. Biotechnol. J.* **2024**, *23*, 3907–3919. [CrossRef]
183. Markovic, M.; Abramov-Harpaz, K.; Regev, C.; Ben-Shabat, S.; Aponick, A.; Zimmermann, E.M.; Miller, Y.; Dahan, A. Prodrug-Based Targeting Approach for Inflammatory Bowel Diseases Therapy: Mechanistic Study of Phospholipid-Linker-Cyclosporine PLA2-Mediated Activation. *Int. J. Mol. Sci.* **2022**, *23*, 2673. [CrossRef] [PubMed]
184. Sun, H.; Sun, K.; Sun, J. Recent Advances of Marine Natural Indole Products in Chemical and Biological Aspects. *Molecules* **2023**, *28*, 2204. [CrossRef]
185. Stránská, M.; Řezanka, T.; Křen, V. Glycosylated mycotoxins: A hidden enemy. *Nat. Prod. Rep.* **2026**. [CrossRef] [PubMed]
186. Al-Shakliah, N.S.; Kadi, A.A.; Abuelizz, H.A.; Al-Salahi, R. In Vitro and Reactive Metabolites Investigation of Metabolic Profiling of Tyrosine Kinase Inhibitors Dubermatinib in HLMs by LC–MS/MS. *Separations* **2023**, *10*, 353. [CrossRef]

https://doi.org/10.3390/targets4010004

187. Wang, Y.; Xie, H.; Ma, J.; Du, T.; Gao, S.; Chen, Y.; Lin, S.-Y.; Liang, D. Drug Metabolism and Pharmacokinetic Evaluation of a Novel RNase H2 Inhibitor for the Treatment of Triple-Negative Breast Cancer. *Pharmaceutics* **2025**, *17*, 1052. [CrossRef]
188. Rongala, G.; Rongala, D.S.; Rongala, A.N. The Future of Precision Medicine: Targeted Therapies, Personalized Medicine and Formulation Strategies. *J. Pharm. BioTech Ind.* **2025**, *2*, 19. [CrossRef]
189. Nagy, M.I.; Darwish, K.M.; Kishk, S.M.; Tantawy, M.A.; Nasr, A.M.; Qushawy, M.; Swidan, S.A.; Mostafa, S.M.; Salama, I. Design, Synthesis, Anticancer Activity, and Solid Lipid Nanoparticle Formulation of Indole- and Benzimidazole-Based Compounds as Pro-Apoptotic Agents Targeting Bcl-2 Protein. *Pharmaceuticals* **2021**, *14*, 113. [CrossRef]
190. Dumitrescu, I.B.; Lupuliasa, D.; Drăgoi, C.M.; Nicolae, A.C.; Pop, A.; Șaramet, G.; Șaramet, G.; Drăgănescu, D. The age of pharmaceutical 3d printing. Technological and therapeutical implications of additive manufacturing. *Farmacia* **2018**, *66*, 365–389. [CrossRef]
191. Drăgoi, C.M.; Nicolae, A.C.; Grigore, C.; Dinu-Pîrvu, C.E.; Arsene, A.L. Characteristics of glucose homeostasis and lipidic profile in a hamster metabolic syndrome model, after the co-administration of melatonin and irbesartan in a multiparticulate pharmaceutical formulation. In *The 2nd International Conference on Interdisciplinary Management of Diabetes Mellitus and Its Complications, INTERDIAB 2016–Diabetes Mellitus As Cardiovascular Disease*; Editura Niculescu Bucureşti: Bucharest, Romania, 2016; pp. 221–229.
192. Shukla, N.; Das, R.; Chanderiya, A.; Cárdenas, C.Y.; Mukhanova, E.; Soldatov, A.; Belbekhouche, S. Harnessing nanotechnology for efficient delivery of indole-based drugs in cancer treatment. *Cancer Treat. Res. Commun.* **2025**, *45*, 101019. [CrossRef]
193. Russo, E.; Grondona, C.; Brullo, C.; Spallarossa, A.; Villa, C.; Tasso, B. Indole Antitumor Agents in Nanotechnology Formulations: An Overview. *Pharmaceutics* **2023**, *15*, 1815. [CrossRef]
194. Yeung, C.H.T.; Sekulich, D.C.; Scott, A.; Nolte, W.M.; Gibson, K.; Su, R.; Alrifai, M.W.; Lopata, S.M.; Lewis, T.; Reese, J.; et al. The Relationship of Indomethacin Exposure With Efficacy and Renal Toxicity Outcomes for Preterm Infants in the Neonatal Intensive Care Unit. *Clin. Transl. Sci.* **2025**, *18*, e70251. [CrossRef]
195. Rendic, S.P.; Guengerich, F.P. Formation of potentially toxic metabolites of drugs in reactions catalyzed by human drug-metabolizing enzymes. *Arch. Toxicol.* **2024**, *98*, 1581–1628. [CrossRef] [PubMed]
196. Pannone, L.; Monaco, C.; Sorgente, A.; Vergara, P.; Calburean, P.; Gauthey, A.; Bisignani, A.; Kazawa, S.; Strazdas, A.; Mojica, J.; et al. Ajmaline-Induced Abnormalities in Brugada Syndrome: Evaluation With ECG Imaging. *J. Am. Heart Assoc.* **2022**, *11*, e024001. [CrossRef]
197. Drăgoi, C.M.; Diaconu, C.C.; Nicolae, A.C.; Dumitrescu, I.B. Redox Homeostasis and Molecular Biomarkers in Precision Therapy for Cardiovascular Diseases. *Antioxidants* **2024**, *13*, 1163. [CrossRef]
198. Diaconu, C.C.; Cozma, M.A.; Dobrică, E.C.; Gheorghe, G.; Jichitu, A.; Ionescu, V.A.; Nicolae, A.C.; Drăgoi, C.M.; Găman, M.-A. Polypharmacy in the Management of Arterial Hypertension—Friend or Foe? *Medicina* **2021**, *57*, 1288. [CrossRef]
199. Mhaske, G.S.; Ganage, A.D.; Satkar, S.S.; Awate, S.S.; Gurav, M.V.; Sribhavani, K.R.; Phadtare, G.; Waghmare, S. High-Throughput Screening of Novel Indole Alkaloids as Potential Tyrosine Kinase Inhibitors for Breast Cancer Therapy. *Curr. Biotechnol.* **2025**, *14*, 181–197. [CrossRef]
200. Ciaglia, T.; Miranda, M.R.; Di Micco, S.; Vietri, M.; Smaldone, G.; Musella, S.; Di Sarno, V.; Auriemma, G.; Sardo, C.; Moltedo, O.; et al. Neuroprotective Potential of Indole-Based Compounds: A Biochemical Study on Antioxidant Properties and Amyloid Disaggregation in Neuroblastoma Cells. *Antioxidants* **2024**, *13*, 1585. [CrossRef]
201. Cortes-Torres, A.G.; López-Castillo, G.N.; Marín-Torres, J.L.; Portillo-Reyes, R.; Luna, F.; Baca, B.E.; Sandoval-Ramírez, J.; Carrasco-Carballo, A. Cymbopogon citratus Essential Oil: Extraction, GC–MS, Phytochemical Analysis, Antioxidant Activity, and In Silico Molecular Docking for Protein Targets Related to CNS. *Curr. Issues Mol. Biol.* **2023**, *45*, 5164–5179. [CrossRef] [PubMed]
202. Sirbu, C.A.; Furdu-Lungut, E.; Plesa, C.F.; Nicolae, A.C.; Drăgoi, C.M. Pharmacological treatment of relapsing remitting multiple sclerosis-where we are? *Farmacia* **2016**, *64*, 651–655.
203. Axente, M.; Mirea, A.; Sporea, C.; Pădure, L.; Drăgoi, C.M.; Nicolae, A.C.; Ion, D.A. Clinical and Electrophysiological Changes in Pediatric Spinal Muscular Atrophy after 2 Years of Nusinersen Treatment. *Pharmaceutics* **2022**, *14*, 2074. [CrossRef]
204. Fagiani, F.; Baronchelli, E.; Pittaluga, A.; Pedrini, E.; Scacchi, C.; Govoni, S.; Lanni, C. The Circadian Molecular Machinery in CNS Cells: A Fine Tuner of Neuronal and Glial Activity With Space/Time Resolution. *Front. Mol. Neurosci.* **2022**, *15*, 937174. [CrossRef]
205. Dragoi, C.M.; Yang, Z.; Fekry, B.; Brenna, A. Editorial: Chronobiology in cardiometabolic health and disease. *Front. Pharmacol.* **2025**, *15*, 1544963. [CrossRef] [PubMed]
206. Huang, X.; Liang, J.; Zhang, J.; Fu, J.; Xie, W.; Zheng, F. Association of cardiovascular-kidney-metabolic health and social connection with the risk of depression and anxiety. *Psychol. Med.* **2024**, *54*, 4203–4211. [CrossRef]
207. Barbu, C.G.; Arsene, A.L.; Florea, S.; Albu, A.; Sirbu, A.; Martin, S.; Nicolae, A.C.; Burcea-Dragomiroiu, G.T.A.; Popa, D.E.; Velescu, B.S.; et al. Cardiovascular risk assessment in osteoporotic patients using osteoprotegerin as a reliable predictive biochemical marker. *Mol. Med. Rep.* **2017**, *16*, 6059–6067. [CrossRef]
208. Roointan, A.; Gheisari, Y.; Hudkins, K.L.; Gholaminejad, A. Non-invasive metabolic biomarkers for early diagnosis of diabetic nephropathy: Meta-analysis of profiling metabolomics studies. *Nutr. Metab. Cardiovasc. Dis.* **2021**, *31*, 2253–2272.

https://doi.org/10.3390/targets4010004

209. Vandana, J.J.; Manrique, C.; Lacko, L.A.; Chen, S. Human pluripotent-stem-cell-derived organoids for drug discovery and evaluation. *Cell Stem Cell* **2023**, *30*, 571–591. [CrossRef]
210. Heinzelmann, E.; Piraino, F.; Costa, M.; Roch, A.; Norkin, M.; Garnier, V.; Homicsko, K.; Brandenberg, N. iPSC-derived and Patient-Derived Organoids: Applications and challenges in scalability and reproducibility as pre-clinical models. *Curr. Res. Toxicol.* **2024**, *7*, 100197. [CrossRef] [PubMed]
211. Ahmed, S.; Zhou, Z.; Zhou, J.; Chen, S.Q. Pharmacogenomics of Drug Metabolizing Enzymes and Transporters: Relevance to Precision Medicine. *Genom. Proteom. Bioinform.* **2016**, *14*, 298–313. [CrossRef] [PubMed]
212. Radosavljevic, M.; Svob Strac, D.; Jancic, J.; Samardzic, J. The Role of Pharmacogenetics in Personalizing the Antidepressant and Anxiolytic Therapy. *Genes* **2023**, *14*, 1095. [CrossRef] [PubMed]
213. Hizbullah Ahmed, S.; Noor Mumtaz, M.; Zulfiqar, Z.; Amir Hamza, S.; Siraj, S.; Jelani, M.; Imran, I.; Khan, A. Genetic variations in drug-metabolizing enzyme CYP2C9 among major ethnic groups of Pakistani population. *Gene* **2020**, *746*, 144659. [CrossRef] [PubMed]

https://doi.org/10.3390/targets4010004

targets

MDPI

Review

Navigating the Molecular and Cellular Landscape of Breast Cancer in India: From Unique Pathogenesis to the Promise of Personalized Medicine and Future Technologies

Anichavezhi Devendran [1,*,†] and Sivasankar Perumal [2]

1 School of Biological Sciences, University of Essex, Colchester CO4 3SQ, UK
2 Norfolk and Suffolk NHS Foundation Trust, Norwich NR6 5BE, UK; shiva3384@gmail.com
* Correspondence: anichavezhi.devendran@gmail.com
† Current address: Cardiovascular Research Institute, Icahn School of Medicine at Mount Sinai, New York, NY 10029, USA.

Abstract

Breast cancer is a substantial and growing public health issue in India, with epidemiological data demonstrating distinct and often severe disease characteristics in contrast to Western countries. Contrary to the global trend, Indian women frequently develop the disease at an earlier age and tend to present with more advanced stages, emphasizing important variations in disease pathophysiology. This review compiles and critically evaluates the current literature to describe the specific pathophysiology of breast cancer in the Indian population. We investigate the unique cellular and molecular landscapes, evaluate the impact of specific Indian demographic and genetic features, and highlight crucial gaps in knowledge, diagnostic tools, and therapeutic approaches. The assessment reveals a molecular landscape determined by the incidence of specific tumor subtypes; triple-negative breast cancer, for instance, is frequently diagnosed in younger women, and genetic profiling research suggests variations in its susceptibility genes and mutation patterns when compared to global populations. While this paper brings together recent advancements, it highlights the challenges of adopting global diagnostic and treatment guidelines in the Indian healthcare system. These challenges are largely due to variances and specific demographic and socioeconomic discrepancies that create substantial hurdles for timely diagnosis and patient care. We highlight significant gaps, such as the need for more complete multi-omics profiling of Indian patient cohorts, an absence of uniform and readily available screening programs, and shortcomings in healthcare infrastructure and qualified oncology experts. Furthermore, the review highlights the crucial need for therapeutic strategies tailored to the distinct genetic and demographic profiles of Indian breast cancer patients. We present significant strategies for addressing these challenges, with a focus on integrating multi-omics data and clinical characteristics to gain deeper insight into the underlying causes of the disease. Promising avenues include using artificial intelligence and advancements in technology to improve diagnostics, developing indigenous and affordable treatment options, and establishing context-specific research frameworks for the Indian population. This review also underlines the necessity for personalized strategies to improve breast cancer outcomes in India.

Keywords: breast cancer; India; molecular landscapes; disease pathophysiology; genetic profiling; multi-omics; artificial intelligence

1. Introduction

Breast cancer (BC) continues to pose a major public health challenge globally, with an estimated 2.3 million new cases and 685,000 deaths reported worldwide in 2020. While the incidence rates are high in developed countries, the burden is growing rapidly in developing nations such as India. In fact, in India, BC has surpassed other type of cancers to become the most common among women, with concerning forecasts for future incidence and mortality [1]. A notable and significant contrast in the Indian demographic is the average age at diagnosis, which is roughly ten years younger than in Western populations. Additionally, greater percentage of patients are diagnosed with more severe or aggressive characteristics, such as triple-negative breast cancer (TNBC) [2]. Effective management of cancer involves more than just treating the physical illness; it also needs to consider the enormous psychosocial burden that patients suffer. Being diagnosed with and treated for BC can be deeply distressing, with a high incidence of psychiatric morbidities, including sadness and anxiety. This aside, unmanaged psychological distress, including depression, has shown to adversely affect a patient's quality of life, adherence to treatment regimens, and survival rates [3].

Breast cancer is an extremely diverse disease; its underlying molecular basis and receptor types primarily determine its prognosis and response to therapy. The global perspective of BC molecular subtypes (luminal A/B, HER2+, and TNBC) often guides therapeutic regimens. Nevertheless, research indicates that the prevalence of these subtypes differs greatly by ethnicity [4]. In India, aggressive subtypes, particularly TNBC, are reported at a higher rate than in Western countries. Additionally, distinct genetic susceptibilities, including particular mutations in *BRCA1* and *BRCA2*, along with other novel gene variants, have been discovered in the diverse Indian gene pool, potentially influencing the early onset and severe progression of the disease. The complete scope and significance of this unique cellular and molecular landscape remains under explored.

This unique Indian BC clinical picture indicates possible variations in the underlying cellular and molecular etiology. Understanding these distinct abnormalities at the genetic and epigenetic levels is critical, as global data and treatment regimens may not be completely applicable to the Indian subcontinent. Regardless of the growing burden and apparent biological disparities, there are considerable gaps in our knowledge and management of BC in India. A key problem is the lack of a comprehensive national cancer registry and large-scale etiological research to determine the causes of early onset and distinct molecular profiles [5]. Setting this aside, diagnostic strategies are often hampered by poor public awareness and limited access to structured screening programs, which results in more advanced diagnoses. In terms of treatment, although there are internationally recognized guidelines, the expensive nature of specialized therapies (such as Trastuzumab), coupled with uneven distribution across the country, reveals significant healthcare inequities, creating a gap in effective treatment for all patients [6]. Therefore, there is an urgent need for research aimed at developing and validating new biomarkers, and for innovative treatment approaches for the Indian population.

Nonetheless, the emergence of frontier technological innovations provides hope for addressing these challenges. For example, programs like the Bharat Cancer Genome Atlas are tailoring advanced techniques, including large-scale genomics and multi-omics profiling to the heterogeneous Indian population [7]. Additionally, the integration of artificial intelligence (AI) and machine learning (ML) is transforming diagnostic options, enabling enhanced image examination and forecasting models tailored to the specific patient demographics of India. These advancements, featuring affordable screening tools, signify an essential progress towards more equitable and efficient cancer treatment.

This review integrates current knowledge of the distinct pathogenesis of breast cancer among the Indian population. It provides an overview of its diverse cellular and molecular landscape in comparison to global understanding and investigates the interplay of demographic, genetic, and environmental factors influencing these discrepancies. Additionally, this paper detects and examines crucial knowledge gaps in the current understanding of the diagnostic procedures and therapeutic strategies in the Indian subcontinent. It also proposes directions for future research and policy changes to improve patient outcomes. Finally, this review will highlight the recent developments and proposes significant future prospects, with an emphasis on Indian-specific innovations in genomics, multi-omics profiling, and artificial intelligence.

2. Breast Cancer Landscape from Indian Epidemiological Context

Breast cancer, as mentioned above, is the most common malignancy in women worldwide. With an estimated 2.3 million new cases, accounting for 11.7% of all cancer cases, it has in fact overtaken lung cancer as the leading cause of cancer incidence worldwide in 2020 [8]. According to epidemiological studies, the global implication and significance of BC is projected to surpass nearly 2 million by 2030 [9]. Especially, looking at the circumstance through an Indian perspective, the prevalence has escalated substantially, by nearly 50% between 1965 and 1985 [10]. To further illustrate this, in 2016 the predicted number of incident cases in India was 118,000 (with 95% confidence interval, 107,000 to 130,000), with women accounting for 98.1% of the total, and prevalent cases accounting for 526,000 (474,000 to 574,000). To highlight the severity, from 1990 to 2016, the age-standardized frequency rate of BC in women augmented by 39.1% (95% CI, 5.1 to 85.5), with an upsurge detected in every state of the country [11]. In fact, according to the Globocan data 2020, BC reckoned for 10% (Figures 1 and 2) [12].

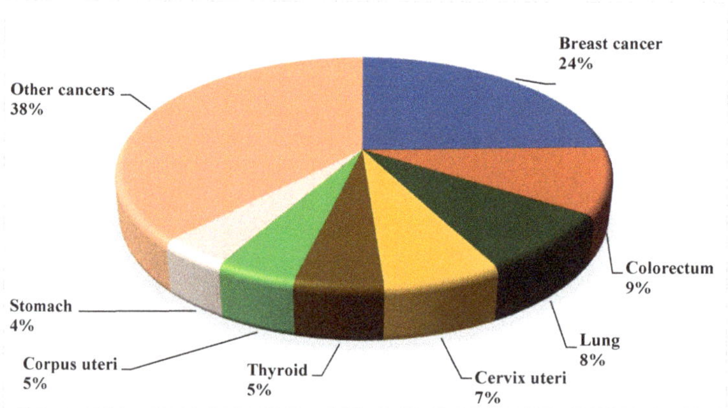

Figure 1. Globally assessed novel cancer cases among women in 2020 (all ages), based on the data from the World Health Organization (WHO, 2020).

In light of the facts presented, and in accordance with the existing trends, a fairly elevated proportion or incidence of the disease was seen to manifest among younger age groups among the Indian women in contrast to their Western counterparts. Furthermore, the National Cancer Registry Program investigated the statistical records from cancer depositories and registries between the years 1998 and 2013 in order to scrutinize the diversity and fluctuation in cancer prevalence. Eventually, the outcomes that ensued from the above examinations in fact revealed a compelling rise in trend of BC. In 1990, the cervix was the prominent site of cancer in India, followed by BC, in Bangalore (23.0% vs.

15.9%), Bhopal (23.2% vs. 21.4%), Chennai (28.9% vs. 17.7%), and Delhi (21.6% vs. 20.3%), respectively, whereas the breast was the prominent site of cancer in Mumbai (24.1% vs. 16.0%), as reported from the above registries. However, from 2000 to 2003 the scenario changed, with breast cancer becoming the prominent cancer site in all registries except the Barshi rural registry (16.9% vs. 36.8%). And noticeably in the case of BC, the registries in Bhopal, Chennai, and Delhi showed a considerable rising trend [13].

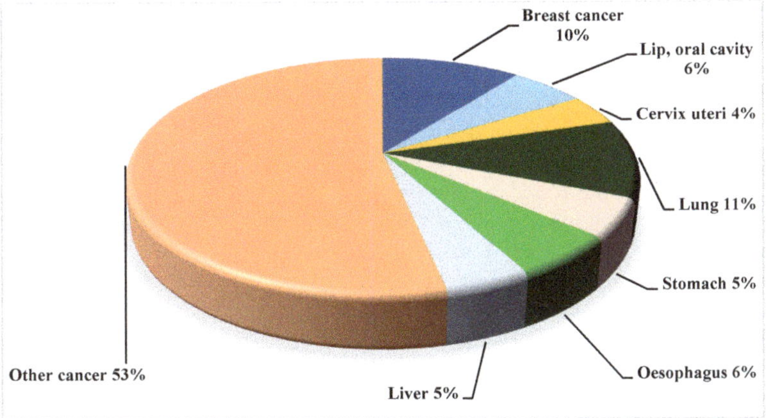

Figure 2. Projected relative cancer burden (based on Disability Adjusted Life Years—DALYs) in India for 2030; data based on the sources such as the Global Cancer Observatory (GLOBOCAN).

To further illustrate the point of conformity, patients with breast cancer have a lower survival rate in India than in Western countries, owing to their early age at the time of onset, the delayed initiation of specific and precise supervision, the late stage of the disease at the time of presentation, and ineffective/imploded treatment regimens [14].

3. Overall Global Understanding of Breast Cancer with the Indian Demographic Characteristics

3.1. Gaps in Knowledge

Compared to Western populations, Indian women have distinct BC characteristics, such as a higher incidence of premenopausal cases, a higher percentage of advanced-stage diagnoses, and a younger average age of diagnosis (40–50 years vs. 60–70 years). Moreover, there is evidence that triple-negative breast cancer (TNBC) is more common in India [15]. In addition, although BRCA1/2 mutations are known risk factors, studies indicate that Indian BC patients may have a higher incidence of these mutations than Western populations, as well as indicating a notable role for non-BRCA genes that warrants additional research.

In India, immense knowledge gaps persist regarding the precise role and function of cancer stem cells (CSCs), particularly in tumor development and progression. These gaps encompass the understanding of the basic attributes and prevalence of CSCs among Indian breast cancer patients. Epigenetic and genetic factors, distinctive to the Indian population, consequently modify CSC function, ultimately impacting therapeutic strategies that target CSCs [16]. Understanding the heterogeneity of cancer stem cells (CSCs) across various breast cancer (BC) subtypes in India and their interaction with the tumor microenvironment is undeniably crucial. This knowledge will enhance therapeutic strategies and improve treatment outcomes. Furthermore, developing these strategies necessitates a deep understanding of how this heterogeneity differs among BC subtypes in India. Therefore, more research is required to pinpoint the precise pathways and mechanisms influencing

CSC behavior in these different subtypes. A collaborative, interdisciplinary approach uniting scientists, healthcare professionals, and experts in cell biology, cancer genomics, and pharmacology can bridge these knowledge gaps and illuminate the complexities of CSC biology and its role in BC in India [16].

Similarly, for precise risk assessment, timely diagnosis, and the development of precision/personalized medicine, it is essential to comprehend the unique molecular and genetic characteristics among the Indian breast cancer patients. In fact, studies are underway to determine distinct mutations (e.g., TP53 and PIK3CA) and their possible effects for personalized treatment approaches [17].

In India, socioeconomic and cultural barriers—including stigma, lack of awareness, and financial hardship—contribute to delayed breast cancer diagnosis and worse outcomes, particularly in rural regions. Therefore, to acquire a thorough understanding about the predisposing factors for BC among Indian women, studies should investigate the relationship between cultural traditions, lifestyle modifications, and genetic predispositions, which are deeply rooted in one another. This aside, with the fair amount of evidence corroborating the implication of environmental factors and lifestyle modifications, research demonstrates possible connections between BC risk and environmental pollutions in air and water, along with urbanization, dietary habits, lack of physical activity, etc., [18].

3.2. Gaps in Diagnostic and Therapeutic Approaches or Methodologies

Despite the fact that early detection of BC makes it treatable, numerous gaps in diagnostic and therapeutic approaches result in higher death rates in India than in Western nations. These disparities are especially apparent when taking into account the distinct demographics and the socioeconomic status of the Indian population as discussed below.

In terms of diagnostics, there is limited access to screening and uptake, regardless of some governmental initiatives. Breast cancer screening program participation is very low, particularly among lower socioeconomic groups and in rural regions. This could be because of a lack of knowledge and awareness, as several women are oblivious about the signs of BC and the significance of early identification [19]. Logistical challenges such as a lack of transportation, a long remote distance to screening amenities, and the necessity for a personal care assistant or attendant further hinder easy access, especially for women from rural areas [20]. This aside, Indian women typically have denser breast tissue which may make the common screening procedures, like mammography, less effective.

In addition, given all the specific and unique BC characteristics among the Indian population, there is an immense need for biomarkers that are specific to the Indian population. Therefore, additional studies and research are needed to detect genetic and epigenetic biomarkers unique to the Indian demography for timely discovery and for further individualized screening approaches [19].

Also given the severity from the genetic and molecular heterogeneity standpoint, the role of tumor microenvironment heterogeneity (TME) has a major impact on treatment resistance, metastasis, and tumor progression. Indeed, research studies focusing on the unique traits and variability of the TME among the Indian BC patients will be vital [1].

Similarly, from limited genetic testing perspective, regardless of the fact that BRCA mutations are more common in India, there is little access to genetic testing and counseling. In addition, further study is needed to determine the clinical significance of variations of unknown significance (VUS) mutations among the Indian population in order to guide treatment selection strategies [21].

However, from therapeutics standpoint, despite the recent medical advances, there is still very few high-quality, multimodal treatment centers, particularly in rural areas. The overall burden is compounded by a severe lack of medical resources, including proper infrastructure and essential equipment like radiotherapy units, and a shortage of qualified oncology specialists. Inadequate care from non-oncologists and financial restrictions lead to high rates of treatment attrition. Furthermore, the scope is complicated by limited access to expensive, novel therapies such as immunotherapy and personalized medicine [22].

On the other hand, with the inadequate studies and reports available on cancer biology from epidemiological and factors unique to India, additional research is crucial to define the frequency of molecular subtypes of BC. This, in turn, will help determine the precise genetic and environmental features that contributes to the disease. In addition, given the severity, and notwithstanding the detailed cancer surveillance initiatives, there is a lack of valid and detailed information on frequency, incidence, and mortality, which in turn impedes the implementation of successful public health interventions.

Therefore, addressing the above gaps implies the essential need to promote better public awareness, and to sensitize the public especially women about the signs and risk factors of BC, as well as the significance of early and prompt detection. Reinforcing and reducing the barriers to screening resources and services through community-centered programs, collaborations with primary healthcare personnel, and mobile screening camps may further help overcome the geographical and socioeconomic obstacles. Improving the BC management landscape requires strategic investment in healthcare personnel training, medical device accessibility, and cancer center expansion. Concurrently, enhancing policy measures—specifically through reinforced national policies, higher healthcare funding, and streamlined financial aid—is vital for addressing current shortcomings.

Ultimately, to devise successful prevention and treatment strategies, further studies and research comprising a comprehensive understanding of the multifaceted nature of BC in terms of genetics, environments, and lifestyle from Indian context, are crucial. This may, in turn, highlight the importance of investing in personalized medicine and tailored treatment approaches.

4. Genetics and Genomics: Implication of Genetic Susceptibility as a Significant Risk Factor for Breast Cancer

Essentially, a risk factor is something that affects the likelihood of developing breast cancer (BC). However, some major risk factors for BC are beyond an individual's control, such as gender and age. For example, the disease is about 100 times more common in women than in men, and the majority of breast cancers are diagnosed in women aged 55 and older, illustrating that aging increases risk. Additionally, having a first-degree relative (mother, sister, or daughter) diagnosed with BC almost doubles a woman's possibility of developing it [23,24].

In general, genetic mutations acquired from a parent are linked to about 5–10% of BC cases. And, in fact, an inherited mutation in the BRCA1 or BRCA2 gene is the most common cause of hereditary BC [25]. This aside, women with a BRCA1 mutation pose a 55–65% lifetime possibility of developing BC. And on the other hand, the lifetime possibility for women with a BRCA2 mutation is about 45%. Furthermore, a woman with both BRCA1 and BRCA2 gene mutation has a 70% possibility of developing BC by the age of 80. BRCA1/2 mutations significantly increase lifetime breast cancer (BC) and ovarian cancer risks, with affected individuals often diagnosed at a younger age. The risk of developing BC in the contralateral breast is also elevated. A positive family history further correlates with increased risk. Specific risks vary by gene: BRCA1 mutations are rarely associated with male breast cancer, while BRCA2 mutations carry a modest lifetime male BC risk of approximately 6.8% [25] (Figures 3 and 4).

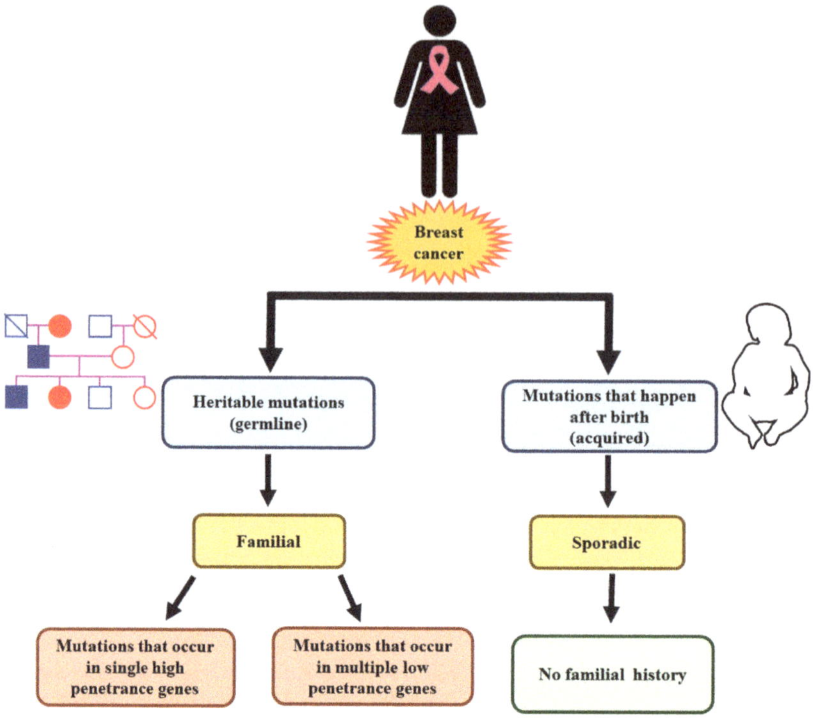

Figure 3. Genetic susceptibility as a significant risk factor for breast cancer.

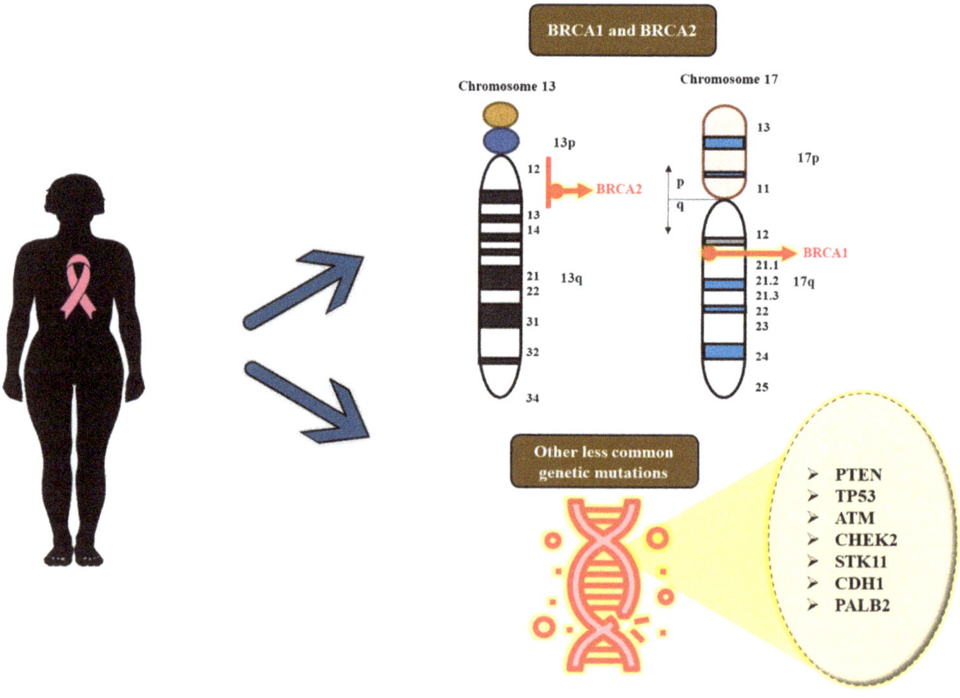

Figure 4. Significant genetic mutations in breast cancer.

This aside, genetic mutations in numerous other genes can also lead to the development of BC, though they are less common than the BRCA mutations and the possibility of developing BC risk is relatively low [25]. A few of these mutated genes include PTEN (mutations acquired from this gene can result in Cowden's syndrome, along with the susceptibility to non-cancerous and cancerous tumors in the breasts, in addition to growths in the thyroid, digestive tract, ovaries and uterus), TP53 (mutations of this gene results in Li-Fraumeni syndrome with an augmented susceptibility to BC, and few other cancers like leukemia, brain tumors, and sarcomas), ATM (acquiring two aberrant copies of this gene causes the disease ataxia–telangiectasia), CHEK2 (mutations in this gene can increase the liability to BC by approximately twofold), STK11 (mutations in this gene can result in Peutz–Jeghers syndrome with an increased susceptibility to several other types of cancer in addition to BC), CDH1 (these genetic mutations can result in the genetic diffuse gastric cancer with high predisposition to the invasive lobular BC), and finally PALB2 (this gene generates a protein that binds with the protein made by the BRCA2 gene, consequently resulting in mutations in this gene with a higher risk of BC) [25].

Genetic screening for mutations in genes like BRCA1, BRCA2, PTEN, and TP53 is beneficial for timely breast cancer (BC) detection and prevention in high-risk women. However, it is crucial to understand the limitations and potential drawbacks of these tests. Furthermore, the high cost of genetic testing means it is not universally applicable, and its practical use should be weighed carefully as not all women require this screening [25].

Moreover, in the context of the Indian scenario, studies indicate that Indian BC patients have a considerably higher frequency and prevalence of *BRCA1* and *BRCA2* mutations when compared to their Western counterparts, with one study indicating a prevalence of 29.1%. However, among the Western populations, the incidence of these mutations is normally between 5 and 10%. Additionally, although BRCA mutations are linked to early-stage BC worldwide, the association is particularly prominent in India, as studies demonstrate a substantial connection between being BRCA-positive and the occurrence of BC in younger women ranging between 20 and 45 years of age [26].

The impact of susceptibility genes like *TP53*, *PTEN*, *CDH1*, *ATM*, *CHEK2*, and *PALB2* in India is indeed substantial for evaluating cancer risk, screening, therapeutic management, and genetic counseling. However, the thorough detailed information tailored to the Indian population is still emerging. Nevertheless, due to India's genetic and socioeconomic variability, the incidence, expression, and the relationships between the genotype and phenotype for these genes may differ from those in the Western population [27].

However, with respect to risk assessment and penetrance, studies suggest that the range of mutations in these genes among the Indian population may show distinct disparities when compared to their Western counterparts. For example, a study in the Kashmir valley discovered a high prevalence of *TP53* mutations in patients with colon cancer, indicating that it is a major contributor to this population's elevated risk [28]. Similarly, penetrance or the risk of developing cancer because of a mutation may vary depending on the particular variant, or other genetic or environmental factors. The range of malignancies in families with Li-Fraumeni syndrome (TP53 mutation) can differ from Western cohorts, with a potentially increased occurrence of particular forms such as sarcomas [29].

This aside, numerous research studies have discovered that the mutation spectrum of cancer susceptibility genes in the Indian population is quite diverse, with notable geographical heterogeneity. In addition, although mutations in BRCA1 and BRCA2 are the most frequent, other genes are also important contributors. Germline pathogenic mutations have a higher incidence and are frequently associated with early-onset cancer.

According to studies, the comprehensive incidence of pathogenic *BRCA1/2* mutations in high-risk Indian breast and ovarian cancer patients is between 20% and 30%, which is greater than that found in Western populations. In some cohorts, *BRCA1* mutations are more common than *BRCA2*, which contrasts with many other Asian ethnicities. This aligns with India's high prevalence of triple-negative breast cancer (TNBC), and early-onset breast cancer, which are frequently associated with *BRCA1* [26,30].

This aside, numerous recurring mutations (as discussed below) have been reported, with diverse frequency across regions.

1. *c.68_69delAG (185delAG)* in *BRCA1*: This mutation, which is a founder mutation in Ashkenazi Jewish communities is recurrent and seems to have originated independently among the South and West Indian populations.
2. *c.1961delA* in *BRCA1*: This is one of the most common mutations discovered in current studies.
3. *c.5074+1G>A* in *BRCA1*: This is most common among West Indian and other populations.
4. *c.5382insC* in *BRCA1*: This is another frequently occurring pathogenic mutation.

Similarly, numerous studies among the Indian population have reported novel *BRCA1/2* mutations not previously cataloged in databases like ClinVar, thus emphasizing the distinct genetic diversity of the Indian subcontinent [26].

Indeed, it will be important to address the issue of specific founder mutations for BC among the Indian population, especially in the *BRCA1* and *BRCA2* genes, that are associated with hereditary breast and ovarian cancer (HBOC) syndrome. India, with its diversified population and endogamous groupings, has distinct regional founder mutations that influence genetic screening and counseling procedures. Founder mutations are essentially genetic changes that occur or originate in a founding (initial) individual and are inherited by their offspring, gradually becoming increasingly prevalent in particular groups or populations that have stayed isolated owing to geographic or cultural reasons. Moreover, owing to the large and genetically diverse population of India, the distribution of BRCA1/2 mutations varies greatly by location and ethnicity. Additionally, the traditional practice of endogamy in certain communities has led to an increase in the frequency of these founder mutations and heritable conditions.

Some of the crucial insights with regard to the founder mutations in India include the following:

1. **Shared mutation (*185delAG*):** The *BRCA1 185delAG* (as mentioned earlier) founder mutation, which is widely linked to the Ashkenazi Jewish population, can also be detected in certain areas of India, especially in the south and west. Indeed, studies indicate that this emerged or originated independently within the South Indian population, instead of originating from a common lineage with Ashkenazi Jews [31].
2. **Regional heterogeneity:** Studies have found a wide range of *BRCA1/2* variants among various Indian populations, showing that the types of mutations vary by geographical locations. This indeed highlights the importance of developing demographically specific genetic profiles [26].
3. **Higher frequency of mutations:** Some studies in India have discovered an increased overall frequency of *BRCA1/2* mutations among breast and ovarian cancer patients when compared to their Western counterparts [26,32].

Additionally, in India, population-specific founder mutations and other socioeconomic factors have a significant impact on genetic screening strategies and counseling, as discussed below:

1. Targeted sequencing–next-generation sequencing (NGS): Instead of expensive, laborious, and lengthy whole gene sequencing, the testing or screening strategies can be highlighted on the most frequent founder mutations that are found in specific regional

or ethnic populations. This strategy provides a faster and less expensive screening tool for high-risk people.

2. Custom-designed gene panel (personalized genetic test): Genetic testing or screening can be performed with panels tailored to contain common founder mutations unique to the Indian population, along with additional pertinent genes associated with high and moderate risks.
3. Overcoming barriers to access: Customized or personalized testing methods can assist in tackling the significant obstacles pertaining to the expense and availability of genetic tests in India.

5. Implication of Non-Genetic Risk Factors for Breast Cancer (Figure 5)

5.1. Familial History of Breast Cancer

Although 15% or less women with BC include a familial member who has the disease, women who have close family members who possess the disease are at an immense risk. Especially given this circumstance, having a first-degree relative like mother, sister, or daughter with BC, for instance, nearly increases the risk by twofold. And, on the other hand, having two first-degree relatives with the disease increases the risk threefold [23,24]. Intriguingly, women who have a father or brother who has BC are more likely to develop the disease themselves. Adding to this, and at an individual level, a woman who has cancer in one breast is more likely to develop cancer in the other breast or another part of the same breast [25].

Familial history and genetic predisposition are more likely to found among the younger BC patients in India. Typically, the average age at which they are diagnosed is about ten years younger than that of patients in Western countries, thus highlighting the connection between genetic predispositions and the early onset of BC, and consequently assessing the importance of genetic counseling [33]. In fact, 19% of women under forty had a first- or second-degree relative who has cancer. However, the actual frequency of inherited breast cancer in India is probably miscalculated. Numerous BC disease manifestations are reported that are not linked to inherited genes or familial predisposition, indicating that familial history-based screening may overlook a large percentage of mutation carriers [26].

5.2. Proliferative Breast Lesions (PBLs)

Certain proliferating lesions with the absence of atypical hyperplasia has demonstrated to increase a woman's susceptibility to BC [23,24]. These include fibroadenoma, sclerosing adenosis, ductal hyperplasia, papillomatosis, and radial scar. Nonetheless, certain proliferative lesions with atypical hyperplasia in the ducts or lobules of the breast tissue will boost the susceptibility to BC by 4–5 fold which comprises the atypical ductal hyperplasia (ADH) and atypical lobular hyperplasia (ALH) [34].

However, studies among the Indian population suggest that proliferative breast lesions are substantial contributing factors to BC, which increasingly aggravates young Indian women more than their Western counterparts. In fact, the diagnosis of PBL indicates the necessity for improved screening. Given this severity among the Indian population, the association between PBLs and breast cancer is especially important because of certain traits in the population such as early onset, lifestyle, and hereditary factors. In addition, owing to this higher risk, studies in India recommend that women with proliferative lesions receive more comprehensive follow-up, like clinical breast examinations and mammography screening. In addition, given the greater likelihood of advanced-stage diagnoses in India, detecting women with PBLs is a fundamental strategy for early diagnosis [1,35].

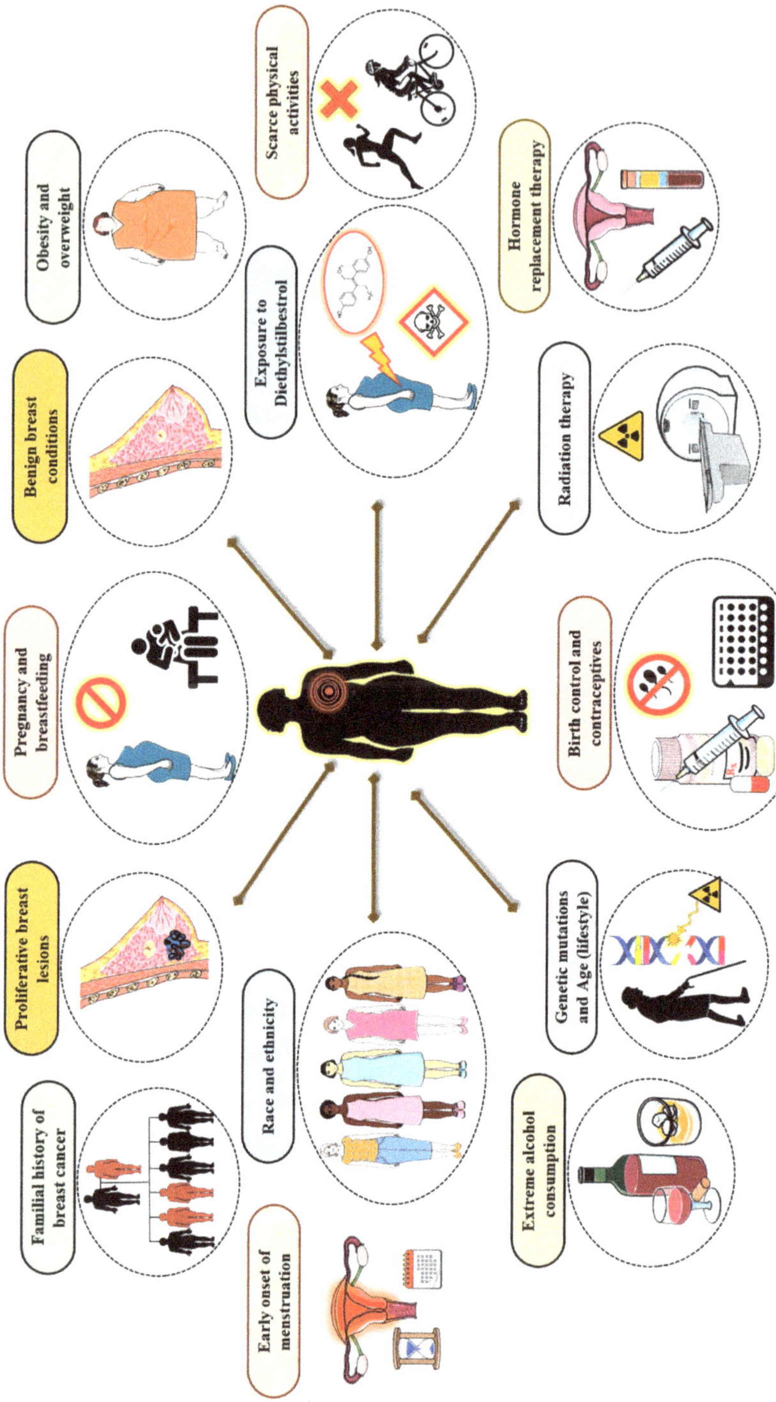

Figure 5. Implication of non-genetic factors in breast cancer.

5.3. Lack of Pregnancy and Breastfeeding

To some extent, women who have never had children or who have had their first child at an age later than 30 years old have an increased risk of BC by and large. Having multiple pregnancies and/or conceiving at an early age, on the other hand, lowers the susceptibility to BC [23,24]. However, pregnancy appears to affect different types of BC in a dissimilar manner, and it in fact appears to increase the risk of triple-negative breast cancer. Breastfeeding may perhaps reduce the risk of BC to some extent, especially if it is carried on for 1.5–2 years. One possible interpretation for this conclusion is that breastfeeding decreases a woman's aggregative number of menstrual cycles over her lifetime.

In India, a lack of pregnancy and breastfeeding, often referred to as nulliparity and never having breastfed, is a substantial and growing risk factor for BC. Although typically Indian women once had timely and numerous pregnancies and breastfed for prolonged durations, these trends are evolving, particularly among the urban, educated population. However, as India undergoes a demographic shift, the prevalence of cancer is rising. In fact, BC has exceeded cervical cancer as the most prevalent malignancy among Indian women predominantly in metropolitan areas. The upsurge in the occurrence of nulliparity and decline of breastfeeding can be attributed as important drivers of this trend, pushing the age of diagnosis to a younger demographic than in Western countries [1].

5.4. Benign Breast Conditions

Women with dense and solid mammograms have an approximately a 1.5- to 2-fold higher susceptibility to BC than women with regular breast density. However, several factors like age, the use of certain medications like menopausal hormone therapy, menopausal status, and pregnancy are found to be involved in defining breast density. This aside, a few non-proliferative lesions may have a minor effect on BC susceptibility. This may comprise fibrosis and/or simple cysts, single papilloma, duct ectasia, mild hyperplasia, adenosis, phyllodes tumor, periductal fibrosis, squamous and apocrine metaplasia, epithelial-related calcifications, and other tumors including lipoma, hamartoma, hemangioma, neurofibroma, adenomyoepithelioma, and mastitis [36].

In the Indian scenario, fibroadenomas and fibrocystic changes are the most frequent benign breast diseases mostly affecting the younger women. Moreover, although the majority of benign conditions pose little to no risk of developing cancer, certain kinds have a marginally increased risk, particularly when combined with other risk factors. Therefore, to differentiate between benign lesions and BC, it is necessary to have an early and precise diagnosis via biopsy, imaging, and clinical testing [37].

5.5. Obesity and Being Overweight

Prior to the onset of menopause, the ovaries produce the majority of the body's estrogen, while the fat tissue produces only a small amount. Nonetheless, when a woman's ovaries cease to produce estrogen after menopause, the majority of her estrogen is derived from the fat tissue. Therefore, harboring extra fat tissue after menopause increases estrogen levels which subsequently escalates the susceptibility to BC [23,24]. Additionally, being overweight tends to result in increased insulin levels in the blood. In turn, the elevated levels of insulin are associated with certain types of cancer, including BC. However, the relationship between weight and susceptibility to BC is complex and not yet completely interpreted.

However, given the severity among the Indian population, obesity and being overweight are intricately linked to BC risk, prognosis, and survival, which differs depending on menopausal state and body fat distribution. High body mass index (BMI) is a substantial risk factor for postmenopausal breast cancer, although central obesity is a major risk factor for both premenopausal and postmenopausal breast cancers. And, in a broader context, owing to the lifestyle modifications, the growing number of BC cases in India can be somewhat linked to trend towards sedentary lifestyle and dietary changes, leading to a rise in the average body mass index [38].

5.6. Early Onset of Menstruation

Women who begin menstruating early, especially prior to the age of 12, will end up having more menstrual cycles which consequently results in a prolonged lifetime exposure to the hormones like estrogen and progesterone, which, to some extent, increase the susceptibility to BC [23,24]. Likewise, women who reach menopause later, especially after the age of 55, will have more menstrual cycles and a longer lifetime exposure to estrogen and progesterone, respectively, which again increases their susceptibility to BC.

Early menstruation (menarche) is a major risk factor for BC in the Indian population. This link is mostly attributed to a woman's prolonged or chronic exposure to hormones such as estrogen and progesterone, which promote breast cell proliferation and raise the risk of cancer. In addition, the landscape of BC in India, characterized by its early onset compared to in Western nations, higher rates of aggressive subtypes like triple-negative BC (TNBC), and substantial discrepancies in detection and treatment, along with lifestyle modifications, contributes its major role [1].

5.7. Race and Ethnicity

On the whole, Caucasian women are to some extent more inclined to develop BC than their African American counterparts, despite the fact that BC is more frequent among the African American women under the of age 45. In addition, they are more prone to die from BC at any age. On the other hand, women from other races, such as those with an Asian, Hispanic, or Native American background, are at a lower risk of developing and dying from BC [23,24].

More specifically, when compared to Western populations, Indian women have a higher prevalence of more aggressive, hormone receptor-negative cancers and are detected at a much younger age. The explanations for these disparities are complex, including genetic predispositions, lifestyle modifications, and socioeconomic background, which contribute to late-stage diagnosis and lower survival rates. Adding to this body of data is a unique clinical presentation with distinct biological characteristics among the Indian BC population. These include early onset, a higher incidence of more invasive tumors (such as TNBC and HER2-positive tumors), and late-stage diagnosis. Such factors make the Indian BC population's profile more complicated than that of its Western counterparts [1].

5.8. Exposure to Environmental Risk Factors

From the 1940s to the early 1970s, a few pregnant women were prescribed to the estrogen-like drug diethylstilbestrol (DES) which was believed to reduce the incidence of miscarriage [23,24]. These women, to some extent, have an increased risk of BC, and women whose mothers used DES during pregnancy possibly will have a considerably increased risk.

On the other hand, environmental and cultural risk factors in India, together with rising urbanization, confers a substantial impact on BC trends and effects. Indeed, these factors lead to its increased prevalence, with early onset compared to in Western populations, and with diagnostic delays resulting in higher fatality rates [39]. This aside, an emerging risk factor for BC among the Indian population is exposure to pollutants like organochlorine pesticides. This is supported by a case–control study in an Indian population, which found that women with BC had considerably higher serum levels of estrogenic OCPs, including

DDT metabolites (DDE) and hexachlorocyclohexane isomers (HCH), compared to healthy control subjects [18].

Similarly, exposure to fine particulate matter like $PM_{2.5}$, has been linked to an increased risk of breast cancer. According to studies, these small particles can enter the bloodstream and permeate into breast tissue, potentially leading to tumor formation [40]. Likewise, another study conducted among the women from northeast India, found that chewing betel quid was a substantial and independent risk factor for the onset of BC, possibly prompting genetic modifications [41].

Other emerging environmental pollutants like phthalates, bisphenol A (BPA), and polyfluoroalkyl substances (PFASs) are being studied for their potential link to breast cancer. The possible impact of these chemicals that interfere with hormones on the progression of BC is still being investigated [42].

5.9. Scarce Physical Activities

An emerging body of evidence and documentation suggests that routine exercise and physical workouts, particularly in women after menopause, could possibly decrease the risk of breast cancer [23,24]. However, it is unclear as to how physical activity may reduce risk of BC, but it could be because of activity levels influencing body weight, inflammation, hormones, and energy balance.

In the Indian scenario, a lack of physical exercise is a key, alterable factor, leading to the rising incidence of BC, especially in younger urban women. This aside, sedentary lifestyle raises the risk of BC owing to its association with obesity, hormonal disruptions, and inflammation, all of which are proven risk factors that contribute to the illness [1].

5.10. Extreme Alcohol Consumption

Consumption of alcohol is undeniably associated with an increased susceptibility to BC, and this increased risk or susceptibility is proportional to the amount of alcohol consumed. For instance, women who consume two to three drinks per day, have a 20% higher risk of BC than women who do not consume alcohol. And, on the other hand, women who consume only one alcoholic drink per day have a minor increased susceptibility [23,24].

High levels of alcohol intake raise the likelihood of BC among Indian women, reflecting global trends. However, unique lifestyle and genetic factors in the population can lead to different risk factors. In addition, although Indian women have a traditionally lower rate of alcohol consumption than their Western counterparts do, the potential for a negative outcome is growing with the advent of urbanization and "westernized" lifestyles [1].

5.11. Genetic Mutations and Age

An extensive proportion of breast cancers is detected in women with no clear familial history. Furthermore, rather than inherited mutations, these cancers may perhaps be brought about by genetic mutations that appear as a result of the aging process and lifestyle-linked risk factors.

In the Indian population, the effects of genetic mutations and age interact significantly, ensuing in a higher prevalence of BC in younger women, with an increased proportion of aggressive subtypes, and a distinct range of genetic mutations compared to Western populations. More specifically, genetic mutations in the *BRCA1* and *BRCA2* genes are major contributing factors for the rise in aggressive and early-onset BC cases in India. The prevalence of these mutations, along with diverse mutation patterns, is significantly higher in India than in Western countries [26].

5.12. Birth Control and Contraceptives

Several birth control approaches contain hormones that can increase the risk of BC [23,24]. In addition, women who use oral contraceptives have a higher risk of BC to some extent than women who have never used them. However, the risk appears to return to normal after the regimen is discontinued. To ascertain this, Depo-Provera, administered as an injectable form of progesterone, has been reported to boost the risk of BC; nevertheless, there was no increased risk in women five years after they stopped receiving the shots.

On the other hand, hormonal contraceptives are linked to a minor but considerable increased risk of BC in Indian women, particularly with long-term use. However, this risk should be examined along with other risk factors prevalent in the Indian population, like being of an advanced maternal age at first childbirth and socioeconomic status. Indeed, some Indian studies have indicated a link between the usage of hormonal contraceptives, especially oral contraceptive pills (OCPs) and an increased risk of BC [43,44].

5.13. Chest Radiation Therapy

Women who have had radiation therapy to the chest for another type of cancer at a young age are more likely to develop BC [23,24]. The effect of this factor is ameliorated in case the individual had received radiation as a teen or young adult, when the breasts were still developing. On the other hand, radiation therapy after the age of 40 did not appear to increase susceptibility to BC.

For Indian BC patients, chest radiation therapy carries the same potential risks as it does to other populations, including heart and lung damage, and the development of a secondary malignancy. However, the overall impact and absolute risk of these complications are intensified by underlying factors more prevalent in India, such as increased baseline cardiovascular risk, late-stage diagnosis, restricted access to modern treatment methods, and major financial constraints [45].

5.14. Hormone Replacement Therapy After Menopause

The hormone estrogen (often in combination with progesterone) has been administered to treat menopausal symptoms and thereupon to prevent osteoporosis [23,24]. Most often, combined hormone therapy is required as the usage of estrogen alone can aggravate the risk of uterine cancer. Nevertheless, estrogen can be used alone by women who have had a hysterectomy. This aside, postmenopausal integrated hormone therapy raises the risk of BC, the likelihood of dying from BC, and the chances of detecting cancer only at a more progressive stage subsequently. This increased risk is typically prominent after 2 years of usage. Nonetheless, the increased risk from concomitant or combined HRT is reversible, and it is only relevant to current and recent users. This is because a woman's BC risk appears to revert to that of the general population after five years of discontinuing HRT. Additionally, the use of bioidentical or "natural" estrogen and/or progesterone is not always safer or more effective and should be regarded as having the same health risks as any other type of HRT. This aside, short-term estrogen usage after menopause does not appear to significantly increase the risk of BC. Yet, long-term treatment (>15 years) has been linked to an increased risk of ovarian and BC, respectively. Therefore, it is up to a woman and her doctor to decide as to whether to use any form of HRT after deliberating the potential risks factors and benefits, and also taking into account other risk factors pertaining to heart disease, BC, and osteoporosis, which occur thereafter.

On the other hand, hormone replacement therapy following menopause offers a complicated risk–benefit scenario for Indian women, which is determined by the type of therapy, the duration it is taken, and the patient's specific health considerations. Especially for Indian women, who reach menopause at a younger age (46 years) than Caucasian women, the necessity of HRT for symptom relief must be balanced against the long-term risks. An Indian Menopause Society study calls for customized risk assessment, highlighting that the advantages may outweigh the modest absolute risks for healthy women in the first decade after menopause [46].

6. Risk Assessment: Breast Cancer Stratification by Way of Pathogenesis, Incidence and Invasiveness

Breast cancers are classified into several types based on where they appear in the breast, for instance in the ducts, lobules, or the tissue in between. The specific cells that are affected define the type of breast cancer. Therefore, they are classified into two distinct types, viz., carcinomas and sarcomas, based on the origin of the cells involved. Principally, carcinomas are breast cancers that develop from the epithelial component of the breast, which includes the cells that outline the lobules and terminal ducts that produce milk. Sarcomas on the other hand, are a much rarer type of breast cancer that develops from the stromal components of the breast, which include myofibroblasts and blood vessel cells (Figure 6). These categories are not always necessary because a single breast tumor can be a mixture of different cell types in some cases [25].

However, the majority of breast cancers are carcinomas. Several different types of breast cancer have been described within the large group of carcinomas depending on their invasiveness corresponding to the primary tumor sites. It is indeed critical to be able to differentiate between the different subtypes since they each have diverse prognoses and treatment implications. Therefore, typically, breast cancers are classified, based on their corresponding pathological features and invasiveness, into three distinct groups: non-invasive (or in situ), invasive, and metastatic breast cancers [25].

6.1. Non-Invasive (In Situ) Breast Cancer

6.1.1. Ductal Carcinoma In Situ (Intraductal Carcinoma; DCIS)

DCIS is a non-invasive or pre-invasive BC that thrives inside from the normal pre-existing ducts, and is one of the most typical types of breast cancer [25]. And, despite the fact that DCIS is not invasive, the in situ carcinomas have a high risk of developing into invasive cancers, and therefore early and appropriate therapeutic regimen is vital in intercepting the patient from developing an invasive cancer.

6.1.2. Invasive/Infiltrating Breast Cancer

Primarily, the cancer cells in invasive breast cancers invade and propagate farther outside of the normal breast lobules and ducts, gradually moving into the adjoining breast stromal tissue [25]. And, when diagnosed with an invasive form of BC, approximately two-thirds of women are 55 or older. In addition, invasive carcinomas have the possibility to progress to other locations of the body, like the lymph nodes or other organs, and form metastases, thus classifying them as metastatic breast cancers [25]. Additionally, the invasive breast cancers are further classified into two types based on the tissue and cell types involved as follows:

6.1.3. Invasive Ductal Carcinoma (IDC)

IDC is the most common type of BC, accounting for approximately 80% of all breast cancers [25]. The IDC group of classification comprises numerous subtypes that include the tubular, medullary, mucinous, papillary, and cribriform carcinomas of the breast [25].

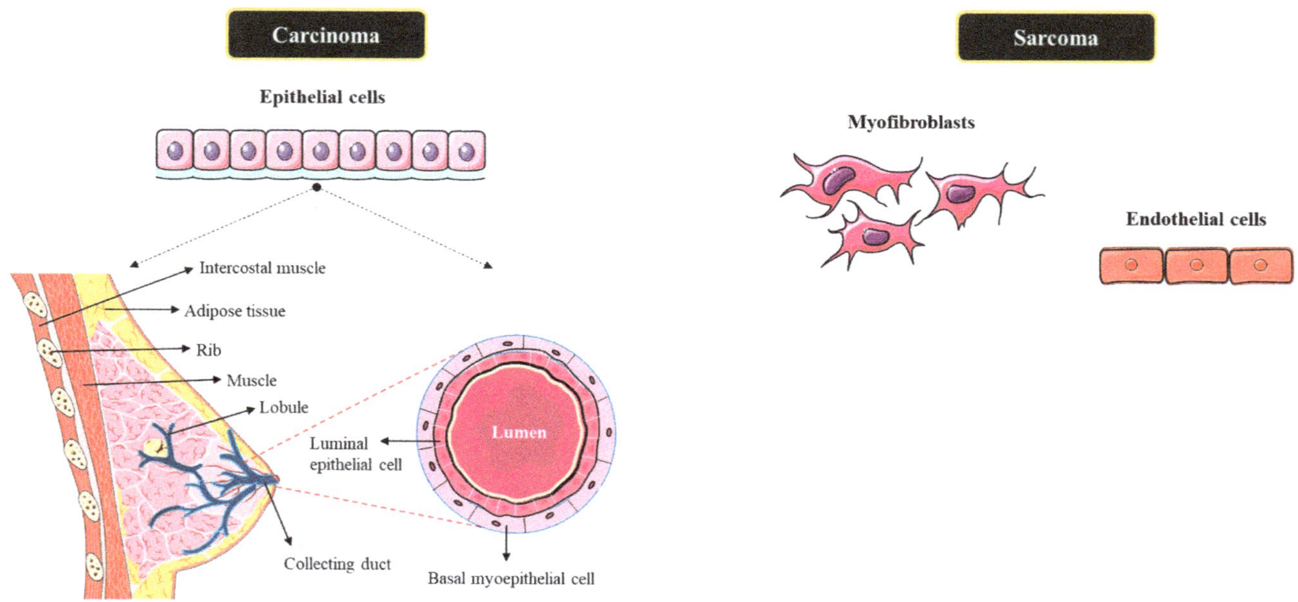

Figure 6. Breast cancer stratification based on specific cell types.

6.1.4. Invasive Lobular Carcinoma (ILC)

ILC is the second most common type of BC, accounting for about 10–15% of all cases [25]. While this subtype of carcinoma can affect women of any age, it is more prevalent in women over the age of 50. Additionally, the invasive lobular carcinoma occur later in life when compared to the invasive ductal carcinoma—for instance, in the early 60s in contrast to the mid-to-late 50s for IDC [25].

Overall, 90–95% of all breast cancers fall into the invasive subcategory. In addition, the invasive ductal carcinoma and the invasive lobular carcinoma have different pathological features. Lobular carcinomas develop as single cells organized individually, in single file, or in sheets, and exhibit distinct molecular and genetic abnormalities. Additionally, both the subtypes distinguish from each other in terms of prognosis and therapeutic regimen (Figure 7).

6.2. Metastatic Breast Cancer

Metastatic breast cancers are late-stage breast cancers that have spread or progressed to other organs in the body. They are also known as stage IV or advanced breast cancers [25]. Breast cancers can spread to lymph nodes in the armpit and/or distant sites like the lung, liver, bone, and brain. Even after eliminating the prime vital tumor, microscopic tumor cells or micro-metastases may still linger around and prevail in the body, thereby enabling the cancer to recur and spread. Clinically, patients may be diagnosed with metastatic disease (or de novo metastatic BC) at the start of treatment, or they may develop metastases months or years later. The prospective likelihood of BC recurrence and metastasis is not well interpreted or foreseen since it differs from person to person, owing to the tumor's unique molecular biology and stage at the time of diagnosis. Unfortunately, given the severity, about 30% of women with early-stage BC will develop a metastatic form of the disease (Figure 7) [25].

6.3. Less Common Types of Breast Cancer

Although IDCs and ILCs account for approximately 90–95% of all BC cases, numerous rare or less common types of BC can be detected in the clinical setting [25].

6.3.1. Inflammatory Breast Cancers (IBCs)

IBC is a rare type of invasive BC that accounts for 1–5% of all breast cancers [25]. The inflammatory breast cancers differ from other group of BC in terms of their manifestations, prognosis, and therapeutic regimen. The characteristic symptoms or manifestations of IBC comprises an inflammation-like breast lump or swelling, purple or red skin discoloration, and the pitting or thickening of the breast skin. All of these are possibly triggered by cancer cells blocking the lymph vessels in the skin. However, IBC is not always associated with a breast lump and may not be detected on mammograms. This aside, IBC is more common among younger women and especially the women of African American origin, along with women who are overweight or obese. In addition, it inclines to be more aggressive than other types of BC, growing and metastasizing much more rapidly. It is invariably detected at a locally progressive stage, when the breast cancer cells have spread into the skin. And, in approximately one-third of cases, the IBCs have been detected already metastasized to distant sites of the body, thereby complicating and making treatment more difficult [25].

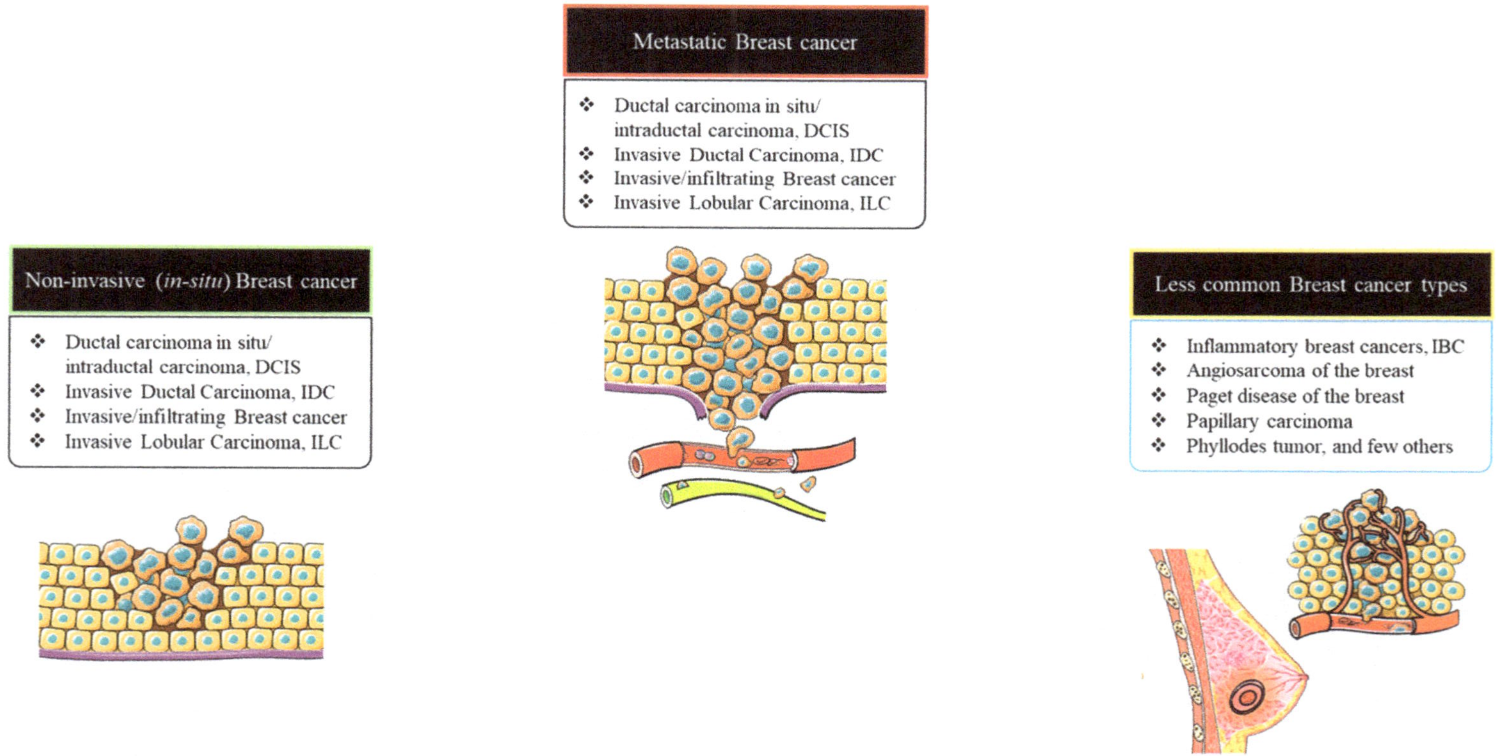

Figure 7. Breast cancer stratification based on pathogenesis and invasiveness.

6.3.2. Angiosarcoma of the Breast

Angiosarcoma is a class of sarcoma that is a very unusual form of cancer that develops from the epithelial cells that line blood or lymph vessels [25]. These angiosarcomas account for 1% of all breast cancer cases and can affect either the breast tissue or the breast skin. A few cases stem as a result of previous radiation therapy in that area. Though uncommon, angiosarcomas grow and spread rapidly and should be treated appropriately.

6.3.3. Paget Disease of the Breast

This uncommon type of BC initiates in the breast ducts, extends to the nipple skin, and then gradually spreads to the areola (the dark circle around the nipple) [25]. Paget disease accounts for 3% of all BC cases and, in addition, the Paget's cells are visually distinct from normal cells and divide rapidly. Around half of the cells express estrogen and progesterone receptors, and the majority express the HER2 protein. Predominantly, a biopsy of the tissue is usually used to diagnose cancer, and this is occasionally trailed by a mammogram, sonogram, or MRI to verify or validate the diagnosis. It is also worth noting that Paget's disease of the breast is not medically related to other clinical settings named after Sir James Paget, such as Paget's disease of the bone.

6.3.4. Papillary Carcinoma

This is another rare form of BC, accounting for less than 3% of all breast cancers [25]. Usually, the cells in papillary carcinoma cancer are typically arranged in finger-like projections. To illustrate further, in certain cases, the cancer cells are relatively minute in dimension and apparently form micropapillary structures. The majority of the papillary carcinomas are invasive, and are treated similarly to IDCs. Nevertheless, the invasive papillary carcinoma typically has a better prognosis than other types of invasive breast cancer. It is worth noting that papillary carcinomas can be identified while still non-invasive. Usually, a non-invasive papillary carcinoma is commonly thought to be a type of DCIS.

6.3.5. Phyllodes Tumor

This is a rare breast tumor that establishes itself in the breast's stromal cells [25]. The majority of these tumors are benign, but approximately one-quarter are malignant. Phyllodes tumors are most common in women in their 40s, as well as in women with Li-Fraumeni syndrome, who are predisposed to this type of tumor.

6.3.6. Breast Cancers in Men, Children, and Juveniles

The male breast carcinomas can be in situ or invasive, and makes up less than 1% of all cases of breast cancer [25]. They can almost resemble as those observed in women, since the majority of cases are invasive ductal carcinomas with estrogen receptor (ER) expression. Gynecomastia (or breast enlargement) is the most common breast lesion in men, and it can affect either the unilateral or bilateral breasts. Likewise, breast lesions are rare in children and adolescents, but they do happen. These rare cases may comprise both benign and malignant lesions, such as juvenile fibro adenoma and malignant carcinoma. On the other hand, pediatric patients have more possibility to develop lymphoma or alveolar rhabdomyosarcoma that has spread or metastasized to the breast.

7. Molecular or Intrinsic Subtyping/Classification of Breast Cancer

Breast cancer entails a complex and phenotypically diverse group of diseases and consists of many biological subtypes with specific mechanisms and therapeutic responses. To elucidate further, gene expression studies have determined numerous specific subtypes of BC that vary notably in prognosis along with the therapeutic targets that exist in the cancer cell types [25]. With the advent of gene expression profiling strategies, the list of intrinsic genes that distinguish these subtypes includes several groups of genes linked with the human epidermal growth factor 2 (HER2) expression, proliferation, the estrogen receptor (ER) expression (the luminal cluster), and a distinct group of genes termed the basal cluster [25]. Accordingly, the BC subtypes are typically classified into five intrinsic or molecular subtypes established on the expression composition or scheme of specific genes using these interpretations (Figure 8).

7.1. Luminal A Breast Cancer

This subtype is essentially positive for the estrogen receptor (ER) and/or progesterone receptor (PR), and negative for HER2. Luminal A cancers make up roughly 40% of all BCs and are of low grade, growing slowly, and have the best prognosis. The treatment strategy usually includes hormonal therapy [25].

Luminal A breast cancer is defined by high estrogen receptor (ER) expression, indicating its dependence on hormonal signaling. While endocrine dependence is prominent, other involved pathways can lead to resistance if disrupted. This subtype exhibits minimal Ki-67 protein levels, suggesting a slow proliferation rate and a favorable prognosis compared to other subtypes. Key genes linked to luminal epithelial cells, such as GATA3 and FOXA1, function as transcription factors for ER alpha and are significantly expressed, playing a vital role in mammary gland differentiation [47]. In addition, based on integrated genomic analysis, common mutations in the luminal A subtype include the following:

1. ***PIK3CA***: Activating mutations are commonly seen in the PI3K/AKT/mTOR pathway.
2. ***GATA3* and *MAP3K1***: Mutations or alterations in these transcription factors are also frequent.
3. ***TP53***: Although not as common as in other subtypes, mutations or alterations in the tumor suppressor gene, TP53, are detected in certain luminal A tumors.

7.2. Luminal B Breast Cancer

This subtype accounts for <20% of all BCs and is ER and/or PR positive, HER2 positive or HER2 negative, and has high Ki-67 levels. Luminal B cancers grow somewhat quicker than luminal A cancers and have a slightly worse prognosis [25].

For patients diagnosed with the intrinsic luminal B subtype of breast cancer, the vital molecular mechanisms and signaling pathways are usually distinguished by an elevated proliferative rate along with increased genomic instability when compared to the luminal A subtype. In addition, despite the fact that these subtypes are estrogen receptor-positive (ER+), these tumors frequently use substitute growth pathways, resulting in a more aggressive phenotype and greater rates of endocrine therapy resistance. Similarly, these subtypes have a higher level of somatic mutations, gene copy number alterations (CNAs), and DNA methylation abnormalities than the luminal A subtype. Moreover, these extensive genetic alterations lead to an increased tumor grade with poor prognosis. Likewise, the tumor suppressor gene *TP53* is more commonly altered in luminal B malignancies. In addition, deactivating this gene impairs cell cycle checkpoints and DNA repair, resulting in unregulated cell division with increased genomic instability [48]. This aside, luminal tumors frequently harbor somatic mutations in the *PIK3CA* gene, which encodes a PI3K subunit. However, the frequency is decreased in luminal B compared to luminal A, indicating that additional factors influence its activation. On the other hand, the deletion or lack of expression of the tumor suppressor gene *PTEN*, which serves as a negative regulator of the *PI3K* pathway, may lead to an increased pathway activity. Similarly, the Akt and mTOR proteins are activated downstream of *PI3K*, triggering cell growth and proliferation, while also increasing resistance to endocrine therapy [49].

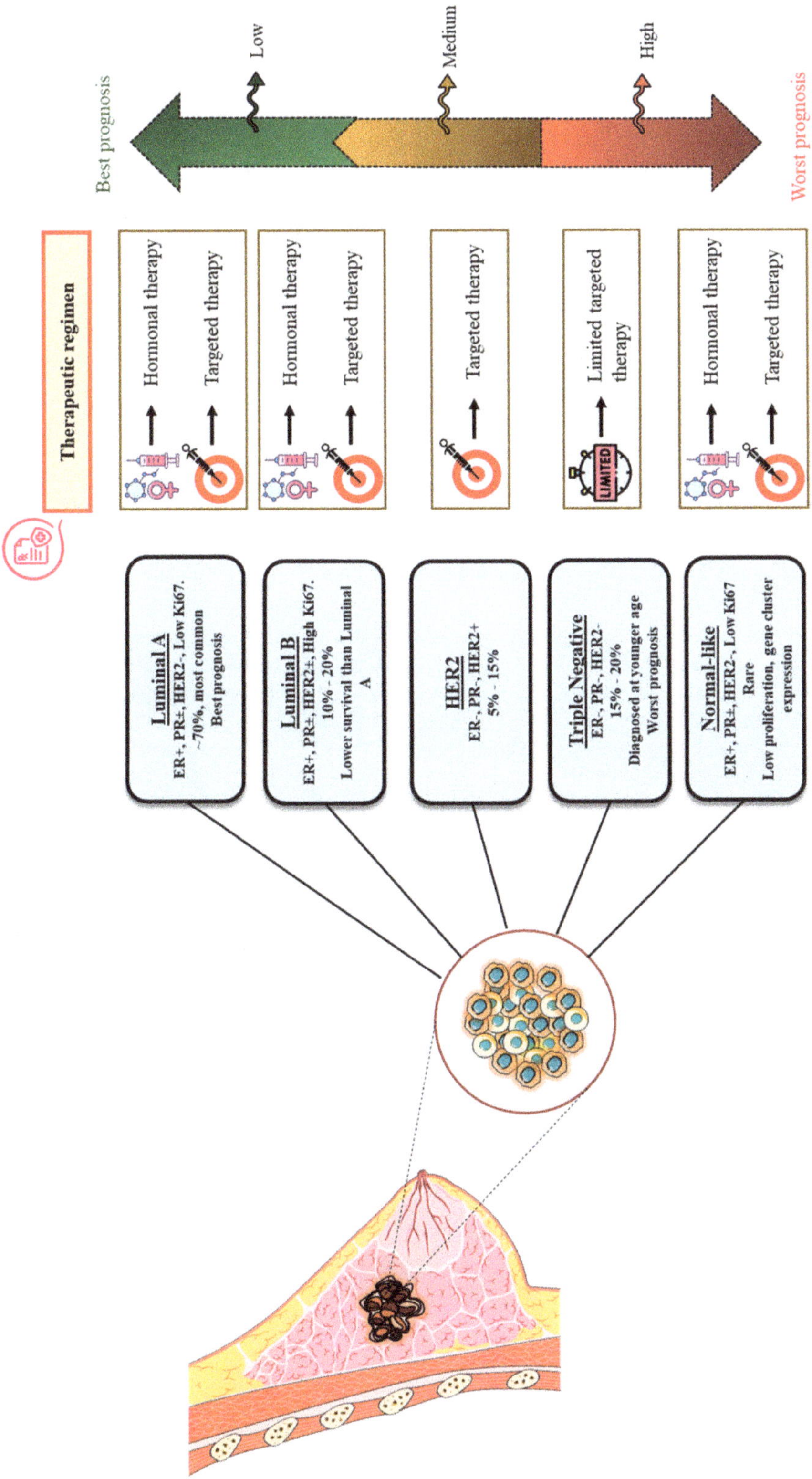

Figure 8. Molecular or intrinsic subtyping/classification of breast cancer.

7.3. HER2-Enriched Breast Cancer

This subtype accounts for 10–15% of all BCs and is defined by the lack of ER and PR expression, high HER2 and proliferation gene cluster expression, and sparse expression of the luminal and basal clusters [25]. This aside, the HER2-supplemented cancers thrive more rapidly than luminal cancers and have a poor prognosis in general. Nevertheless, they can be effectively managed with HER2-targeted therapies like Tykerb (or lapatinib), Herceptin (or trastuzumab), Kadcyla (or T-DM1 or ado-trastuzumab emtansine), and Perjeta (or pertuzumab). Additionaaly, it should be noted that the HER2-supplemented subtype does not imply clinically HER2-positive BC. Despite the fact that approximately 50% of clinical HER2-positive BCs are HER2 supplemented, the residual 50% can be of any molecular subtype but are predominantly HER2-positive luminal subtypes. Yet, approximately 30% of HER2-enriched tumors are clinically HER2-negative [25].

The main molecular mechanism underlying the HER2-enriched (HER2-E) intrinsic subtype of BC is the overexpression or amplification of the *ERBB2* gene, which encodes the HER2 receptor. As a result, there is an overactivation of the HER2 signaling pathway, which triggers a number of downstream cascade of events that regulate uncontrolled cell proliferation, growth, and survival. The primary mechanism driving HER2-E BC is the occurrence of too many copies of the ERBB2 gene, which results in a surplus of the HER2 protein on the cell surface [50]. The dimerization of HER2 receptors stimulates a number of critical downstream signaling pathways including the following:

1. **PI3K/Akt/mTOR pathway**: This signaling pathway is one of the most commonly activated downstream of HER2 and is an important regulator of cell survival, growth, metabolism, and proliferation. Mutations in the *PIK3CA* gene, which encodes the p110α subunit of PI3K, along with the loss of the tumor suppressor gene *PTEN*, are prevalent in HER2+ BC, leading to constitutive PI3K/Akt activation and treatment resistance.
2. **MAPK/ERK pathway**: The MAPK signaling pathway is yet another important modulator of HER2 signaling, which stimulates cell proliferation. Certain metastatic HER2-positive BCs exhibit genetic modifications that trigger the MAPK pathway, including the deletion of NF1. This can lead to a "pathway switch", rendering tumors resistant to Akt obstruction, yet more reliant on and susceptible to MEK/ERK inhibition.
3. Likewise, other related signaling pathways like HIF-1, Wnt/β-catenin, and STAT signaling pathways are also known to be activated in some HER2-enriched tumors, thus contributing to the resistance of anti-HER2 therapies.

7.4. Triple-Negative/Basal-like Breast Cancer (TNBC)

Triple-negative/basal-like breast cancer are ER, PR, and HER2-negative and account for approximately 20% of all BCs [25]. TNBC is more prevalent among women with BRCA1 gene mutations, women under the age of 40, and African American women. It is a high-grade BC since it acts more destructively than other types of BC. TNBC is most commonly diagnosed with infiltrating ductal carcinoma, though an uncommon histologic subtype, medullary carcinoma, is also triple negative. Additionally, TNBC's non-surgical treatment/management has been restricted to usual chemotherapy, in contrast to other breast cancer subtypes, which have an arsenal of targeted regimens such as ER antagonists and HER2 monoclonal antibodies, until the validation of the PARP inhibitor Olaparib for BRCA1 and BRCA2 mutation carriers, who are prone to develop TNBC [25].

Likewise, TNBC is more invasive and has a greater rate of early recurrence than other forms of BC. The prognosis is often very bad, and most patients relapse five years following surgery. Not to mention, TNBC is resistant to both targeted and endocrine therapies since ER, PR, and HER2 are negatively expressed. In addition, TNBC can only be treated with a few, often ineffective therapies, and thus there is a pressing, urgent need for novel therapies [51,52].

Additionally, targeting the mechanisms that underpin genomic instability may be required for such cancers, in which chromosomal instability is a driving factor. It is still unclear how significant many genetic dependencies and anomalies are across the various forms of breast cancer. In order to unravel the essential molecular pathways and interconnected nodes, a thorough functional screening such as CRISPR-Cas9 technology coupled with deeper analyses and data integration would be needed [53–55].

Furthermore, TNBC is a diverse and aggressive BC subtype, characterized by unique molecular mechanisms and signaling cascades. Certain genetic mutations and dysregulated pathways define this subtype, influencing its proliferation, metastasis, and prognosis [56]. The landscape of TNBC is intricate and can be categorized into molecular subtypes, each presenting distinct treatment-responsive deficiencies and featuring numerous significant pathways discussed below, which play a role among them.

1. **PI3K/Akt/mTOR pathway**: The PI3K/AKT/mTOR (PAM) signaling pathway is among the most frequently modified and excessively activated in TNBC. This pathway regulates cell proliferation, survival, metabolism, and angiogenesis. Mutations in the PIK3CA gene or the deletion of the tumor suppressor PTEN are common causes of hyperactivation. This pathway may also play a role in contributing to chemoresistance and relapse [57].
2. **DNA damage response (DDR) pathways**: A considerable proportion of TNBC tumors are distinguished by deficiencies in DNA repair processes, a feature known as "BRCAness", due to its similarities to tumors with inherited BRCA1/2 mutations. This aside, mutations in *BRCA1* are especially known to be linked to the basal-like type of TNBC.
3. **Wnt/β-catenin signaling pathway**: The Wnt/β-catenin pathway is essential for embryonic development and adult tissue homeostasis, but it is abnormally active in certain malignancies, including TNBC. In TNBC, the Wnt/β-catenin pathway is frequently overactivated because of the high levels of Wnt receptors such Frizzled (FZD7) and the Wnt co-receptor LRP6. As a result, β-catenin is stabilized and subsequently moves into the nucleus.

7.5. Normal-like Breast Cancer

This subtype resembles to the luminal A disease. They are ER and/or PR positive, HER2 negative, and have low levels of protein Ki-67 [25]. While normal-like breast cancer has a noble prognosis, it is still somewhat worse than luminal breast cancer.

Due to the inherent cellular and molecular diversity of breast cancers (BCs), simultaneous investigation of multiple genetic alterations is essential [27]. Advances in next-generation genomics and transcriptomics have facilitated this, leading to the identification of numerous BC subtypes with distinct prognoses and therapeutic targets [58]. These intrinsic subtypes naturally divide into two main groups based on hormone receptor gene expression, reflecting the clinically understood biological differences between ER-positive and ER-negative cancers, which originate from different progenitor cells.

To sum up, the histopathological and molecular/intrinsic classification of BC are two diverse but complementary approaches used in understanding and categorizing BC. Indeed, each provides a distinctive perspective on the nature of the disease's manifestation. While histopathology concentrates on the microscopic features of cancer cells and tissue architecture, molecular subtyping focuses on the patterns of gene expression to categorize tumors based on their biological functions. Invasive ductal carcinoma (IDC), invasive lobular carcinoma (ILC), mucinous carcinoma, tubular carcinoma, medullary carcinoma, papillary carcinoma, and others are among the most prevalent histological forms. In addition, HER2-enriched, basal-like/triple-negative, luminal A, and luminal B are the most well-known molecular subtypes. In addition, molecular subtyping counts on the condition of important biomarkers like estrogen receptors (ER), progesterone receptors (PR), and human epidermal growth factor receptor 2 (HER2), as well as proliferation markers like Ki-67.

To elaborate, histopathological classification (e.g., ILC, IDC) examines cancer cells' microscopic organization and features, while molecular subtyping analyzes attributes like gene expression signatures to categorize tumors. Both methods have prognostic significance: specific histological subtypes like tubular carcinoma indicate favorable prognosis, while others like metaplastic cancer suggest a negative one. Similarly, the likelihood of recurrence and response to treatment vary by molecular subtype; for example, luminal A tumors generally have a better prognosis than luminal B tumors [1].

In addition, although prognostic information is provided by both the classifications, molecular subtyping might enable a more accurate evaluation of risk and possible outcomes. Similarly, while there is not always a one-to-one correlation, some histological types are more likely to be linked to particular molecular subtypes. For example, luminal A is a common presentation of invasive lobular cancer. Essentially, histopathology lays the groundwork for comprehending the visual aspects of BC, whereas molecular subtyping sheds light on the disease's underlying biology and behavior, which in turn influences prognosis and therapeutic approaches.

8. Prevalence and Clinical Significance of Molecular Subtypes in the Indian Setting

However, in the Indian context, there is a well-established correlation between the molecular subtypes of BC, particularly those defined by immunohistochemistry (IHC), their histological attributes, and their respective clinical outcomes. In addition, although there is considerable variation across studies, recent research suggests that luminal A is the most prevalent subtype, followed by triple-negative breast cancer (TNBC) and luminal B, both of which are observed at high rates in India [59–61]. Luminal A, which is frequently regarded as the most prevalent subtype among Western populations, may be less common in certain Indian studies, or differences in Ki-67 evaluation may have an impact on its diagnosis. In addition, corresponding to the global trends, invasive ductal carcinoma (IDC) is the most prevalent histological form of BC among all molecular subtypes in India. Moreover, while lower histological grades (Grade I) are often linked to luminal A cancers, higher histological grades (Grade III) are frequently associated with TNBC, HER2-enriched, and luminal B subtypes.

Moreover, in terms of lymphovascular invasion (LVI), some Indian studies suggest a strong association between molecular subtypes and LVI. While the highest prevalence of LVI has been observed in various subtypes depending on the study, it is generally considered higher in aggressive subtypes, such as HER2-positive BC and TNBC compared to luminal A tumors [59–61]. However, some studies from India demonstrate a higher rate of lymph node positivity in TNBC, in comparison to some other regions [58]. Additionally, these studies also suggest that HER2 and TNBC-enriched subtypes might present with bigger tumors than that of luminal A.

Given the current oncological landscape of India, molecular profiling is becoming more popular, since it can help guide personalized therapy choices, especially when combined with techniques like next-generation sequencing (NGS). Although promising, a number of limitations in local data solutions and access to the gene expression assays hinder its extensive and fair implementation [21].

In terms of its positive impact on treatment decisions, it undeniably helps gain a deeper insight into tumor heterogeneity. By identifying each tumor's distinct genetic composition, molecular profiling goes beyond conventional classification systems and enables a more specialized and prospective therapeutic regime. It advances personalized and precision medicine by diagnosing precise molecular subtypes and predicting treatment outcomes [62]. Oncotype DX and similar tests can predict recurrence risk, allowing clinicians to identify individuals who can safely evade chemotherapy, saving unnecessary toxicity exposure and expenses, thus refining prognosis and abating overtreatment [15]. Similarly, gene expression profiling (GEP) aids in the identification of appropriate chemotherapy regimens, predicting therapeutic response, and tracking disease progression [15]. Indeed, clinical studies or trials aimed at specific genetic profiles can accelerate the discovery and regulatory approval of novel therapies.

In addition, given the inadequacy of local data infrastructure and gene expression assay availability, current genomic classifiers, such as Oncotype DX and MammaPrint, are often exorbitant for most Indian patients. The fact that most genomic classifiers were established with data predominantly from Western populations raises questions about their applicability and efficacy within the heterogeneous Indian population [63]. Indeed, there is a significant scarcity of comprehensive and thorough research investigating the molecular heterogeneity of BC subtypes among the Indian population. In fact, ongoing RNA and DNA profiling studies on Indian patients are demonstrating distinct genetic and expression patterns when compared to data from Western populations [64]. In addition, due to India's insufficient government investment in healthcare and inadequate coverage plans, a substantial amount of healthcare spending comes directly from the patients/individuals, thus making expensive examinations like GEP unaffordable to many.

Furthermore, limited access to diagnostic and treatment technologies, especially in rural areas, insufficient resources and equipment, and a scarcity of experienced specialists in molecular testing make access more challenging. Discrepancies in molecular technologies like microarray, sequencing, RNA extraction, and reverse transcription procedures can result in variations. Therefore, standardized operating procedures or guidelines across the testing centers are essential for consistency [65]. Therefore, overcoming these hurdles, and leveraging focused research, with improved governmental measures, prioritizing healthcare equity initiatives, are critical to realize the prospects of molecular profiling for breast cancer patients in India.

9. Clinical Phasing/Staging and Characteristic Survival Percentage in BC Cases

When a patient is detected with breast cancer, routine assessments are implemented to define the stage of the disease, which influences the treatment they receive. According to the American Joint Committee on Cancer (AJCC) and the International Union for Cancer Control's (UICC) Tumor, Node, and Metastasis (TNM) breast cancer staging system, the clinical staging of breast cancer is alike across breast cancer subtypes: Stage 0, Stage I, Stage II, Stage III, and Stage IV, as detailed in Figure 9.

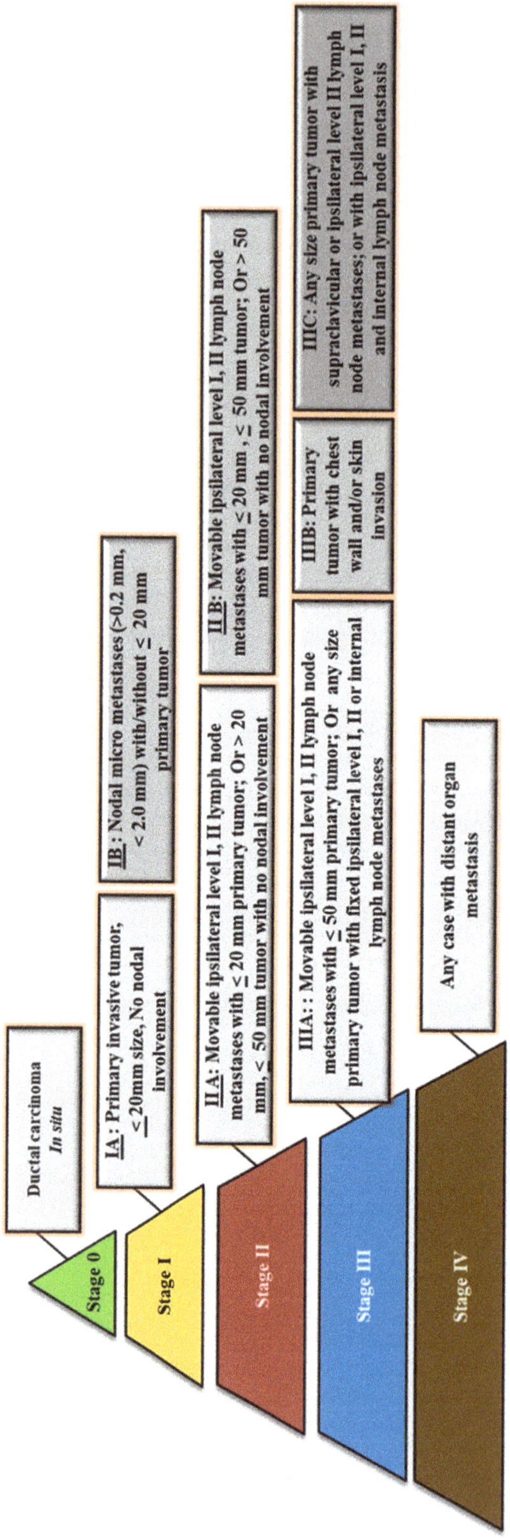

Figure 9. Clinical phasing/staging and characteristic survival percentage in BC cases.

10. Treatment Options for Breast Cancer

Treatment options for BC are often determined by numerous factors, such as the specific type, progression of the disease, the patient's genetic composition, and their clinical status. Treatment may consist of a single approach or be used in conjunction with various therapies, such as surgery, radiation therapy, hormone therapy, chemotherapy, immunotherapy, and targeted therapies. While some treatment strategies focus on the specific area surrounding the tumor, others are systemic, affecting the entire body with cancer-fighting agents. Table 1 below outlines and explains the mechanisms of action for various cancer therapies and their application in the Indian context, taking into account factors like population demographics, disease burden, and healthcare infrastructure.

In India, the use of breast cancer therapies is increasing, despite challenges like resource disparities, high costs, and a need for better infrastructure. These obstacles have spurred new research efforts aimed at producing more cost-effective and holistic therapeutic options, as well as gaining a deeper insight into cancer with the resident population. The most prominent challenges and the ongoing research for each of these BC treatment options in the Indian context are discussed below.

10.1. Surgical Therapy

A late-stage diagnosis presents with advanced malignancy, making surgical treatment more challenging and requiring more resources. This is further complicated by a lack of specialized oncologists along with the disproportionate distribution of cancer facilities or centers, which results in many procedures being conducted in less than ideal conditions. Moreover, in terms of research investigations, Indian guidelines are progressively promoting a multidisciplinary team (MDT) strategy for complicated surgeries, particularly for advanced tumors. Studies are also being carried out to perceive the high out-of-pocket expenditure associated with challenging cancer surgeries, including at government institutions [66].

10.2. Radiation Therapy

Indeed, there is a substantial disparity between the demand for radiation therapy and its actual application. Approximately around 50–64% of new cancer patients require radiation therapy globally, yet only 28.5% obtain it. This is due to a significant lack of radiation machines, with most treatment centers located in urban areas, resulting in challenges for rural patients who face extended wait times with limited access [67]. In the Indian context, research investigations using epidemiological data are in full swing to make a more accurate evaluation of the overall national radiotherapy requirements. Additionally, there are also initiatives to build and implement locally produced low-cost radiotherapy devices to expand access and lessen the dependence on expensive imports [68].

10.3. Hormone Therapy

About one in four individuals with hormone-sensitive breast cancer become unresponsive to these medications, resulting in a recurrence of the disease. This indeed underscores a significant obstacle in managing therapeutic effectiveness over time. In addition, most recently, a study from the Tata Memorial Centre and the National Institute of Biomedical Genomics, India, has detected critical genetic markers that contribute to hormone therapy resistance among Indian BC patients. This finding may undoubtedly pave the way for individualized or personalized treatment approaches and for remodeling the current medications to combat resistance [69].

10.4. Chemotherapy

Chemotherapy plays a fundamental role in the therapeutic regimen of BC in India, applied both before (neoadjuvant) and after (adjuvant) surgery. The standard regimen frequently includes anthracyclines and taxanes, along with additional therapies to alleviate the side effects. However, the expensive nature and harmful effects of certain advanced medications, alongside the restricted availability of newer treatments, limit the choices of treatment, especially within public healthcare [20].

10.5. Immunotherapy

Immunotherapy, which enhances the body's immune system to combat cancer, is a recent therapeutic option available in India for specific cases. This aside, the Drug Controller General of India (DCGI) authorized atezolizumab for metastatic TNBC in combination with nab-paclitaxel in 2020, following the conclusions from the Phase III IMpassion130 trial. Indeed the DCGI made its conclusion based on the results from the Phase III IMpassion130 study, which demonstrated that the combined therapeutic regimen significantly improved progression-free survival (PFS) in patients with PD-L1-positive malignancies [70]. However, the high price of immunotherapy treatment is a significant obstacle, with each cycle costing several lakh rupees and making the treatment unaffordable for many patients. Indian organizations are actively involved in clinical trials focused on immunotherapy. To improve accessibility, personalized treatments, such as dendritic cell-based therapy (Denvax), are being pursued as more affordable alternatives to expensive, imported immunotherapies [71].

10.6. Targeted Therapies

The recent advancement in the development of Trastuzumab Deruxtecan (Enhertu), the antibody conjugate for the metastatic HER2+ BC, was introduced in India in January 2024, marking a significant breakthrough [72]. In addition, in terms of Indian generics, the availability of affordable generic drugs and biosimilars has been vital for lowering costs and increasing treatment accessibility. However, the quality of some Indian-made drugs has been subject to debate due to varying manufacturing and regulatory standards. Moreover, in terms of research context, research investigations focusing on refining targeted therapy combinations and investigating innovative drug delivery technologies are ongoing.

Table 1. Mechanism of action for various cancer therapies.

Therapy	Mechanism of Action	References
Surgery	The primary mechanism of action is the physical removal of a tumor, along with the adjacent healthy tissue (often referred to as the margin). It can be effective in treating early-stage, localized cancers. For more advanced diseases, surgery can be utilized to remove as much of the tumor as possible (debulking), thereby improving the efficacy of other treatments such as chemotherapy or radiation.	[73]
Radiation Therapy	High-energy radiation destroys the DNA of cancer cells, stopping them from proliferating and replicating. Cancer cells are more vulnerable to DNA damage than healthy cells because they divide faster and are less effective at DNA repair. The two most common methods are external beam radiation and internal radiation (brachytherapy), both of which are localized treatments.	[74]
Hormone Therapy	Also known as endocrine therapy, this approach functions by inhibiting the synthesis or activity of hormones that promote the growth of hormone-sensitive malignancies, such as breast and prostate cancer. This can be accomplished with medications that inhibit hormone synthesis, the blocking of hormone receptors, or the surgical removal of hormone-producing glands.	[75]
Chemotherapy	Cytotoxic chemotherapy medications aim to attack and eliminate rapidly dividing cells, including cancer cells, as well as healthy cells in the bone marrow, digestive tract, and hair follicles. These drugs disrupt the process of cell division through several mechanisms like causing DNA damage, blocking essential enzymes, or interrupting the cell's mitotic machinery.	[76]

Table 1. *Cont.*

Therapy	Mechanism of Action	References
Immunotherapy	This therapy activates or strengthens the body's immune system to identify and fight cancer cells. Examples of mechanisms are as follows: 1. *Checkpoint inhibitors*: These drugs suppress proteins (such as PD-1 and CTLA-4) that cancer cells utilize to hide from the immune system. 2. *CAR T-cell treatment*: This involves genetically engineering a patient's own T-cells to express chimeric antigen receptors (CARs), which target specific proteins on cancer cells. 3. *Cancer vaccines*: This involves presenting tumor antigens into the immune system to elicit an anti-tumor response.	[77]
Targeted Therapies	Unlike chemotherapy, targeted therapies focus on particular molecular targets that are essential for the growth, development, and survival of cancer cells, while reducing damage to normal cells. They can inhibit signals that stimulate cancer cell development, trigger cell death, deprive the tumor by preventing the formation of new blood vessels (angiogenesis inhibitors), and transport toxic compounds directly to cancer cells.	[76]

11. Cellular Genesis of Breast Cancer

11.1. Establishment of Mammary Gland and Mammary Stem Cells

The development of the human mammary gland is a continual process that commences during embryonic life. The mammary gland is made up of a distinctly branched network of epithelial tubes that are enclosed in a complex stroma. During the process of embryonic development, an epidermal placode gives rise to the mammary epithelium. The placodes eventually descend into the underlying mesenchyme and give rise to the rudimentary ductal structure of the gland that is present at birth which is chiefly promoted by the epithelial–mesenchymal interactions [78]. The breast rudiment is shaped at birth by 10–12 nascent ductal elements that are positioned underneath the nipple–areola complex [78]. In response to steroid hormone and growth factor receptor signaling, the breast goes through dramatic alterations in size, shape, and function during puberty, pregnancy, and lactation, respectively. This aside, the mammary epithelium is extremely receptive to local and systemic signals, and exhibits profound morphological alterations in the ductal tree during puberty and pregnancy.

Transplantation and lineage tracing investigation studies have demonstrated that the mammary epithelium has a grading of stem and progenitor cells. To elucidate furthermore, lineage tracing has indeed revealed the presence of bipotent mammary stem cells (MaSCs) in situ, in addition to long-lasting unipotent cells that steer the ductal tree morphogenesis and homeostasis. A fair amount of evidence indicates the presence of a heterogeneous MaSC section that includes the fetal MaSCs, slow-cycling cells, along with the long- and short-term repopulating cells. In addition, various luminal progenitor subtypes have been described in both normal mouse and normal human mammary tissue. This in turn suggests that understanding the "cells of origin" for BC requires interpreting the normal cellular hierarchy (Figure 10) [79].

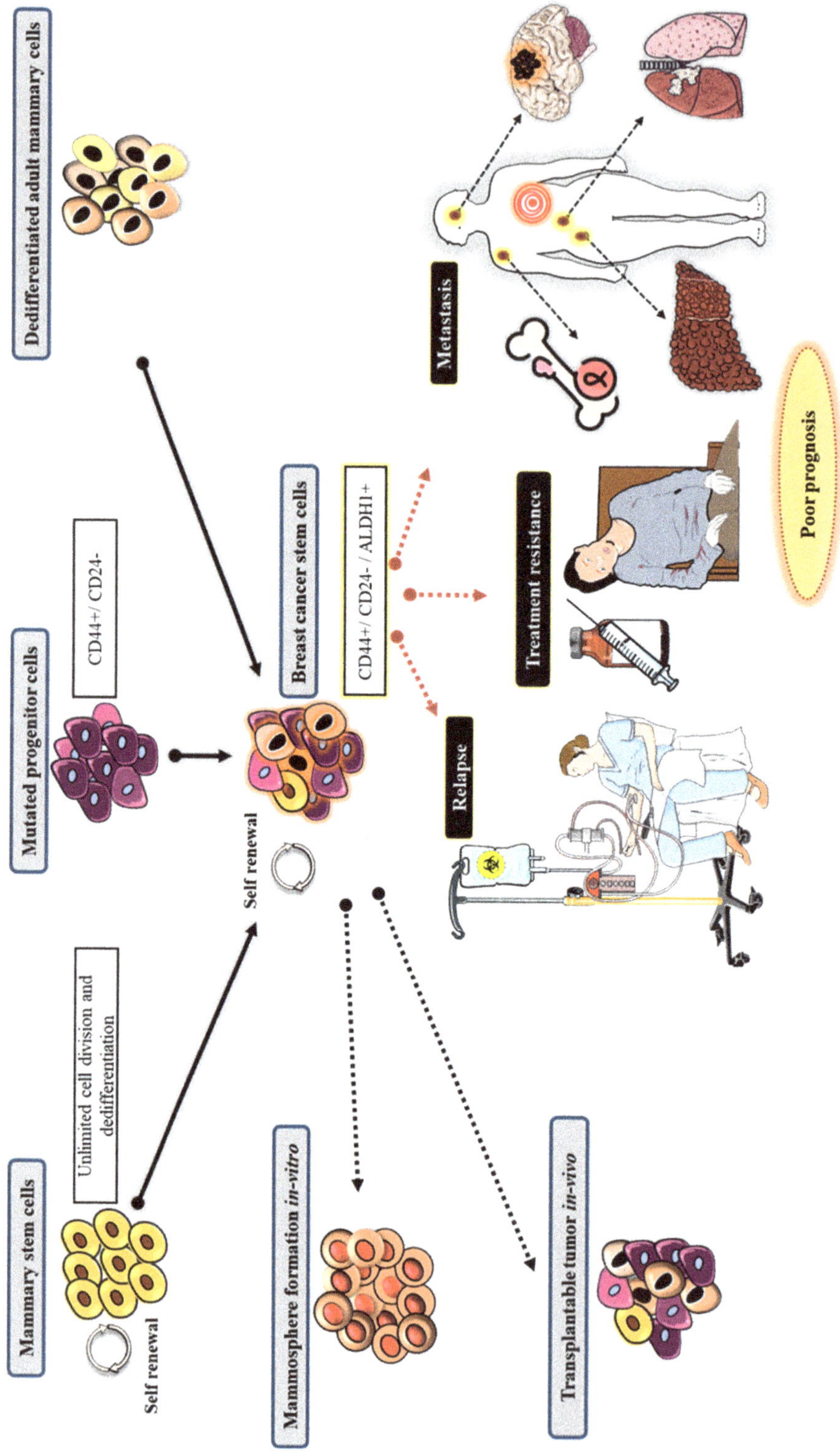

Figure 10. Cellular genesis of breast cancer.

Mammary stem cells (MaSCs) occur as a small section of undifferentiated cells in the mammary gland that can self-renew and generate diverse types of differentiated cells through symmetric and asymmetric divisions. Basically, the progenitor cells are thought to be produced by asymmetric divisions. In addition, the durability and self-renewal properties of the stem cells in a healthy body are well-known for their role in maintenance, repair, and tissue regeneration [79,80]. Under normal conditions, the MaSCs are thought to have functional roles that involve responding to cellular requirements during the progression of reproductive life. The MaSCs are believed to operate through close communication with their distinct cellular microenvironment, also known as the mammary stem cell niche. [79,80]. Notwithstanding, the precise location of these cells, with their distinct self-renewal and differentiation attributes, is still being debated, along with their role in BC. The MaSCs are slow-operating and long-existing stem cells, or progenitor cells, that prevail as a single, role-committed population. Additionally, there is still much discussion about the cells' ability to differentiate into one of two types of mammary epithelial cell lineages, along with the impact of cell plasticity [81].

11.2. Contribution of Mammary Stem Cells in BC Progression

The umbrella concept or the cancer stem cell (CSC) theory is one effort to elucidate MaSCs in relation to breast cancer, as these cells do have a normal role in the body [82]. Despite the fact that the CSC theory has a long history, the emphasis herein will be on the current developments and interpretations of the theory related to breast cancer (Figure 10) [79]. The main point of disagreement in this theory is regardless of whether CSCs are the cell of derivation or origin from which cancer cells progresses, a theory or hypothesis based on explanations of resemblances between tissue replenishment and tumorigenesis, or whether malignant cells acquire stem cell potential [80,83]. The discussion commendably revolves around the significance of bipotent and unipotent luminal and myoepithelial stem cells during normal cell development and carcinogenesis [79].

One of the obstacles in accruing data to back-up or disprove this theory is the inaccuracy and variability between human and mouse models of selected biomarkers, such as CD49f, EpCAM, CD10, ALDH1A1, SSEA4, CD44, and CK12/CK19 66. Understanding cellular heterogeneity, which CSC theory attempts to describe, is critical to developing more specific treatment regimens for breast cancer, where cell heterogeneity is a major barrier. The CSC theory is based on a broadly accepted cell hierarchy structure both in normal and carcinogenic cells, with stem cells at the top and cells differentiating from that point [66]. As a result of this structure, MaSCs, as opposed to mammary repopulating cells, are bi- or multipotent, which is a well-supported conclusion. Under regenerative conditions as perceived through transplantation, these mammary repopulating cells can be recognized as adopting multipotent characteristics [84].

The MaSCs are thought to prevail in a specific microenvironment known as a stem cell niche, which emphasizes the significance of mutual response interactions between MaSC-derived epithelial cells, neighboring stromal cells, and the extracellular matrix. Nevertheless, the signaling pathways and how they impart to the normal and carcinogenic roles of MaSCs remain unknown. In terms of breast CSC origin, two major cell lineages are at the heart of this debate, viz., basal cells and luminal cells [79]. Basal cells are one of the two major cell lines in the mammary gland, the other being luminal cells. In the structure of the mammary gland, basal cells border the luminal cell layer. Similarly, the lumen of the mammary ducts and alveoli is lined with luminal cells. The basal cells usually express cytokeratins 5 and 14, as well as smooth muscle actin, whereas luminal cells express cytokeratin-8 and can be positive or negative to hormone receptors [79]. Few studies have demonstrated that patient-derived breast cancer organoids can outperform mouse models for cellular lineage tracing [69]. One such project is the creation of a living breast cancer organoid biobank, which could be used as a more precise model for research on cancer cell heterogeneity and drug discovery [85,86].

12. The Role of Genetics in the Effect of Breast Cancer Drug Resistance: Requirement for Predictive Biomarkers to Address the Outstanding Questions

The treatment regimens that are extensively presented for BC are human epidermal growth factor receptor 2 (HER2)-targeted therapy, endocrine therapy, and systemic chemotherapy. To elaborate further, based on the tumor's features or attributes, and its hormone receptor status, and the presence of HER2, progesterone receptors, and estrogen receptors, pharmacotherapy is selected accordingly so as to yield a better disease prognosis [87]. Pharmacotherapy for breast cancer frequently has variable outcomes (inter-patient heterogenicity) in different populations, which may result in therapeutic failure. This can be brought about by several forms of congenital medication resistance among individuals. Congenital drug resistance is a phenomenon that is influenced by genetic variation. The link between genetic variations and treatment resistance for breast cancer will be briefly discussed herein.

Even while the prognosis for the disease has ameliorated through the positive outcomes of chemotherapy and adjuvants, a few designated sets of populations frequently respond differently to breast cancer treatment. Such variations or differences are brought about by the intrinsic resistance to certain prescribed medications or drugs [88]. Drug resistance is one of the chief causes for the cancer treatment failure [89]. These individual differences in drug response are mostly caused by DNA mutations that can change treatments' or drugs' efficacy [90]. Different mechanisms like the altered drug pharmacokinetics [91], increased or decreased cell signaling [92], altered pharmacodynamic-related receptor counts [93], etc., may all contribute to resistance. In order to identify drug resistance and provide information that enables the development of personalized medicine, it is indeed extremely pertinent to investigate additional gene polymorphisms that may influence therapeutic responses in breast cancer.

Hence, this segment will specifically be focused on the effects of pharmacogenetics in breast cancer drug resistance, and will discuss the reports or studies demonstrating the relationship of breast cancer drug resistance for significant genes like *ABC1*, *BARDI*, *CD24*, *CYP19A1*, *CY2C19*, *CY2C9*, *CYP2D6*, and a few others.

12.1. *ABCB1*

The ATP-binding cassette (ABC) is a transporter for diverse drug molecules, including chemotherapy drugs. This gene transporter aids and facilitates the passage of drug molecules through cellular membranes by expelling energy. The genes *ABCB*, *ABCG*, *ABCD*, *ABCF*, *ABCCI*, and *ABCCII* make up the ABC subfamily [94]. The *ABCB1* gene is found in chromosome 7 which produces a 45-kB mRNA [95]. It encrypts a drug-active transporter that helps cells secrete cytotoxic chemicals. In addition, it encodes for P-glycoprotein (P-gp), an ATP-dependent drug efflux pump which transfers or delivers numerous hydrophobic substrates like anti-cancer drugs such as anthracyclines (doxorubicin), epipodophyllotoxins (etoposide), vinca alkaloids (vincristine), dactinomycin, and paclitaxel. The P-gp is usually expressed in tissues with excretory function, like the jejunum, kidney, liver, and adrenal gland [96]. The overexpression of MDR1 will result in the augmented drug efflux activity of P-gp, resulting in multi-drug resistance (MDR).

According to a study by Vulsteke et al. (2014) [97], patients with early breast cancer treated with 5-fluorouracil (FU), eirenicon, and cyclophosphamide (FEC) had better therapeutic outcomes when they had the *ABCB1* GT genotype than when they had the *ABCB1* GG/GA genotype. Similarly, based on the findings from a different study, the *ABCB1* 2677 GG genotype polymorphisms showed resistance to paclitaxel and anthracycline therapy. This gene polymorphism may contribute to the cross-resistance between paclitaxel and anthracycline in the treatment of metastatic breast cancer. Additionally, cases of *ABCB1* 3435 CT genotype resistance were discovered, which led to a poorer disease control rate and shorter overall survival (OS) when anthracycline treatment was administered [98].

Similarly another study by Vishnu et al. (2013) [99] on P-gp expression as a determinant factor in predicting response to NACT (Neoaduvant chemotherapy) proved that the detection of P-gp expression status before the initiation of chemotherapy can be used as a predictive marker for NACT response and which will eventually help in evading the toxic side effects of NACT (neoadjuvant chemotherapy) among non-responders.

12.2. eNOS

Numerous intracellular and extracellular factors, such as reactive oxygen and nitrogen species (RONS), have an impact on the development of cancers, especially breast cancer. In fact, the RONS's accumulation within the cell is thought to contribute to the development and spread of cancer by altering the gene expression profile, dysregulating signaling pathways, and thereby inducing aberrant protein modifications [100]. As a crucial prerequisite for the physiological process, nitric oxide (NO), a water-soluble signaling molecule, is widely distributed and endogenously generated throughout the body [101]. In cancer, nitric oxide renders pleiotropic effects, some of which may promote tumor development and others may inhibit it, depending on the local concentration [102]. Reports on the function of *eNOS* in breast cancer revealed that it is in fact manifested in the breast tumors [100]. This indeed has a positive correlation with ER expression, which in turn has negative association with tumor grade and lymph node status. Additionally, the BC cell lines [103] and advanced breast tumors have been shown to express NOS [104]. This aside, the concentration of NO in the tumor microenvironment determines whether or not NO plays a role in tumor development or regression. While a higher NO concentration has cytotoxic effects and inhibits tumor growth and apoptosis, a lower NO concentration accelerates tumor progression by boosting tumor invasion, proliferation, migration, and chemoresistance [105]. Similarly, the results from a recent meta-analysis study described the link between the eNOS 894G>T polymorphism and the likelihood of developing cancer and stroke. Indeed, the conclusion from this study reveals a number of significant details, advancing the knowledge of NO's role in these illnesses. The results pinpoint a possible association between the eNOS polymorphisms and a higher risk of developing cancer, especially in those who have the TT or GT genotype as opposed to the GG. This correlation seems to hold true for different ethnic groups (African, Asian, and European). However, the possibility that other variables could be at play was brought to light by other studies that found no correlation. It is thus inferred from this study that the eNOS polymorphisms, especially 894G>T, have a significant risk of cancer among the overall population based on the GG vs. GT+TT genetic model. Nevertheless, in light of these conclusions, these findings should be reassessed for further prospective validation with a larger sample size [106].

12.3. BARD1

Proteins that can communicate with breast cancer genes 1 (BRCA1) and 2 (BRCA2) and thereupon cause DNA and tumor suppressor damage may be encoded by a number of genes. The BRCA1-associated RING domain 1 (BARD1) gene is one such gene [107]. In terms of structure and function, the BARD1 gene generates a protein that is comparable to the BRCA1 protein [108]. Both BRCA1 and BARD1 have the ability to form homodimer and heterodimer structures. The former can be formed by linking with the an interesting new gene (RING) finger domain in the N-terminal portion, while the latter is stabilized by bonds between the amino acid residues of 26–119 in BARD1 and 1–109 in BRCA1respectively [109]. These interactions play a significant role in determining how breast cancer tumor suppression may manifest [108]. The stability of genotype and phenotype is generally controlled by BARD1, which also participates in DNA repair and ubiquitination. However, it was demonstrated that there is a significant association, with patients who carry single nucleotide polymorphism (SNP) in *BARD1* rs2070096 with minor alleles having worse relapse-free survival rates compared to patients who have received non-TCH treatment. Nonetheless, another study found no substantial association between the polymorphism of *BARD1* in rs2229571 and the response of patients using docetaxel, carboplatin, and trastuzumab (TCH) treatment compared with non-TCH treatment. Therefore, this may indicate chemoresistance [88].

12.4. CD24

The cluster of differentiation 24 (CD24) gene, which encodes for sialoglycoprotein and can bring about cell proliferation and signal changes, is found at chromosome locus 6q21 [110]. It is a protein located on the cell surface that offers links to the cell membrane via glycosylphosphatidylinositol [111]. Breast cancer cases have been linked to the overexpression of the CD24 protein. The *CD24* expression was typically found in luminal and HER2-positive breast cancer cells [112]. In addition, the CD24 expression has been linked to breast cancer patients' prognoses since it can control tumor cell proliferation [87] and escalate the prospects of metastasis [113].

There are currently no strong associations between genotype differentiation and CD24 expression. According to studies, *CD24* Val is involved in the response to chemotherapy and has been linked to increased sensitivity, increased autoreactive immunity, and faster disease development. In a previous study on primary breast cancer, *CD24* Val showed a significant sensitivity to anthracycline- and taxane-based therapy. According to the study, *CD24* Ala/Val is the only single-factor indicator of post-neoadjuvant chemotherapy (NCT) positive complete response (pCR) in breast cancer patients. This aside, the anti-tumor immune response of the host may be able to be controlled by *CD24* Ala/Val to make it more autoreactive. Differences in the CD24 genotype affect how patients respond to NCT therapy, as shown by the substantial association between CD24 Val/Val and intratumoral lymphocytic [114]. However, a study by Zhou et al. [115] revealed that pCR in breast cancer patients who had received NCT therapy was not anticipated by CD24 polymorphisms in rs3838646 and rs52812045.

12.5. Cytochrome P450 (CYP450), CYP

Cytochrome P450 (CYP450) is an enzyme that acts as a xenobiotic metabolizer by engaging xenobiotics into an oxidation reaction that converts the drug into its metabolites [116]. This mechanism is vital in BC therapy because drugs like tamoxifen must be transformed further into active and effective metabolites like 4-hydroxy-tamoxifen (4-HT) and endoxifen by CYP450 3A4 and CYP2D6, resulting in an elevated binding connection for the ER [117].

In one of the studies reported by Artigalás et al. [118], it is described that the rs4646 polymorphism in the *CYP19A1* gene could possibly be a prognostic determinant in the aromatase inhibitor (AI) therapy. The SNP rs4646 was concomitant with the increased time to progression (TTP) compared to the wild-type gene in metastatic BC patients treated with AI (hazard ratio (HR) = 0.51 [95% confidence interval (CI), 0.33–0.78], $p = 0.002$). Additionally, Liu et al. [119] discovered a statistically significant link amongst the rs4646

T alleles (G/T or T/T) and improved overall survival in women with metastatic breast cancer (HR, 2.37 [95% CI, 1.20–4.65], $p = 0.001$). Miron et al. [120], on the other hand, found no compelling association with OS in the same SNP. Similarly, Henry et al. [121] found no statistically significant association between 127 SNPs in CYP19A1 associated with estrogen metabolism and breast density modulation. These findings suggest that CYP19A1 genotypes may be related to OS in BC patients receiving AIs. This association, however, seems to vary significantly between patients.

This aside, in a retrospective cohort study to describe chemoresistance brought about by *CYP3A4* polymorphisms, Gor et al. [122] discovered that patients with at least one *CYP3A4 *1B* variant allele had a substantial relation with worse disease-free survival (DFS) in comparison to those with a wild-type **1A/*1A*. The causal mechanism for this chemoresistance is that the **1B* polymorphism decreases Phase I enzyme activity, resulting in a suboptimal 4-hydroxy-cyclophosphamide concentration. Because of the nature of cyclophosphamide pharmacokinetics, it must be activated to 4-hydroxy-cyclophosphamide before it can be distributed into the cancer cells via Phase I CYP enzymes, one of which is 3A4. Preceding studies by Beelen et al. [123] described a substantial link between *CYP2C19* variant alleles and time to treatment failure (TTF) among tamoxifen patients, with *CYP2C19*2* carriers having a longer TTF and those with the *CYP2C19*17* allele having a shorter TTF, but not statistically significantly. Tamoxifen metabolism is effectively influenced by *CYP2C19* inhibition, with transition to active metabolites like endoxifen having been detected. Tamoxifen resistance may be linked to lower tamoxifen and trans-4-OH-tamoxifen levels stimulated by isomerization to the cis isomer, and this isomerization may also happen for endoxifen [124–126]. Likewise, Vulsteke et al. [97] proposed that the *CYP2C9* rs1057910 polymorphism produced resistance in his study, where there was a considerably worse RFI, but the C-allele variant carrier was only observed in three subjects, thereupon signifying additional investigation.

Furthermore, Regan et al. [127] found no link between tamoxifen therapy and alterations in the *CYP2D6* phenotype metabolism. Endoxifen, a tamoxifen metabolite with a greater affinity for the ER, is thought to be linked to disease control, whereas *CYP2D6* polymorphism, an enzyme that can metabolize tamoxifen into its active metabolite, was thought to be linked to lower endoxifen levels, and hence poorer disease control and higher side effects. According to Regan et al. [127], the *CYP2D6* metabolism phenotype did not predict tamoxifen efficacy, implying that more research into tamoxifen metabolism and its mechanism of disease control is necessary. However, Dezentjé et al. [128] spotted an incorrect interpretation in that study and repeated it while investigating whether loss of heterozygozity (LOH) could explain a Hardy–Weinberg equilibrium deviation that may exclude false genotype by LOH in tumor tissues. Nevertheless, their study did not find any link between *CYP2D6* genotype differences and the efficacy of tamoxifen. Further investigations by Neven et al. [129] and Sanchez-Spitman et al. [130] backed these findings, addressing no associations between low-activity *CYP2D6* genotypes and low endoxifen levels and clinical outcomes.

13. Pharmacogenetics of Breast Cancer Therapeutics in India

The concept of pharmacogenetics, the study of how genetic variants influence drug response, is surfacing as a critical element of personalized and precision medicine, particularly for BC in India. Indeed, this discipline has a great potential for improving therapeutic options and dose regimen, thereby refining efficacy and eventually reducing side effects among the Indian BC patients, as corroborated by the studies discussed in Table 2.

Identifying the significance of genetic polymorphisms in determining BC treatment response and toxicity is crucial for personalized medicine, particularly in diversified populations like India [131]. Although there is a substantial body of research conducted worldwide, studies that focus exclusively on the Indian population are emerging, bringing to light specific concerns.

In the context of clinical implications, comprehending the genetic profiles of Indian BC patients, especially with the discussed gene polymorphisms, warrants targeted screening and treatment approaches improving drug dosage, and possibly reducing side effects. The gene polymorphisms in *UGT1A1* and *CYP2D6* can indicate an elevated risk of adverse events from certain drugs thus streamlining for preventive and strategic management of the drug dose modifications [132]. This aside, genetic polymorphisms in *ABCB1* and *GSTs* can facilitate the efficacy of chemotherapy and the possibility of developing drug resistance, thus aiding in therapy escalation or the selection of alternate therapies. Likewise, comprehending the frequency and the scale of BRCA1/2 mutations in the Indian population is critical for genetic counseling and for strategies in managing hereditary BC.

At present, India does not have dedicated specific and regional pharmacogenetic guidelines for BC. Nonetheless, acknowledging the need and significance of personalized and precision medicine to the distinct genetic characteristics of Indian patients, initiative attempts are underway to incorporate pharmacogenomics into oncology practice in India [21,133].

Below are some suggestions and recommendations on how this information can be incorporated into oncology practice in India:

13.1. Evidence Generation and Dissemination

- ***Funding research on Indian populations:*** Additional investigation is required to explore and examine the genetic polymorphisms unique to the Indian population and their effect on treatment and drug response in BC.
- ***Implementing nationwide data repositories:*** Developing a complete and an extensive database of pharmacogenomic variations and drug–gene interactions related to the Indian population would be tremendously beneficial.
- ***Utilize real-world evidence, implementing observational studies:*** Corroborating results from real-world research or evidence-based studies in India, demonstrating the influence of pharmacogenomics testing on BC therapeutic response, is critical for initiating therapeutic value and increasing uptake.

13.2. Practical Implementation Strategies

- ***Clinically actionable drug–gene interactions (DGIs):*** Focus on employing pharmacogenomics tests for well-recognized drug–gene interactions in BC; for example, *CYP2D6* for tamoxifen.
- ***Establish accessible testing policies or programs:*** Capitalizing on commercial, affordable and effective genetic testing tools like next-generation sequencing (NGS), is critical for increasing access.
- ***Proof of concept programs and step-by-step implementation:*** Initiate proof of concept or pilot programs at all oncology centers of excellence or specialized cancer hospitals to assess the feasibility and impact of integrating pharmacogenomics testing into regular BC care.

Table 2. Details of some key genes and their associated polymorphisms linked to Breast cancer among the Indian population along with their potential clinical implications and the drugs affected.

	Polymorphisms	Drugs Affected	Clinical Implications Relevant to Indian Population
CYP2D6	Multiple variant alleles (e.g., *4 null allele, *10 reduced function allele)	Tamoxifen	**Increased risk of adverse effects:** A combination of specific SNPs (e.g., 2988G>A, -1584C, and 2850C>T, suggestive of *41 reduced functional allele) may predict a higher risk of fatty liver with tamoxifen therapy [134]. **Altered tamoxifen metabolism:** Individuals with reduced function alleles (*10) may show reduced benefit from tamoxifen and poorer disease-free survival [134].
UGT1A1	UGT1A1*28*, *UGT1A16*	Irinotecan	**Increased irinotecan toxicity:** Both UGT1A1*28 and UGT1A16* (more common in Asians) are associated with decreased UGT1A1 activity, increasing the risk of severe neutropenia and diarrhea from irinotecan. Double heterozygosity (UGT1A1*6/*28) is also associated with increased toxicity [135].
ABCB1 (MDR1)	1236C>T, 2677G>T/A, 3435C>T	Doxorubicin, cyclophosphamide, paclitaxel, and other MDR1 substrates	**Drug resistance:** Polymorphisms affecting the function and expression of ABCB1 (encoding P-glycoprotein) can lead to reduced intracellular drug levels and resistance to chemotherapy agents like anthracyclines and taxanes. Some studies suggest a higher risk of non-response to FAC chemotherapy with specific ABCB1 genotypes [136].
GSTM1	Null genotype	Anthracyclines (and potentially other chemotherapy drugs metabolized by GSTM1)	**Worse chemotherapy responsiveness:** The null genotype has been associated with worse responsiveness to anthracycline-based chemotherapy. Some studies suggest an increased risk of breast cancer susceptibility and shorter overall survival [137].
GSTT1	Null genotype	Cyclophosphamide, doxorubicin, 5-fluorouracil (CAF regimen)	**Better DFS and OS with certain chemotherapy regimens:** The null genotype, in some contexts, has been associated with better disease-free survival (DFS) and overall survival (OS) when treated with CAF (cyclophosphamide, doxorubicin, and 5-fluorouracil) chemotherapy compared to other variants. It may be associated with an increased frequency of p53 mutations [136].
GSTP1	105Val/Val genotype, GG genotype	Cyclophosphamide (CTX)	Worse DFS with CTX: The 105Val/Val genotype showed a significant association with worse DFS in patients treated with cyclophosphamide. No significant association with cyclophosphamide outcome: Another study found no significant association with the GG genotype and cyclophosphamide outcome [136].
BRCA1/2	Pathogenic mutations and variants of uncertain significance (VUS)	PARP inhibitors	Increased hereditary breast and ovarian cancer risk: Mutations in BRCA1/2 are a major cause of hereditary breast and ovarian cancer. Predictive marker for PARP inhibitors: BRCA1/2 mutations are a strong indicator for PARP inhibitor efficacy [138].
HER2	Amplification	Trastuzumab, Pertuzumab, Lapatinib, Neratinib (HER2 inhibitors and targeted agents)	Predictive marker for HER2-targeted therapies: HER2 amplification leads to overexpression, which makes tumors responsive to anti-HER2 therapies like trastuzumab and pertuzumab. Resistance mechanisms can develop over time [139,140].
Other genes			
TP53	Polymorphisms		Increased breast cancer risk [141,142].
PTEN	Germline mutations		Increased breast cancer risk [27].
MTHFR	C677T, A1298C		Association with increased homocysteine levels. A link with breast cancer risk is not clearly established [143,144].
NBN, RAD51, XRCC3	Polymorphisms	Radiotherapy	Potential biomarkers for radiotherapy toxicity (e.g., cardiac toxicity with RAD51 and skin adverse events with XRCC3) in HER2-positive breast cancer [145].

13.3. Policy and Screening Guidelines

- ***Establish and formulate guidelines specific for India:*** Developing national guidelines for pharmacogenomic testing and interpretation in BC is crucial for standardizing the reliability and interpretation of results for therapeutic regimens.
- ***Review compliance with legal and ethical guidelines:*** Acknowledging concerns about patient confidentiality, sharing data resources, and possible discernment based on genetic information are vital for cultivating transparency and accountability, and eventually maintaining integrity.
- ***Developing a standardized curriculum for precision medicine in pharmacotherapy:*** Incorporating pharmacogenomics into medical training programs for both undergraduate and postgraduate students will prepare the next generation of oncologists to use this discipline proficiently.

By concentrating on these aspects, India can advance pharmacogenomics in BC treatment, leading to more accurate, efficient, and safer therapies for patients.

14. The Role of Tumor Microenvironment in Breast Cancer

The basic principle of the tumor microenvironment (TME) is that a tumor is an intricate, dynamic ecosystem with both malignant and non-malignant components, rather than an isolated mass of cancer cells. This environment plays an active role in and promotes cancer initiation, growth, progression, metastasis, and resistance to therapy. The non-cancerous parts of the TME are not inactive spectators, but are "hijacked" and reprogrammed by cancer cells to enhance their existence and proliferation [146].

As discussed below, the TME is made up of several key cellular and non-cellular components that interact via complex signaling networks.

14.1. Cellular Components

1. ***Cancer cells:*** These primary cancerous cells comprise the tumor.
2. ***Cancer-associated fibroblasts:*** These essentially reprogrammed fibroblasts are key producers of extracellular matrix (ECM) and numerous signaling molecules. They stimulate tumor development, angiogenesis, and metastasis.
3. ***Immune cells:*** A varied group of immune cells, comprising the following:
    i. *Tumor-associated macrophages (TAMs):* Tumor-associated macrophages (TAMs) are usually adopted to an M2-like phenotype, which usually suppresses anti-tumor responses, stimulating angiogenesis, and facilitating metastasis.
    ii. *Regulatory T-cells (Tregs):* These are a subclass of T-cells that inhibit the activity of other immune cells, resulting in an immunosuppressive environment.
    iii. *Myeloid-derived suppressor cells (MDSCs):* These underdeveloped myeloid cells inhibit anti-tumor immune responses and promote tumor development.
4. ***Endothelial cells:*** These cells form the inner layer of blood vessels and are engaged and stimulated by tumors to generate new, often defective, blood vessels that provide oxygen and nutrients to the tumor (angiogenesis).
5. ***Pericytes:*** These are cells that encase the blood vessels and communicate with endothelial cells, prompting the development and durability of the tumor's vasculature.

14.2. Non-Cellular Components

1. ***Extracellular matrix (ECM):*** The ECM is a complex network of proteins and large molecules (such as collagen, fibronectin, and proteoglycans) that offers structural and biochemical assistance. The ECM in tumors frequently becomes rigid and un-

dergoes modifications, thus facilitating the spread of cancer cells and impeding the effectiveness of therapeutic delivery [147].

2. ***Signaling molecules:*** Several TME cells release a network of cytokines, chemokines, and growth factors to regulate their interactions. However, depending on the circumstances, these signals might stimulate or prevent tumor growth.
3. ***Extracellular vesicles (EVs):*** These are minute vesicles, such as exosomes, that are secreted by cells to transport substances such as proteins and RNA. The tumor-derived EVs have the potential to reprogram other cells in the TME, promoting tumor development and prime distant locations for metastasis [148].

Indeed, the interplay and interaction within the tumor microenvironment (TME) are critical factors in cancer development and resistance to treatment regimen.

In terms of reciprocal communication, cancer cells are not passive in their environment; they actively regulate it by secreting cytokines and growth factors (signaling molecules) that in turn attract and alter the behavior of nearby healthy cells. As a result, the reprogrammed or modified stromal and immune cells produce signals that in turn regulate cancer cell proliferation and survival [149].

In the context of angiogenesis, because of the reduced oxygen (hypoxia) levels within the tumor, the TME produces pro-angiogenic substances like vascular endothelial growth factor (VEGF). This results in the development of a defective vascular network that delivers oxygen and nutrients to the tumor while simultaneously functioning as pathway for metastasis [150]. However, it is noteworthy that in therapeutic resistance, the TME can potentially obstruct a treatment's efficacy. Additionally, immunosuppressive cells can deactivate the body's inherent defenses against cancer, thus reducing the effectiveness of immunotherapies [151].

The importance of TME for Indian BC patients is significant because of unique clinical and epidemiological features, including a higher prevalence of aggressive subtypes in younger, premenopausal patients compared to Western populations. Furthermore, the unique genomic profiles observed in these may be driven by or connected to distinct TME architectures, which could play a role in both cancer progression and therapy response [152].

15. Recent Developments and Key Prospective Directions for Breast Cancer Research in India

Recent advancements in Indian breast cancer research are determined by the disease's increasing prevalence and the significant number of late-stage detections. Key research objectives should therefore focus on improving diagnostic tools, including the use of artificial intelligence (AI) and local technology (indigenous), and creating affordable, scalable screening methods for neglected rural communities. Additionally, the research focus should prioritize India-specific challenges, such as a younger average age of diagnosis, a higher prevalence of aggressive cancer subtypes, late detection, dense breast tissue, and socioeconomic hurdles affecting care. Furthermore, key directions for the future may have to focus on utilizing advanced technologies such as AI, nanotechnology, and genomics to address the specific issues of the Indian population as discussed below.

15.1. Generating Comprehensive Genomic Data Specific for Indian Population (India-Specific Genome Atlas)

Breast cancer in India frequently affects younger individuals and is often more severe than in Western populations, resulting in lower survival rates. In addition, since Indian populations are inadequately represented in global databases, large-scale genomic studies are needed to generate comprehensive data, examine the distinct genomic landscape, and create customized predictive models for India, along with nationwide policies to address systemic challenges. These studies would focus on identifying and generating Indian-specific genetic resources, classify population-specific mutations, comprehending the younger average age of onset, and better understanding the higher prevalence of aggressive subtypes like TNBC, and eventually developing customized treatments and diagnostics strategies [15].

- *Genomic and multi-omics profiling:*

Large-scale genomic studies in India are aimed at producing comprehensive, high-quality multi-omics data that cover the entire spectrum of breast cancer. The Indian Cancer Genome Atlas (ICGA), a pan-India consortium spearheaded by the Council of Scientific and Industrial Research (CSIR) along with other governmental bodies has initiated a non-profit foundation for creating nationwide datasets, and genetic profiles of the cancers prevalent in India. In September 2024, it launched its initial multi-omics data platform specifically for BC patients in India, and is in fact the first multi-omics data portal for cancer in India [153]. On the other hand, the Bharat Cancer Genome Atlas (BCGA), another initiative launched by organizations like IIT Madras in conjunction with Karkinos Healthcare, is also aimed at creating a complete genetic database for cancers predominant in India. In fact, the BCGA was launched in early 2025, and its first public release is aimed at breast cancer [7]. The ultimate goal of these initiatives is to provide a multi-omics database that will manage and provide a specific genetic landscape of Indian cancer patients while also reducing current inequities in worldwide cancer research.

India's heterogeneous population necessitates sequencing initiatives across different zones and socioeconomic classes to construct genetic classifiers that are clinically relevant throughout the country. Moreover, through the initiatives like ICGA and BCGA, researchers are examining genomic, transcriptomic, epigenetic, and proteomic data from Indian cancer patients. The initial cohort of 1000 BC patients is anticipated to create an extensive dataset that will be ethically collected and standardized. In addition, through these resources it will be possible to identify genetic traits and mutations unique to the Indian population, which may vary from those found in Western populations. This in turn may pave way to identify distinct mutation patterns (for example, genes like *TP53* and *ZFHX3*), and detect specific or unique gene fusions and drivers of metastasis, and eventually the incidence of diverse molecular subtypes of BC in India, which may differ from those found in Western populations.

Gathering representative genomic data from the Indian population may help in developing more accurate, India-specific predictive tools. This will enable the use of specific SNPs to improve polygenic scores, which will provide insight into the unique mutational landscapes and pathways that can lead to more personalized medicine for patients. Eventually, the integration of genomic, clinical, and pathological data improves the overall understanding of the relationship between genetic profiles and patient outcomes.

15.2. Addressing High Breast Density Among Indian Breast Cancer Population

The recent developments and advances in BC research in India have indeed focused on using artificial intelligence (AI) and genomics to tackle the rising incidence, by advancing early diagnosis, addressing the distinct biological and socioeconomic issues specific to the Indian subcontinent. Indeed the recent research areas include harnessing technology for early detection, leveraging India's unique genetic diversity for genomic investigations, enhancing treatment accessibility, and establishing therapeutic protocols tailored to specific needs. Moreover, this widens the scope of key directions, which aids in developing more accessible screening tools, implementing personalized medicine, and adopting public health policies to lessen inequities in care between urban and rural areas [153].

Undeniably, new research directions for BC in India combine deep learning (DL) with digital tomosynthesis (DBT) and genomic profiling. These efforts are aimed at overcoming healthcare and population-specific obstacles to increase screening accuracy and early detection. By integrating these advanced technologies, researchers hope to address the limitations of traditional approaches such as mammography, especially given the severity of higher breast density, which are particularly common among younger Indian women [154]. Nonetheless, effectively establishing such a system would necessitate considerable funding and validation to overcome significant resource and logistical obstacles within the Indian healthcare context.

- *Using Artificial Intelligence (AI), deep learning (DL), and genomic profiling to refine tomosynthesis:*

Indian researchers are indeed integrating these technologies to enhance breast cancer detection. Deep learning facilitates the analysis of massive amounts of image and genetic data, DBT provides more comprehensive 3D imaging than traditional mammography, and genomic profiling provides a better understanding of the disease's biology.

Indian women generally exhibit higher breast density, which complicates the identification of tumors using the standard mammography since both dense tissue and tumors show up as white on the imaging. However, the deep learning algorithms can evaluate the more intricate, 3D visuals obtained from DBT to distinguish between dense tissue and possible anomalies, considerably increasing accuracy. This aside, deep learning and generative adversarial networks (GANs) can produce synthetic images to train detection algorithms, possibly refining the presentation in women with dense breast tissue, by generating bigger and more diversified training datasets [155].

On the other hand, genomic profiling enhances precision by examining the biological features of the tumor, which can subsequently be paired or integrated with imaging data. It can indeed help in forecasting a patient's specific risk of developing BC as well as the chance of recurrence. When combined with tomosynthesis imaging, it offers a more comprehensive and precise evaluation of risk [156]. Similarly, genomic testing detects specific gene mutations and molecular features of tumors, thereby guiding towards personalized medicine. Indeed, genomic profiling can help identify biomarkers, track disease progression, and the onset of drug resistance by examining circulating tumor DNA via a "liquid biopsy" [157]. With the facts presented, AI models may integrate a wide range of data types, including tomosynthesis imaging and genomic data. This integrated strategy, known as radiogenomics, links the visual landscapes of a tumor to its molecular features, offering a deeper insight into the disease.

15.3. Adapting Screening Technologies for Early Breast Cancer Detection and Screening for Indian Needs

Breast cancer research and policies in India are prioritizing several key directions to address late diagnosis and enhance treatment availability, including the use of technology, reinforcing health systems, and culturally tailoring interventions. Therefore, to acclimatize BC screening for India, novel technologies must identify and overcome constraints like high costs, inadequate facilities, a lack of public awareness, and higher breast density among Indian women. Moreover, in place of depending only on exorbitant, foreign technology, India is slowly exploring low-cost, portable, and AI-powered BC screening approaches as discussed below, for rural and resource-limited settings.

- *Technological innovation and research adapted for India:*

Various startups and organizations in India are working on artificial intelligence(AI) technologies and solutions for BC screening, including NiramAI, an AI-powered diagnostic.

It employs thermal imaging and AI (Thermalytix) to detect temperature fluctuations produced by tumors. The method is non-invasive, radiation-free, portable, and involves little training to use, making it ideal for rural areas and places with limited resources [158]. Likewise, Indian researchers are also working on non-invasive biomarkers, like blood-based diagnostics (immunoassays) and serum biomarkers for early cancer diagnosis, which could enhance sensitivity over mammography in younger Indian women [155].

This aside, AI-powered technologies such as "MammoAssist" are being investigated to improve diagnostic accuracy and lessen the burden on radiologists. Additionally, studies are exploring AI's application in personalized therapies and treatment strategies [159]. Similarly, the internet is being used to offer easily accessible and trustworthy details about cancer in various languages. Telemedicine and digital health serve as an important study topic for delivering remote consultations and follow-up care, particularly in rural areas in India [160].

It is noteworthy that Indian researchers have effectively created novel and sophisticated drug delivery systems that improve therapeutic efficacy for breast cancer. To benchmark this, in February 2025, researchers at IIT Madras created and secured a patent for a drug delivery system using nanoparticles, which encapsulates anti-cancer drugs and specifically targets cancer cells, thus reducing adverse effects [161]. Correspondingly, in January 2025, researchers from the Bose Institute Kolkata and the IIT, Guwahati, created an injectable hydrogel for targeted therapy. This hydrogel delivers anti-cancer medications in a regulated manner to tumor areas [162].

In addition, in terms of utilizing local technology or indigenous treatment options, India is working on developing its own innovative remedies to boost accessibility. For example, In September 2025, a study reported that diselenide 7, a newly synthesized organoselenium compound, shows potential in addressing aggressive TNBC in animal models by lowering in invasiveness and metastasis [163].

15.4. Integrating Socio-Cultural Factors for Improving Program Implementation and Access for BC Research in India

In India, breast cancer screening uptake is chiefly influenced by the socioeconomic status and deeply rooted cultural beliefs. Therefore, BC research in India should focus on incorporating socio-cultural dynamics to improve the implementation and accessibility of breast cancer programs by overcoming challenges like stigma, lack of knowledge (ignorance), and patriarchal norms. Consequently, the key directions should include developing culturally tailored outreach strategies, incorporating socio-cultural aspects that enhance the capabilities of community health workers, addressing financial and systemic barriers, and mitigating economic disparities to effectively address the diverse Indian landscape. In addition, by establishing context-specific initiatives as discussed below, researchers can enhance screening rates, promote early diagnosis, and thereby lower the disproportionately high mortality among Indian women. Studies are underway evaluating the efficiency of introducing early detection strategies or programs into India's healthcare system to enhance outreach and reduce late-stage diagnosis.

- *Community-based solutions; a culturally adapted approach:*

Successful BC screening in India would necessitate a nationwide plan that should combine various technologies and leverage collaborations between public and private sectors. Therefore, to improve coverage and lower the incidence of late-stage diagnosis, studies are evaluating the efficacy of implementing early detection programs into India's public health system [155]. Predominantly, the primary research should aim to develop and test awareness programs that are sensitive to traditions, societal expectations and education levels, leveraging community channels.

Indeed, through community-based education, public health initiatives can reduce stigma and raise understanding of screening benefits and risk factors. This fosters women's confidence in their healthcare choices. Studies are exploring the efficiency of disseminating knowledge through local resources. These resources include community leaders (such as religious frontrunners, political figures), media networks (including local talk shows and podcasts), and social media. This aside, in terms of community-based screening models, studies are underway to determine the practicality of deploying frontline health workers, such as Accredited Social Health Activists (ASHAs), to provide breast health education and clinical breast examinations in the community. These employees are trustworthy and can assist in breaking down cultural boundaries [164]. This in turn would enable women, especially those with a familial history or other high-risk factors, to benefit from accessible and affordable genetic testing or screening for mutations like *BRCA*, which would allow for targeted and early surveillance. Further studies are required to build successful community engagement models through local healthcare personnel and self-help groups. Investigations in urban slums across India have already demonstrated that engaging these groups is crucial to the long-term viability of new screening procedures for breast cancer [165].

Another key direction would be to incorporate BC screening into existing programs like the National Program for the Prevention and Control of Cancer, Diabetes, Cardiovascular Diseases, and Stroke (NPCDCS), which currently targets women aged 30 years and older. Undeniably, additional investigation is required to determine why increasing socioeconomic status and women's empowerment characteristics may not consistently correlate with increased screening uptake in certain regions. This includes reviewing the implementation of the present and future screening programs. Similarly, studies are also underway to determine how mobile screening camps can deliver services directly to the communities, especially in urban and rural areas with limited infrastructure [166].

16. Conclusions

The intricate landscape of breast cancer in India, as discussed in this comprehensive review, highlights specific challenges rooted in the country's diversified socioeconomic and genetic factors. This review presents integrated information on the molecular, cellular, and developmental biology of BC, with a particular emphasis on India's distinct epidemiological and clinical landscape. The findings from this review highlight a serious public health concern, with a considerable incidence of aggressive, triple-negative breast cancers and a younger median age of initiation than Western nations. The pathophysiological and molecular subtypes of BC are outlined, highlighting disease heterogeneity and the significance of personalized therapies. The mechanisms that drive BC progression are explained, demonstrating current studies that interpret complex signaling pathways. It also identifies specific knowledge gaps as well as diagnostic and therapeutic procedures or methodologies, indicating an urgent need for multi-faceted and comprehensive interventions.

Indeed, the global data on BC provides a basis, but it is evident that simply applying Western models to the Indian setting is inadequate to tackle the high mortality rates and the current inequities in treatment regimens. Therefore, the implications of this review are clear: the effective BC control in India demands culturally sensitive awareness campaigns, the development of effective national screening programs using suitable technologies (like ultrasound for younger women), and an increased investment in genetic counseling and treatment infrastructure.

In addition, to bridge the knowledge and diagnostic gaps, significant future strategies should emphasize generating India-specific molecular data and leveraging technological innovation. Specific strategies should focus on the development of the proposed "India-specific Genome Atlas" to uncover population-specific biomarkers, and should utilize AI-driven solutions to address issues like high breast density in screening and the need for early detection tools.

Finally, a comprehensive and integrated approach that combines research, education, early diagnosis, and accessible, high-quality healthcare services is not only necessary, but is also morally required, in order to reduce the growing burden of BC and increase survival rates across India.

Author Contributions: Conceptualization, writing, reviewing, and editing: A.D. and S.P. All authors have read and agreed to the published version of the manuscript.

Funding: This research received no external funding.

Institutional Review Board Statement: Not applicable.

Informed Consent Statement: Not applicable.

Data Availability Statement: No new data were created or analyzed in this study. Data sharing is not applicable to this article.

Conflicts of Interest: The authors declare no conflicts of interest.

References

1. Roy, P. Breast cancer in young Indian women: Factors, challenges in screening, and upcoming diagnostics. *J. Cancer Res. Clin. Oncol.* **2023**, *149*, 14409–14427. [CrossRef]
2. Kulkarni, A.; Kelkar, D.A.; Parikh, N.; Shashidhara, L.S.; Koppiker, C.B.; Kulkarni, M. Meta-Analysis of Prevalence of Triple-Negative Breast Cancer and Its Clinical Features at Incidence in Indian Patients With Breast Cancer. *JCO Glob. Oncol.* **2020**, *6*, 1052–1062. [CrossRef]
3. Sivasankar Perumal, A.D. Breast cancer and mental health: Understanding the link to depressioninanindianscenario. *Glob. J. Res. Anal.* **2025**, *14*, 1–4.
4. Samad, M.A.; Ahmad, I.; Khan, M.R.; Suhail, M.; Zughaibi, T.A.; Al-Abbasi, F.A.; Alhosaini, K.A.; Khan, M.S.; Kumer, A.; Tabrez, S. Breast Cancer: Molecular Pathogenesis and Targeted Therapy. *MedComm* **2025**, *6*, e70404. [CrossRef]
5. Srivastava, T.P.; Goel, I.; Gogia, A.; Parshad, R.; Monga, S.; Talukdar, J.; Rai, A.; Dhar, R.; Karmakar, S. Comprehensive Overview of Breast Cancer in India. *JCO Glob. Oncol.* **2025**, *11*, e2500083. [CrossRef]
6. Gupta, N.; Verma, R.K.; Gupta, S.; Prinja, S. Cost Effectiveness of Trastuzumab for Management of Breast Cancer in India. *JCO Glob. Oncol.* **2020**, *6*, 205–216. [CrossRef]
7. Mahalingam, S.; Scaria, V.; Sivasubbu, S. The Bharat Cancer Genome Atlas: Charting India's Unique Cancer Landscape for Precision Oncology. *Technol. Cancer Res. Treat.* **2025**, *24*, 15330338251381404. [CrossRef]
8. Sung, H.; Ferlay, J.; Siegel, R.L.; Laversanne, M.; Soerjomataram, I.; Jemal, A.; Bray, F. Global Cancer Statistics 2020: GLOBOCAN Estimates of Incidence and Mortality Worldwide for 36 Cancers in 185 Countries. *CA Cancer J. Clin.* **2021**, *71*, 209–249. [CrossRef]
9. DeSantis, C.; Siegel, R.; Bandi, P.; Jemal, A. Breast cancer statistics, 2011. *CA Cancer J. Clin.* **2011**, *61*, 409–418. [CrossRef] [PubMed]
10. Saxena, S.; Szabo, C.I.; Chopin, S.; Barjhoux, L.; Sinilnikova, O.; Lenoir, G.; Goldgar, D.E.; Bhatanager, D. BRCA1 and BRCA2 in Indian breast cancer patients. *Hum. Mutat.* **2002**, *20*, 473–474. [CrossRef]
11. India State-Level Disease Burden Initiative Cancer Collaborators. The burden of cancers and their variations across the states of India: The Global Burden of Disease Study 1990–2016. *Lancet Oncol.* **2018**, *19*, 1289–1306. [CrossRef]
12. Kulothungan, V.; Sathishkumar, K.; Leburu, S.; Ramamoorthy, T.; Stephen, S.; Basavarajappa, D.; Tomy, N.; Mohan, R.; Menon, G.R.; Mathur, P. Burden of cancers in India—Estimates of cancer crude incidence, YLLs, YLDs and DALYs for 2021 and 2025 based on National Cancer Registry Program. *BMC Cancer* **2022**, *22*, 527. [CrossRef]
13. Takiar, R.; Srivastav, A. Time trend in breast and cervix cancer of women in India—(1990–2003). *Asian Pac. J. Cancer Prev.* **2008**, *9*, 777–780.
14. Maurya, A.P.; Brahrnachari, S. Current Status of Breast Cancer Management in India. *Indian J. Surg.* **2021**, *83* (Suppl. S2), S316–S321. [CrossRef]
15. Ranganathan, S.; Dee, E.C.; Debnath, N.; Patel, T.A.; Jain, B.; Murthy, V. Access and barriers to genomic classifiers for breast cancer and prostate cancer in India. *Int. J. Cancer* **2024**, *154*, 1335–1339. [CrossRef]
16. Talukdar, J.; Srivastava, T.P.; Sahoo, O.S. Cancer stem cells: Signaling pathways and therapeutic targeting. *MedComm—Oncology* **2023**, *2*, e62. [CrossRef]

17. Weerarathna, I.N.; Luharia, A.; Uke, A.; Mishra, G. Challenges and Innovations in Breast Cancer Screening in India: A Review of Epidemiological Trends and Diagnostic Strategies. *Int. J. Breast Cancer* **2024**, *2024*, 6845966. [CrossRef] [PubMed]
18. Kaur, N.; Swain, S.K.; Banerjee, B.D.; Sharma, T.; Krishnalata, T. Organochlorine pesticide exposure as a risk factor for breast cancer in young Indian women: A case-control study. *South. Asian J. Cancer* **2019**, *8*, 212–214. [CrossRef] [PubMed]
19. Pillai, R.N.; Alex, A.; M.S., N.; Verma, V.; Shaji, A.; Pavithran, K.; Vijaykumar, D.K.; John, D. Economic burden of breast cancer in India, 2000–2021 and forecast to 2030. *Sci. Rep.* **2025**, *15*, 1323. [CrossRef]
20. Wadasadawala, T.; Mohanty, S.K.; Sen, S.; Kanala, T.S.; Maiti, S.; Puchali, N.; Gupta, S.; Sarin, R.; Parmar, V. Out-of-pocket payment and financial risk protection for breast cancer treatment: A prospective study from India. *Lancet Reg. Health Southeast Asia* **2024**, *24*, 100346. [CrossRef]
21. Pemmasani, S.K.; Raman, R.; Mohapatra, R.; Vidyasagar, M.; Acharya, A. A Review on the Challenges in Indian Genomics Research for Variant Identification and Interpretation. *Front. Genet.* **2020**, *11*, 753. [CrossRef] [PubMed]
22. Kumar, N.; Kapoor, R. Challenges and Opportunities in Oncology in India: Is There a Paradigm Shift? *Indian J. Surg.* **2025**, *87*, 976–977. [CrossRef]
23. Sun, Y.S.; Zhao, Z.; Yang, Z.N.; Xu, F.; Lu, H.J.; Zhu, Z.Y.; Shi, W.; Jiang, J.; Yao, P.-P.; Zhu, H.-P. Risk Factors and Preventions of Breast Cancer. *Int. J. Biol. Sci.* **2017**, *13*, 1387–1397. [CrossRef]
24. Ozsoy, A.; Barça, N.; Dolek, B.A.; Aktaş, H.; Elverici, E.; Araz, L.; Ozkaraoğlu, O. The Relationship Between Breast Cancer and Risk Factors: A Single-Center Study. *Eur. J. Breast Health* **2017**, *13*, 145–149. [CrossRef]
25. Allison, K.H. Molecular pathology of breast cancer: What a pathologist needs to know. *Am. J. Clin. Pathol.* **2012**, *138*, 770–780. [CrossRef]
26. Chikkala, R.; Bhayal, D.; Rani, N.; Modali, R.; Bhatia, K.; Dubey, B. Mutational landscape of BRCA gene mutations in Indian breast cancer patients: Retrospective insights from a diagnostic lab. *Egypt. J. Med. Hum. Genet.* **2024**, *25*, 101. [CrossRef]
27. Pal, M.; Das, D.; Pandey, M. Understanding genetic variations associated with familial breast cancer. *World J. Surg. Oncol.* **2024**, *22*, 271. [CrossRef]
28. Rehman, S.; Sameer, A.; Zahoor, L.; Abdullah, S.; Shah, Z.; Afroze, D.; Hussain, I.; Shaffi, S.M.; Syeed, N.; Rizvi, M.A.; et al. Distinct pattern of mutations of conserved regions of TP53 in colorectal cancer patients in the Kashmir population: An emerging high-risk area. *Ecancermedicalscience* **2009**, *3*, 129. [CrossRef] [PubMed]
29. Lalli, E.; Zambetti, G.P.; Ribeiro, R.C.; Figueiredo, B.C. Cancer associated to low-penetrance TP53 variants: Insights from p.R337H. *Trends Cancer* **2025**, *11*, 499–502. [CrossRef]
30. Daly, M.B.; Pal, T.; Berry, M.P.; Buys, S.S.; Dickson, P.; Domchek, S.M.; Elkhanany, A.; Friedman, S.; Goggins, M.; Hutton, M.L.; et al. Genetic/Familial High-Risk Assessment: Breast, Ovarian, and Pancreatic, Version 2.2021, NCCN Clinical Practice Guidelines in Oncology. *J. Natl. Compr. Canc Netw.* **2021**, *19*, 77–102. [CrossRef]
31. John, A.O.; Singh, A.; Yadav, P.; Joel, A.; Thumaty, D.B.; Ninan, K.F.; Georgy, J.T.; Cherian, A.J.; Thomas, S.; Thomas, A.; et al. The BRCA mutation spectrum among breast and ovarian cancers in India: Highlighting the need to screen BRCA1 185delAG among South Indians. *Eur. J. Hum. Genet.* **2024**, *32*, 1319–1326. [CrossRef]
32. Patra, A.; Ali, S.S.; Devi, N.M.; Qadeer, A.S.; Kamalakannan, S.; Nag, S.; Kulkarni, S.S.; Rajappa, S.; Hariharan, N.; Pant, H.B.; et al. Prevalence of BRCA mutation in breast and ovarian cancer among women in India: A systematic review and meta-analysis protocol. *PLoS ONE* **2024**, *19*, e0306612. [CrossRef]
33. Goyal, J.; Yadav, A.; Malhotra, H. Association of Positive Family History and Clinicopathological Features in Breast Cancer in Young Indian Females—A Pilot Study. *J. Radiat. Cancer Res.* **2025**, *16*, 23–26. [CrossRef]
34. Dupont, W.D.; Parl, F.F.; Hartmann, W.H.; Brinton, L.A.; Winfield, A.C.; Worrell, J.A.; Schuyler, P.A.; Plummer, W.D. Breast cancer risk associated with proliferative breast disease and atypical hyperplasia. *Cancer* **1993**, *71*, 1258–1265. [CrossRef] [PubMed]
35. Liebermann, E.; Patwardhan, V.; Usmanova, G.; Aktar, N.; Agrawal, S.; Bhamare, P.; McCarthy, M.; Ginsburg, O.; Kumar, S. Barriers to Follow-Up of an Abnormal Clinical Breast Examination in Uttar Pradesh, India: A Qualitative Study. *JCO Glob. Oncol.* **2024**, *10*, e2400001. [CrossRef] [PubMed]
36. Hartmann, L.C.; Sellers, T.A.; Frost, M.H.; Lingle, W.L.; Degnim, A.C.; Ghosh, K.; Vierkant, R.A.; Maloney, S.D.; Pankratz, V.S.; Hillman, D.W.; et al. Benign breast disease and the risk of breast cancer. *N. Engl. J. Med.* **2005**, *353*, 229–237. [CrossRef] [PubMed]
37. Kumar, N.; Babu, A.; Singh, S.; Lakhera, K.K.; Patel, P.; Gurjar, B.; Kumar, A.; Gora, B.S.; Sharma, R.G. Patterns and Prevalence of Benign Breast Diseases: An Epidemiological Chart Review Analysis of 5965 Patients from North Western India. *Indian J. Surg. Oncol.* **2025**, *16*, 1264–1269. [CrossRef]
38. Sagiraju, H.K.R.; Gogia, A.; Goel, A.; Batra, A.; Sharma, D.; Saini, S.K.; Mishra, A.; Deo, S.; Prasad, C.P.; Tanwar, P.; et al. Effect of obesity on the survival outcomes among Indian women with breast cancer. *J. Clin. Oncol.* **2025**, *43*, e12558. [CrossRef]
39. Sen, S.; Khan, P.K.; Wadasadawala, T.; Mohanty, S.K. Socio-economic and regional variation in breast and cervical cancer screening among Indian women of reproductive age: A study from National Family Health Survey, 2019–2021. *BMC Cancer* **2022**, *22*, 1279. [CrossRef]

40. Chatterjee, D.; McDuffie, E.E.; Smith, S.J.; Bindle, L.; van Donkelaar, A.; Hammer, M.S.; Venkataraman, C.; Brauer, M.; Martin, R.V. Source Contributions to Fine Particulate Matter and Attributable Mortality in India and the Surrounding Region. *Environ. Sci. Technol.* **2023**, *57*, 10263–10275. [CrossRef]
41. Rajbongshi, N.; Mahanta, L.B.; Nath, D.C. Evaluation of Female Breast Cancer Risk Among the Betel Quid Chewer: A Bio-Statistical Assessment in Assam, India. *Nepal. J. Epidemiol.* **2015**, *5*, 494–498. [CrossRef]
42. Das, A.M.; Gogia, A.; Garg, M.; Elaiyaraja, A.; Arambam, P.; Mathur, S.; Babu-Rajendran, R.; Deo, S.; Kumar, L.; Das, B.C.; et al. Urinary concentration of endocrine-disrupting phthalates and breast cancer risk in Indian women: A case-control study with a focus on mutations in phthalate-responsive genes. *Cancer Epidemiol.* **2022**, *79*, 102188.
43. Maurya, A.P.; Brahmachari, S. Association of hormonal and reproductive risk factors with breast cancer in Indian women: A systematic review of case-control studies. *Indian J Cancer* **2023**, *60*, 4–11. [CrossRef]
44. Vishwakarma, G.; Ndetan, H.; Das, D.N.; Gupta, G.; Suryavanshi, M.; Mehta, A.; Singh, K.P. Reproductive factors and breast cancer risk: A meta-analysis of case-control studies in Indian women. *South Asian J. Cancer* **2019**, *8*, 80–84. [CrossRef]
45. Parker, S.A.; Weygand, J.; Bernat, B.G.; Jackson, A.M.; Mawlawi, O.; Barreto, I.; Hao, Y.; Khan, R.; Yorke, A.A.; Swanson, W.; et al. Assessing Radiology and Radiation Therapy Needs for Cancer Care in Low-and-Middle-Income Countries: Insight From a Global Survey of Departmental and Institutional Leaders. *Adv. Radiat. Oncol.* **2024**, *9*, 101615. [CrossRef]
46. Meeta, M.; Tanvir, T. Menopause Hormone Therapy in the Changing Trends of Breast Cancer in India. *J. Midlife Health* **2020**, *11*, 196–198. [CrossRef] [PubMed]
47. Clusan, L.; Ferriere, F.; Flouriot, G.; Pakdel, F. A Basic Review on Estrogen Receptor Signaling Pathways in Breast Cancer. *Int. J. Mol. Sci.* **2023**, *24*, 6834. [CrossRef]
48. Rodriguez-Casanova, A.; Costa-Fraga, N.; Castro-Carballeira, C.; González-Conde, M.; Abuin, C.; Bao-Caamano, A.; García-Caballero, T.; Brozos-Vazquez, E.; Rodriguez-López, C.; Cebey, V.; et al. A genome-wide cell-free DNA methylation analysis identifies an episignature associated with metastatic luminal B breast cancer. *Front. Cell Dev. Biol.* **2022**, *10*, 1016955. [CrossRef] [PubMed]
49. Ulaganathan, K.; Puranam, K.; Mukta, S.; Hanumanth, S.R. Expression profiling of luminal B breast tumor in Indian women. *J. Cancer Res. Clin. Oncol.* **2023**, *149*, 13645–13664. [CrossRef]
50. Prat, A.; Pascual, T.; De Angelis, C.; Gutierrez, C.; Llombart-Cussac, A.; Wang, T.; Cortés, J.; Rexer, B.; Paré, L.; Forero, A.; et al. HER2-Enriched Subtype and ERBB2 Expression in HER2-Positive Breast Cancer Treated with Dual HER2 Blockade. *J. Natl. Cancer Inst.* **2020**, *112*, 46–54. [CrossRef] [PubMed]
51. Yin, L.; Duan, J.J.; Bian, X.W.; Yu, S.C. Triple-negative breast cancer molecular subtyping and treatment progress. *Breast Cancer Res.* **2020**, *22*, 61. [CrossRef]
52. Prat, A.; Solovieff, N.; André, F.; O'SHaughnessy, J.; Cameron, D.A.; Janni, W.; Sonke, G.S.; Yap, Y.-S.; Yardley, D.A.; Partridge, A.H.; et al. Intrinsic Subtype and Overall Survival of Patients with Advanced HR+/HER2- Breast Cancer Treated with Ribociclib and ET: Correlative Analysis of MONALEESA-2, -3, -7. *Clin. Cancer Res.* **2024**, *30*, 793–802. [CrossRef] [PubMed]
53. Nolan, E.; Lindeman, G.J.; Visvader, J.E. Deciphering breast cancer: From biology to the clinic. *Cell* **2023**, *186*, 1708–1728. [CrossRef] [PubMed]
54. Li, W.; Zou, L.; Huang, S.; Miao, H.; Liu, K.; Geng, Y.; Liu, Y.; Wu, W. The anticancer activity of bile acids in drug discovery and development. *Front. Pharmacol.* **2024**, *15*, 1362382. [CrossRef]
55. Wang, J.; Li, B.; Luo, M.; Huang, J.; Zhang, K.; Zheng, S.; Zhang, S.; Zhou, J. Progression from ductal carcinoma in situ to invasive breast cancer: Molecular features and clinical significance. *Signal Transduct. Target. Ther.* **2024**, *9*, 83. [CrossRef] [PubMed]
56. Kim, E.S. Molecular targets and therapies associated with poor prognosis of triple-negative breast cancer (Review). *Int. J. Oncol.* **2025**, *66*, 52. [CrossRef]
57. Hassan, A.; Aubel, C. The PI3K/Akt/mTOR Signaling Pathway in Triple-Negative Breast Cancer: A Resistance Pathway and a Prime Target for Targeted Therapies. *Cancers* **2025**, *17*, 2232. [CrossRef]
58. Pereira, C.; Martis, M.; D'Souza, R.; Tauro, L.F. Correlation of Clinicopathological Features of Breast Cancer with Molecular Subtypes Taking Ki-67 into Consideration: Single Institution Experience Over 5 Years. *Curr. Health Sci. J.* **2021**, *47*, 348–352. [CrossRef]
59. Jonnada, P.K.; Sushma, C.; Karyampudi, M.; Dharanikota, A. Prevalence of Molecular Subtypes of Breast Cancer in India: A Systematic Review and Meta-analysis. *Indian J. Surg Oncol.* **2021**, *12* (Suppl. S1), 152–163. [CrossRef]
60. Siddiqui, B.; Ahmed, S.; Sinha, D.; Sharma, A.V.L. Molecular Classification of Breast Carcinoma in a Tertiary Hospital of India: The Recent Trends. *Indian. J. Surg. Oncol.* **2023**, *14*, 176–180. [CrossRef]
61. Jain, S.; Narang, V.; Jain, K.; Paul, D.; Singh, J.; Sohi, A.S.; Sood, S.; Aggarwal, R.; Sood, N.; Brar, G.S. Prevalence of Molecular Subtypes in Operated Cases of Breast Cancer and Its Clinicopathological Correlation: A Single Institute Study from a Tertiary Cancer Centre in North India. *Indian. J. Surg. Oncol.* **2021**, *12*, 538–544. [CrossRef]
62. Rathnasamy, N.; Mullapally, S.; Sirohi, B. Precision oncology in Low and Middle income countries: A word of caution. *Int. J. Cancer Care Deliv.* **2021**, *1*, 27–31. [CrossRef]

63. Suresh AVSS, M.; Dadireddy, P.K. Cost-effectiveness analysis of gene expression profiling for breast cancer treatment decisions in India. *Int. J. Community Med. Public Health* **2025**, *12*, 2100–2102. [CrossRef]
64. Gulwe, A.B.; Dawkhar, S.S. Role of SNP in the Breast Cancer Development of Indian Women. *Int. J. Res. Innov. Appl. Sci.* **2025**, *10*, 1160–1173. [CrossRef]
65. Nirgude, S.; Desai, S.; Khanchandani, V.; Nagarajan, V.; Thumsi, J.; Choudhary, B. Integration of exome-seq and mRNA-seq using DawnRank, identified genes involved in innate immunity as drivers of breast cancer in the Indian cohort. *PeerJ* **2023**, *11*, e16033. [CrossRef] [PubMed]
66. Sengar, M.; Pramesh, C.S.; Mehndiratta, A.; Shah, S.; Munshi, A.; Vijaykumar, D.K.; Puri, A.; Mathew, B.; Arora, R.S.; T, P.K.; et al. Ensuring quality in contextualised cancer management guidelines for resource-constraint settings: Using a systematic approach. *BMJ Glob. Health* **2022**, *7*, e009584. [CrossRef]
67. Zhu, H.; Chua, M.L.K.; Chitapanarux, I.; Kaidar-Person, O.; Mwaba, C.; Alghamdi, M.; Mignola, A.R.; Amrogowicz, N.; Yazici, G.; Bourhaleb, Z.; et al. Global radiotherapy demands and corresponding radiotherapy-professional workforce requirements in 2022 and predicted to 2050: A population-based study. *Lancet Glob. Health* **2024**, *12*, e1945–e1953. [CrossRef]
68. Sankarapillai, J.; Ramamoorthy, T.; Sarveswaran, G.; Venugopal, G.; Mathur, P. Epidemiological analysis of radiation therapy utilization and its implications for cancer care in India-insights from National Cancer Registry Programme. *BMC Cancer* **2025**, *25*, 1062. [CrossRef]
69. Ghosh, A.; Chaubal, R.; Das, C.; Parab, P.; Das, S.; Maitra, A.; Majumder, P.P.; Gupta, S.; Biswas, N.K. Genomic hallmarks of endocrine therapy resistance in ER/PR+HER2- breast tumours. *Commun. Biol.* **2025**, *8*, 207. [CrossRef] [PubMed]
70. Kang, C.; Syed, Y.Y. Atezolizumab (in Combination with Nab-Paclitaxel): A Review in Advanced Triple-Negative Breast Cancer. *Drugs* **2020**, *80*, 601–607. [CrossRef]
71. Laureano, R.S.; Sprooten, J.; Vanmeerbeerk, I.; Borras, D.M.; Govaerts, J.; Naulaerts, S.; Berneman, Z.N.; Beuselinck, B.; Bol, K.F.; Borst, J.; et al. Trial watch: Dendritic cell (DC)-based immunotherapy for cancer. *Oncoimmunology* **2022**, *11*, 2096363. [CrossRef]
72. Pandey, P.; Chaudhary, R.; Tripathi, D.; Lavudi, K.; Dua, K.; Weinfeld, M.; Lavasanifar, A.; Rajinikanth, P.S. Personalized treatment approach for HER2-positive metastatic breast cancer. *Med. Oncol.* **2024**, *41*, 252. [CrossRef] [PubMed]
73. Deepa, K.V.; Gadgil, A.; Lofgren, J.; Mehare, S.; Bhandarkar, P.; Roy, N. Is quality of life after mastectomy comparable to that after breast conservation surgery? A 5-year follow up study from Mumbai, India. *Qual. Life Res.* **2020**, *29*, 683–692. [CrossRef]
74. Bhattacharya, S.; Asaithamby, A. Repurposing DNA repair factors to eradicate tumor cells upon radiotherapy. *Transl. Cancer Res.* **2017**, *6* (Suppl. S5), S822–S839. [CrossRef]
75. Khan, A.; Sisodiya, S.; Aftab, M.; Tanwar, P.; Hussain, S.; Gupta, V. Mechanisms and Therapeutic Strategies for Endocrine Resistance in Breast Cancer: A Comprehensive Review and Meta-Analysis. *Cancers* **2025**, *17*, 1653. [CrossRef] [PubMed]
76. Dey, A.; Chandel, A.K.S.; Sanyal, R.; Mishra, A.; Pandey, D.K.; De Falco, V.; Upadhyay, A.; Kandimalla, R.; Chaudhary, A.; Dhanjal, J.K.; et al. Cancer chemotherapy and beyond: Current status, drug candidates, associated risks and progress in targeted therapeutics. *Genes. Dis.* **2023**, *10*, 1367–1401.
77. Rai, A.; Deshpande, S.G.; Vaidya, A.; Shinde, R.K. Advancements in Immunotherapy for Breast Cancer: Mechanisms, Efficacy, and Future Directions. *Cureus* **2024**, *16*, e68351. [CrossRef] [PubMed]
78. Macias, H.; Hinck, L. Mammary gland development. *Wiley Interdiscip. Rev. Dev. Biol.* **2012**, *1*, 533–557. [CrossRef]
79. Lloyd-Lewis, B.; Harris, O.B.; Watson, C.J.; Davis, F.M. Mammary Stem Cells: Premise, Properties, and Perspectives. *Trends Cell Biol.* **2017**, *27*, 556–567. [CrossRef]
80. Soteriou, D.; Fuchs, Y. A matter of life and death: Stem cell survival in tissue regeneration and tumour formation. *Nat. Rev. Cancer* **2018**, *18*, 187–201. [CrossRef]
81. Mascré, G.; Dekoninck, S.; Drogat, B.; Youssef, K.K.; Brohée, S.; Sotiropoulou, P.A.; Simons, B.D.; Blanpain, C. Distinct contribution of stem and progenitor cells to epidermal maintenance. *Nature* **2012**, *489*, 257–262. [CrossRef]
82. Papaccio, F.; Paino, F.; Regad, T.; Papaccio, G.; Desiderio, V.; Tirino, V. Concise Review: Cancer Cells, Cancer Stem Cells, and Mesenchymal Stem Cells: Influence in Cancer Development. *Stem Cells Transl. Med.* **2017**, *6*, 2115–2125. [CrossRef] [PubMed]
83. Brooks, M.D.; Burness, M.L.; Wicha, M.S. Therapeutic Implications of Cellular Heterogeneity and Plasticity in Breast Cancer. *Cell Stem Cell* **2015**, *17*, 260–271. [CrossRef] [PubMed]
84. van Amerongen, R.; Bowman, A.N.; Nusse, R. Developmental stage and time dictate the fate of Wnt/beta-catenin-responsive stem cells in the mammary gland. *Cell Stem Cell* **2012**, *11*, 387–400. [CrossRef]
85. Sachs, N.; de Ligt, J.; Kopper, O.; Gogola, E.; Bounova, G.; Weeber, F.; Balgobind, A.V.; Wind, K.; Gracanin, A.; Begthel, H.; et al. A Living Biobank of Breast Cancer Organoids Captures Disease Heterogeneity. *Cell* **2018**, *172*, 373–386.e10. [CrossRef]
86. Pauli, C.; Hopkins, B.D.; Prandi, D.; Shaw, R.; Fedrizzi, T.; Sboner, A.; Sailer, V.; Augello, M.; Puca, L.; Rosati, R.; et al. Personalized In Vitro and In Vivo Cancer Models to Guide Precision Medicine. *Cancer Discov.* **2017**, *7*, 462–477. [CrossRef] [PubMed]
87. McDonald, E.S.; Clark, A.S.; Tchou, J.; Zhang, P.; Freedman, G.M. Clinical Diagnosis and Management of Breast Cancer. *J. Nucl. Med.* **2016**, *57* (Suppl. S1), 9S–16S. [CrossRef]

88. Coté, D.; Eustace, A.; Toomey, S.; Cremona, M.; Milewska, M.; Furney, S.; Carr, A.; Fay, J.; Kay, E.; Kennedy, S.; et al. Germline single nucleotide polymorphisms in ERBB3 and BARD1 genes result in a worse relapse free survival response for HER2-positive breast cancer patients treated with adjuvant based docetaxel, carboplatin and trastuzumab (TCH). *PLoS ONE* **2018**, *13*, e0200996. [CrossRef]
89. Polgar, O.; Bates, S.E. ABC transporters in the balance: Is there a role in multidrug resistance? *Biochem. Soc. Trans.* **2005**, *33 Pt 1*, 241–245. [CrossRef]
90. Housman, G.; Byler, S.; Heerboth, S.; Lapinska, K.; Longacre, M.; Snyder, N.; Sarkar, S. Drug resistance in cancer: An overview. *Cancers* **2014**, *6*, 1769–1792. [CrossRef]
91. Gandhi, N.; Das, G.M. Metabolic Reprogramming in Breast Cancer and Its Therapeutic Implications. *Cells* **2019**, *8*, 89. [CrossRef]
92. Raghav, D.; Sharma, V. An In Silico Evaluation of Deleterious Nonsynonymous Single Nucleotide Polymorphisms in the ErbB3 Oncogene. *Biores. Open Access* **2013**, *2*, 206–211. [CrossRef]
93. Thussbas, C.; Nahrig, J.; Streit, S.; Bange, J.; Kriner, M.; Kates, R.; Ulm, K.; Kiechle, M.; Hoefler, H.; Ullrich, A.; et al. FGFR4 Arg388 allele is associated with resistance to adjuvant therapy in primary breast cancer. *J. Clin. Oncol.* **2006**, *24*, 3747–3755. [CrossRef]
94. Mackenzie, P.I.; Owens, I.S.; Burchell, B.; Bock, K.W.; Bairoch, A.; Belanger, A.; Gigleux, S.F.; Green, M.; Hum, D.W.; Iyanagi, T.; et al. The UDP glycosyltransferase gene superfamily: Recommended nomenclature update based on evolutionary divergence. *Pharmacogenetics* **1997**, *7*, 255–269. [CrossRef]
95. Fojo, A.; Lebo, R.; Shimizu, N.; Chin, J.E.; Roninson, I.B.; Merlino, G.T.; Gottesman, M.M.; Pastan, I. Localization of multidrug resistance-associated DNA sequences to human chromosome 7. *Somat. Cell Mol. Genet.* **1986**, *12*, 415–420. [CrossRef] [PubMed]
96. Tashbaeva, R.E.; Hwang, D.N.; Song, G.S.; Choi, N.H.; Lee, J.H.; Lyoo, Y.S.; Lee, S.J.; Jung, D.I.; Kim, H.Y.; Sur, J.H. Cellular characterization of multidrug resistance P-glycoprotein, alpha fetoprotein, and neovascular endothelium-associated antigens in canine hepatocellular carcinoma and cirrhotic liver. *Vet. Pathol.* **2007**, *44*, 600–606. [CrossRef]
97. Vulsteke, C.; Pfeil, A.M.; Schwenkglenks, M.; Pettengell, R.; Szucs, T.D.; Lambrechts, D.; Peeters, M.; van Dam, P.; Dieudonné, A.S.; Hatse, S.; et al. Impact of genetic variability and treatment-related factors on outcome in early breast cancer patients receiving (neo-) adjuvant chemotherapy with 5-fluorouracil, epirubicin and cyclophosphamide, and docetaxel. *Breast Cancer Res. Treat.* **2014**, *147*, 557–570. [CrossRef] [PubMed]
98. Chang, H.; Rha, S.Y.; Jeung, H.-.-C.; Im, C.-.-K.; Ahn, J.B.; Kwon, W.S.; Yoo, N.C.; Roh, J.K.; Chung, H.C. Association of the ABCB1 gene polymorphisms 2677G>T/A and 3435C>T with clinical outcomes of paclitaxel monotherapy in metastatic breast cancer patients. *Ann. Oncol.* **2009**, *20*, 272–277. [CrossRef] [PubMed]
99. Vishnukumar, S.; Umamaheswaran, G.; Anichavezhi, D.; Indumathy, S.; Adithan, C.; Srinivasan, K.; Kadambari, D. P-glycoprotein expression as a predictor of response to neoadjuvant chemotherapy in breast cancer. *Indian J. Cancer* **2013**, *50*, 195–199. [CrossRef]
100. Mishra, D.; Patel, V.; Banerjee, D. Nitric Oxide and S-Nitrosylation in Cancers: Emphasis on Breast Cancer. *Breast Cancer* **2020**, *14*, 1178223419882688. [CrossRef]
101. Devendran, A.; Nampoothiri, S.; Shewade, D.G.; Chatterjee, S.; Jayaraman, B.; Chandrasekharan, A. Allele, Genotype and Haplotype Structures of Functional Polymorphic Variants in Endothelial Nitric Oxide Synthase (eNOS), Angiotensinogen (ACE) and Aldosterone Synthase (CYP11B2) Genes in Healthy Pregnant Women of Indian Ethnicity. *J. Reprod. Infertil.* **2015**, *16*, 180–192.
102. Coutinho, L.L.; Femino, E.L.; Gonzalez, A.L.; Moffat, R.L.; Heinz, W.F.; Cheng, R.Y.S.; Lockett, S.J.; Rangel, M.C.; Ridnour, L.A.; Wink, D.A. NOS2 and COX-2 Co-Expression Promotes Cancer Progression: A Potential Target for Developing Agents to Prevent or Treat Highly Aggressive Breast Cancer. *Int. J. Mol. Sci.* **2024**, *25*, 6103. [CrossRef]
103. Choudhari, S.K.; Chaudhary, M.; Bagde, S.; Gadbail, A.R.; Joshi, V. Nitric oxide and cancer: A review. *World J. Surg. Oncol.* **2013**, *11*, 118. [CrossRef]
104. Loibl, S.; von Minckwitz, G.; Weber, S.; Sinn, H.; Schini-Kerth, V.B.; Lobysheva, I.; Nepveu, F.; Wolf, G.; Strebhardt, K.; Kaufmann, M. Expression of endothelial and inducible nitric oxide synthase in benign and malignant lesions of the breast and measurement of nitric oxide using electron paramagnetic resonance spectroscopy. *Cancer* **2002**, *95*, 1191–1198. [CrossRef]
105. Liang, F.; Wang, M.; Li, J.; Guo, J. The evolution of S-nitrosylation detection methodology and the role of protein S-nitrosylation in various cancers. *Cancer Cell Int.* **2024**, *24*, 408. [CrossRef] [PubMed]
106. Tualeka, A.R.; Jalaludin, J.; Gasana, J.; Sopian, N.A.; Chao, H.R.; Yusmaidie, M.; Perumal, V.; Zurimi, S.; Rahmawati, P.; Ahsan, A.; et al. Association between eNOS rs1799983 (894G>T) Polymorphism with Cancer and Stroke Risk: A meta-analysis. *F1000Research* **2024**, *12*, 1467. [CrossRef]
107. Suszynska, M.; Kluzniak, W.; Wokolorczyk, D.; Jakubowska, A.; Huzarski, T.; Gronwald, J.; Debniak, T.; Szwiec, M.; Ratajska, M.; Klonowska, K.; et al. BARD1 is A Low/Moderate Breast Cancer Risk Gene: Evidence Based on An Association Study of the Central European p.Q564X Recurrent Mutation. *Cancers* **2019**, *11*, 740. [CrossRef]
108. Hashizume, R.; Fukuda, M.; Maeda, I.; Nishikawa, H.; Oyake, D.; Yabuki, Y.; Ogata, H.; Ohta, T. The RING heterodimer BRCA1-BARD1 is a ubiquitin ligase inactivated by a breast cancer-derived mutation. *J. Biol. Chem.* **2001**, *276*, 14537–14540. [CrossRef]

109. Meza, J.E.; Brzovic, P.S.; King, M.C.; Klevit, R.E. Mapping the functional domains of BRCA1. Interaction of the ring finger domains of BRCA1 and BARD1. *J. Biol. Chem.* **1999**, *274*, 5659–5665. [CrossRef] [PubMed]
110. National Library of Medicine N-CHs. CD24 Molecule [Homo Sapiens (Human)]. 2023. Available online: https://www.ncbi.nlm.nih.gov/gene?Db=gene&Cmd=DetailsSearch&Term=100133941 (accessed on 22 November 2025).
111. Schabath, H.; Runz, S.; Joumaa, S.; Altevogt, P. CD24 affects CXCR4 function in pre-B lymphocytes and breast carcinoma cells. *J. Cell Sci.* **2006**, *119 Pt 2*, 314–325. [CrossRef]
112. Kwon, M.J.; Han, J.; Seo, J.H.; Song, K.; Jeong, H.M.; Choi, J.-S.; Kim, Y.J.; Lee, S.-H.; Choi, Y.-L.; Shin, Y.K. CD24 Overexpression Is Associated with Poor Prognosis in Luminal A and Triple-Negative Breast Cancer. *PLoS ONE* **2015**, *10*, e0139112. [CrossRef]
113. Baumann, P.; Cremers, N.; Kroese, F.; Orend, G.; Chiquet-Ehrismann, R.; Uede, T.; Yagita, H.; Sleeman, J.P. CD24 expression causes the acquisition of multiple cellular properties associated with tumor growth and metastasis. *Cancer Res.* **2005**, *65*, 10783–10793. [CrossRef]
114. Marmé, F.; Werft, W.; Walter, A.; Keller, S.; Wang, X.; Benner, A.; Burwinkel, B.; Sinn, P.; Hug, S.; Sohn, C.; et al. CD24 Ala57Val polymorphism predicts pathologic complete response to sequential anthracycline- and taxane-based neoadjuvant chemotherapy for primary breast cancer. *Breast Cancer Res. Treat.* **2012**, *132*, 819–831. [CrossRef]
115. Zhou, X. CD24 polymorphisms cannot predict pathologic complete response to anthracycline- and taxane-based neoadjuvant chemotherapy in breast cancer. *Clin. Breast Cancer* **2014**, *14*, e33–e40. [CrossRef]
116. Rao, U.S.C.; Devendran, A.; Satyamoorthy, K.; Shewade, D.G.; Krishnamoorthy, R.; Chandrasekaran, A. Functional characterization of promoter region polymorphisms of human CYP2C19 gene. *Mol. Biol. Rep.* **2011**, *38*, 4171–4179.
117. Crewe, H.K.; Ellis, S.W.; Lennard, M.S.; Tucker, G.T. Variable contribution of cytochromes P450 2D6, 2C9 and 3A4 to the 4-hydroxylation of tamoxifen by human liver microsomes. *Biochem. Pharmacol.* **1997**, *53*, 171–178. [CrossRef] [PubMed]
118. Artigalas, O.; Vanni, T.; Hutz, M.H.; Ashton-Prolla, P.; Schwartz, I.V. Influence of CYP19A1 polymorphisms on the treatment of breast cancer with aromatase inhibitors: A systematic review and meta-analysis. *BMC Med.* **2015**, *13*, 139. [CrossRef] [PubMed]
119. Liu, L.; Bai, Y.X.; Zhou, J.H.; Sun, X.W.; Sui, H.; Zhang, W.J.; Yuan, H.H.; Xie, R.; Wei, X.L.; Zhang, T.T.; et al. A polymorphism at the 3′-UTR region of the aromatase gene is associated with the efficacy of the aromatase inhibitor, anastrozole, in metastatic breast carcinoma. *Int. J. Mol. Sci.* **2013**, *14*, 18973–18988. [CrossRef]
120. Miron, L.; Negura, L.; Peptanariu, D.; Marinca, M. Research on aromatase gene (CYP19A1) polymorphisms as a predictor of endocrine therapy effectiveness in breast cancer. *Rev. Med. Chir. Soc. Med. Nat. Iasi* **2012**, *116*, 997–1004.
121. Henry, N.L.; Chan, H.P.; Dantzer, J.; Goswami, C.P.; Li, L.; Skaar, T.C.; Rae, J.M.; Desta, Z.; Khouri, N.; Pinsky, R.; et al. Aromatase inhibitor-induced modulation of breast density: Clinical and genetic effects. *Br. J. Cancer* **2013**, *109*, 2331–2339. [CrossRef]
122. Gor, P.P.; Su, H.I.; Gray, R.J.; A Gimotty, P.; Horn, M.; Aplenc, R.; Vaughan, W.P.; Tallman, M.S.; Rebbeck, T.R.; DeMichele, A. Cyclophosphamide-metabolizing enzyme polymorphisms and survival outcomes after adjuvant chemotherapy for node-positive breast cancer: A retrospective cohort study. *Breast Cancer Res.* **2010**, *12*, R26. [CrossRef]
123. Beelen, K.; Opdam, M.; Severson, T.M.; Koornstra, R.H.T.; Vincent, A.D.; Hauptmann, M.; van Schaik, R.H.N.; Berns, E.M.J.J.; Vermorken, J.B.; van Diest, P.J.; et al. CYP2C19 2 predicts substantial tamoxifen benefit in postmenopausal breast cancer patients randomized between adjuvant tamoxifen and no systemic treatment. *Breast Cancer Res. Treat.* **2013**, *139*, 649–655. [CrossRef]
124. Johnston, S.; Dowsett, M.; Smith, I.; Sacks, N.; Haynes, B.; Jarman, M.; Ebbs, S. Acquired tamoxifen resistance in human breast cancer and reduced intra-tumoral drug concentration. *Lancet* **1993**, *342*, 1521–1522. [CrossRef]
125. Satyanarayana, C.R.; Devendran, A.; Sundaram, R.; Gopal, S.D.; Rajagopal, K.; Chandrasekaran, A. Genetic variations and haplotypes of the 5′ regulatory region of CYP2C19 in South Indian population. *Drug Metab. Pharmacokinet.* **2009**, *24*, 185–193. [CrossRef]
126. Anichavezhi, D.; Chakradhara Rao, U.S.; Shewade, D.G.; Krishnamoorthy, R.; Adithan, C. Distribution of CYP2C19*17 allele and genotypes in an Indian population. *J. Clin. Pharm. Ther.* **2012**, *37*, 313–318. [CrossRef] [PubMed]
127. Regan, M.M.; Leyland-Jones, B.; Bouzyk, M.; Pagani, O.; Tang, W.; Kammler, R.; Dell'Orto, P.; Biasi, M.O.; Thürlimann, B.; Lyng, M.B.; et al. CYP2D6 genotype and tamoxifen response in postmenopausal women with endocrine-responsive breast cancer: The breast international group 1-98 trial. *J. Natl. Cancer Inst.* **2012**, *104*, 441–451. [CrossRef] [PubMed]
128. Dezentjé, V.O.; Van Schaik, R.H.N.; Vletter-Bogaartz, J.M.; Van Der Straaten, T.; Wessels, J.A.M.; Kranenbarg, E.M.-K.; Berns, E.M.; Seynaeve, C.; Putter, H.; Van De Velde, C.J.H.; et al. CYP2D6 genotype in relation to tamoxifen efficacy in a Dutch cohort of the tamoxifen exemestane adjuvant multinational (TEAM) trial. *Breast Cancer Res. Treat.* **2013**, *140*, 363–373. [CrossRef]
129. Neven, P.; Jongen, L.; Lintermans, A.; Van Asten, K.; Blomme, C.; Lambrechts, D.; Poppe, A.; Wildiers, H.; Dieudonné, A.-S.; Brouckaert, O.; et al. Tamoxifen Metabolism and Efficacy in Breast Cancer: A Prospective Multicenter Trial. *Clin. Cancer Res.* **2018**, *24*, 2312–2318. [CrossRef]
130. Sanchez-Spitman, A.; Dezentjé, V.; Swen, J.; Moes, D.J.A.r.; Böhringer, S.; Batman, E.; Van Druten, E.; Smorenburg, C.; Van Bochove, A.; Zeillemaker, A.; et al. Tamoxifen Pharmacogenetics and Metabolism: Results From the Prospective CYPTAM Study. *J. Clin. Oncol.* **2019**, *37*, 636–646. [CrossRef]

131. Kadri, M.S.N.; Patel, K.M.; Bhargava, P.A.; Shah, F.D.; Badgujar, N.V.; Tarapara, B.V.; Patel, P.S.; Shaikh, M.I.; Shah, K.; Patel, A.; et al. Mutational Landscape for Indian Hereditary Breast and Ovarian Cancer Cohort Suggests Need for Identifying Population Specific Genes and Biomarkers for Screening. *Front. Oncol.* **2020**, *10*, 568786. [CrossRef] [PubMed]
132. Kumar, R.M.R.; Joghee, S. Enhancing breast cancer treatment through pharmacogenomics: A narrative review. *Clin. Chim. Acta* **2024**, *562*, 119893. [CrossRef]
133. Jadeja, N.; Rajakumar, N.; Reddy, N.; Ali, N.; Lichten, L. Reflections on my international genetic counseling rotations: Contrasts in practice between India and the United States. *Genet. Med. Open* **2024**, *2* (Suppl. S2), 101871. [CrossRef]
134. Damodaran, S.E.; Pradhan, S.C.; Umamaheswaran, G.; Kadambari, D.; Reddy, K.S.; Adithan, C. Genetic polymorphisms of CYP2D6 increase the risk for recurrence of breast cancer in patients receiving tamoxifen as an adjuvant therapy. *Cancer Chemother. Pharmacol.* **2012**, *70*, 75–81. [CrossRef] [PubMed]
135. Bandyopadhyay, A.; Sharma, S.; Behera, D.; Singh, N. UGT1A1 Gene Polymorphisms in Patients with Small Cell Lung Cancer Treated with Irinotecan-Platinum Doublet Chemotherapy and Their Association with Gastrointestinal Toxicity and Overall Survival. *Oncologist* **2021**, *26*, 701–713. [CrossRef]
136. Gudur, R.A.; Bhosale, S.J.; Gudur, A.K.; Kale, S.R.; More, A.L.; Datkhile, K.D. Influence of (C1236T and C3435T) Polymorphisms of ABCB1 Gene on Chemotherapy Treatment Outcome and Toxicity in Breast Cancer Patients. *Asian Pac. J. Cancer Prev.* **2024**, *25*, 1567–1577. [CrossRef]
137. Behlam, I.; Sudershan, A.; Bharti, S.; Saini, A.K.; Manhas, S.; Kumar, P. Association of GSTT1 & GSTM1 Null Mutations with Breast Cancer Risk in North Indian Population: A Case-Control Study. *Afr. J. Biomed. Res.* **2024**, *27*, 4016–4022.
138. Bhaskaran, S.P.; Huang, T.; Rajendran, B.K.; Guo, M.; Luo, J.; Qin, Z.; Zhao, B.; Chian, J.; Li, S.; Wang, S.M. Ethnic-specific BRCA1/2 variation within Asia population: Evidence from over 78 000 cancer and 40 000 non-cancer cases of Indian, Chinese, Korean and Japanese populations. *J. Med. Genet.* **2021**, *58*, 752–759. [CrossRef] [PubMed]
139. Veeraraghavan, J.; Gutierrez, C.; Sethunath, V.; Mehravaran, S.; Giuliano, M.; Shea, M.J.; Mitchell, T.; Wang, T.; Nanda, S.; Pereira, R.; et al. Neratinib plus trastuzumab is superior to pertuzumab plus trastuzumab in HER2-positive breast cancer xenograft models. *NPJ Breast Cancer* **2021**, *7*, 63. [CrossRef] [PubMed]
140. Mathew, S.P.; Avaronnan, M.; Devi, N.; Praveen Kumar Shenoy, V.P. Her2 positive metastatic breast cancer treated with low dose lapatinib in a resource-constrained setting in South India: A retrospective audit. *Ecancermedicalscience* **2024**, *18*, 1758. [CrossRef]
141. Ghosh, M.; Saha, S.; Bettke, J.; Nagar, R.; Parrales, A.; Iwakuma, T.; van der Velden, A.W.; Martinez, L.A. Mutant p53 suppresses innate immune signaling to promote tumorigenesis. *Cancer Cell* **2021**, *39*, 494–508.e5. [CrossRef]
142. Akhter, N.; Dar, S.A.; Chattopadhyay, S.; Haque, S.; Anwer, R.; Wahid, M.; Jawed, A.; Lohani, M.; Mandal, R.K.; Shukla, N.K.; et al. Impact of p53 arg72pro SNP on Breast Cancer Risk in North Indian Population. *Curr. Genom.* **2018**, *19*, 395–410. [CrossRef]
143. Waseem, M.; Hussain, S.R.; Kumar, S.; Serajuddin, M.; Mahdi, F.; Sonkar, S.K.; Bansal, C.; Ahmad, M.K. Association of MTHFR (C677T) Gene Polymorphism with Breast Cancer in North India. *Biomark. Cancer* **2016**, *8*, 111–117. [CrossRef]
144. Kuras, S.; Pence, M.E.; Arac, E.; Yildiz, A.; Erdogan, B.; Pence, S. Investigation of the effects of MTHFR gene variations and homocysteine levels in hypertensive patients. *North. Clin. Istanb.* **2025**, *12*, 55–61. [CrossRef]
145. Rajagopal, T.; Seshachalam, A.; Rathnam, K.K.; Talluri, S.; Venkatabalasubramanian, S.; Dunna, N.R. Homologous recombination DNA repair gene RAD51, XRCC2 & XRCC3 polymorphisms and breast cancer risk in South Indian women. *PLoS ONE* **2022**, *17*, e0259761.
146. Wang, Q.; Shao, X.; Zhang, Y.; Zhu, M.; Wang, F.X.C.; Mu, J.; Li, J.; Yao, H.; Chen, K. Role of tumor microenvironment in cancer progression and therapeutic strategy. *Cancer Med.* **2023**, *12*, 11149–11165. [CrossRef]
147. Prakash, J.; Shaked, Y. The Interplay between Extracellular Matrix Remodeling and Cancer Therapeutics. *Cancer Discov.* **2024**, *14*, 1375–1388. [CrossRef] [PubMed]
148. Li, J.; Yu, S.; Rao, M.; Cheng, B. Tumor-derived extracellular vesicles: Key drivers of immunomodulation in breast cancer. *Front. Immunol.* **2025**, *16*, 1548535. [CrossRef]
149. Neophytou, C.M.; Panagi, M.; Stylianopoulos, T.; Papageorgis, P. The Role of Tumor Microenvironment in Cancer Metastasis: Molecular Mechanisms and Therapeutic Opportunities. *Cancers* **2021**, *13*, 2053. [CrossRef] [PubMed]
150. Obeagu, E.I. Hypoxia-driven angiogenesis in breast cancer mechanisms and therapeutic targets: A narrative review. *Ann. Med. Surg.* **2025**, *87*, 4246–4254. [CrossRef]
151. Gill, G.S.; Kharb, S.; Goyal, G.; Das, P.; Kurdia, K.C.; Dhar, R.; Karmakar, S. Immune checkpoint inhibitors and immunosuppressive tumor microenvironment: Current challenges and strategies to overcome resistance. *Immunopharmacol. Immunotoxicol.* **2025**, *47*, 485–507. [CrossRef]
152. McDonald, J.A.; Rao, R.; Gibbons, M.; Janardhanan, R.; Jaswal, S.; Mehrotra, R.; Pandey, M.; Radhakrishnan, V.; Ramakant, P.; Verma, N.; et al. Symposium report: Breast cancer in India-trends, environmental exposures and clinical implications. *Cancer Causes Control* **2021**, *32*, 567–575. [CrossRef]
153. ICGA. Teams—Indian Cancer Genome Atlas: Mapping India's Cancer Genome SD, Ankita Singh. The Indian Cancer Genome Atlas: A Multi-Omics Resource for Advancing Cancer Research. *bioRxiv* **2025**.

154. Bhattacharya, S.; Varshney, S.; Heidler, P.; Tripathi, S.K. Expanding the horizon for breast cancer screening in India through artificial intelligent technologies -A mini-review. *Front. Digit. Health* **2022**, *4*, 1082884. [CrossRef]
155. Mangayarkarasi, V.; Durairaj, E.; Ramanathan, V. Enhancing Cancer Screening and Early Diagnosis in India: Overcoming Challenges and Leveraging Emerging Technologies. *Cureus* **2025**, *17*, e78808. [CrossRef] [PubMed]
156. Demetriou, D.; Lockhat, Z.; Brzozowski, L.; Saini, K.S.; Dlamini, Z.; Hull, R. The Convergence of Radiology and Genomics: Advancing Breast Cancer Diagnosis with Radiogenomics. *Cancers* **2024**, *16*, 1076. [CrossRef]
157. Adapa, K.; Gupta, A.; Singh, S.; Kaur, H.; Trikha, A.; Sharma, A.; Rahul, K. A real world evaluation of an innovative artificial intelligence tool for population-level breast cancer screening. *npj Digit. Med.* **2025**, *8*, 2. [CrossRef]
158. Bhattacharya, S.; Saleem, S.M.; Singh, A.; Singh, S.; Tripathi, S. Empowering precision medicine: Regenerative AI in breast cancer. *Front. Oncol.* **2024**, *14*, 1465720. [CrossRef]
159. Das, S.K.; Dasgupta, R.K.; Roy, S.D.; Shil, D. AI in Indian healthcare: From roadmap to reality. *Intell. Pharm.* **2024**, *2*, 329–334. [CrossRef]
160. Haleem, A.; Javaid, M.; Singh, R.P.; Suman, R. Telemedicine for healthcare: Capabilities, features, barriers, and applications. *Sens. Int.* **2021**, *2*, 100117. [CrossRef] [PubMed]
161. Basha, S.S.; Sasmal, A.; Ariraman, S.B.M.; Raj, T.; Selvaraj, V.; Arockiarajan, A.; Sudhakar, S. Cobalt-nickel metal-organic frameworks (CNMs) as drug delivery agents for triple-negative breast cancer. *Nanoscale Adv.* **2025**, *7*, 5361–5376. [CrossRef] [PubMed]
162. Halder, S.; Das, T.; Kushwaha, R.; Misra, A.K.; Jana, K.; Das, D. Targeted and precise drug delivery using a glutathione-responsive ultra-short peptide-based injectable hydrogel as a breast cancer cure. *Mater. Horiz.* **2025**, *12*, 987–1001. [CrossRef] [PubMed]
163. Press Information Bureau (PIB), G.o.I. New Compound Synthesised Can Combat Aggressive Breast Cancer: Press Information Bureau (PIB), Government of India; 2025 [Cited Government of India]. Available online: https://www.pib.gov.in/PressReleasePage.aspx?PRID=2162654 (accessed on 22 November 2025).
164. Sangwan, R.K.; Huda, R.K.; Panigrahi, A.; Toteja, G.S.; Sharma, A.K.; Thakor, M.; Kumar, P. Strengthening breast cancer screening program through health education of women and capacity building of primary healthcare providers. *Front. Public Health* **2023**, *11*, 1276853. [CrossRef] [PubMed]
165. Mehrotra, R.; Yadav, K. Breast cancer in India: Present scenario and the challenges ahead. *World J. Clin. Oncol.* **2022**, *13*, 209–218. [CrossRef] [PubMed]
166. Gupta, G.; Jamwal, N.; Gupta, R. The Chiraiya project: A retrospective analysis of breast cancer detection gaps addressed via mobile mammography in Jammu Province, India. *BMC Public Health* **2024**, *24*, 2087. [CrossRef]

targets

MDPI

Review

The Expanding E3 Ligase-Ligand Landscape for PROTAC Technology

Zhenzhen Li [1], Xiaoli Huang [2], Xuchi Zhao [2], Yunxiu Zhang [2] and Ping Li [2,*]

[1] College of Life Sciences, Northwest Normal University, Lanzhou 730070, China; lizhenzhen@nwnu.edu.cn
[2] College of Chemistry & Chemical Engineering, Northwest Normal University, Lanzhou 730070, China; 15293613899@163.com (X.H.); 18293351215@163.com (X.Z.); yxiu_zhang@nwnu.edu.cn (Y.Z.)
* Correspondence: lip@sdnu.edu.cn

Abstract

Proteolysis-targeting chimeras (PROTACs) are a transformative therapeutic modality that co-opts the ubiquitin-proteasome system for selective protein degradation. To date, the development of PROTACs has been overwhelmingly dominated by the recruitment of four canonical E3 ligases: CRBN, VHL, MDM2, and IAP. This limited repertoire represents a critical bottleneck, restricting the scope of degradable proteins and potential therapeutic applications. Addressing this challenge, recent years have witnessed a surge in the successful recruitment of novel E3 ligases. This review provides a dedicated and comprehensive summary of this progress, focusing exclusively on the emerging E3 ligases and their cognate ligands reported for PROTAC technology outside of the well-established quartet. We detail their discovery and strategic application, highlighting how this rapidly expanding toolbox promises to overcome existing limitations and unlock the full potential of targeted protein degradation.

Keywords: targeted protein degradation; PROTACs; E3 ubiquitin ligase; E3 ligase ligand; drug discovery

1. Introduction

The traditional "occupancy-driven" paradigm of pharmacology, which relies on competitive inhibitors or antagonists to modulate protein function, has been a cornerstone of drug discovery for decades. While immensely successful, this approach is fundamentally limited to proteins with well-defined, functional binding pockets and often requires high, sustained drug concentrations to achieve therapeutic efficacy. In recent years, a paradigm shift has occurred with the advent of event-driven pharmacology, most prominently represented by targeted protein degradation (TPD) [1]. This innovative strategy does not merely inhibit a target protein but eliminates it entirely from the cellular environment by hijacking the cell's own protein disposal machinery [2,3].

At the forefront of TPD is the proteolysis-targeting chimera (PROTAC) technology [4,5]. PROTACs are heterobifunctional molecules comprising three key components: a ligand that binds to a protein of interest (POI), a ligand that recruits an E3 ubiquitin ligase, and a flexible linker connecting the two. By simultaneously engaging the POI and an E3 ligase, a PROTAC induces the formation of a productive ternary complex (POI-PROTAC-E3 ligase). This proximity-induced event triggers the transfer of ubiquitin from a charged E2 conjugating enzyme to lysine residues on the surface of the POI. The resulting polyubiquitinated POI is then recognized and degraded by the 26S proteasome, releasing the PROTAC

 https://doi.org/10.3390/targets3040030

molecule to engage in further catalytic cycles of degradation (Figure 1A). This catalytic nature allows for potent, sub-stoichiometric activity and can lead to a more profound and durable pharmacological response compared to traditional inhibitors. Furthermore, by targeting proteins for degradation rather than inhibition, PROTACs have opened the door to addressing the vast portion of the proteome previously considered "undruggable", including scaffolding proteins and transcription factors that lack enzymatic activity [2,6–9].

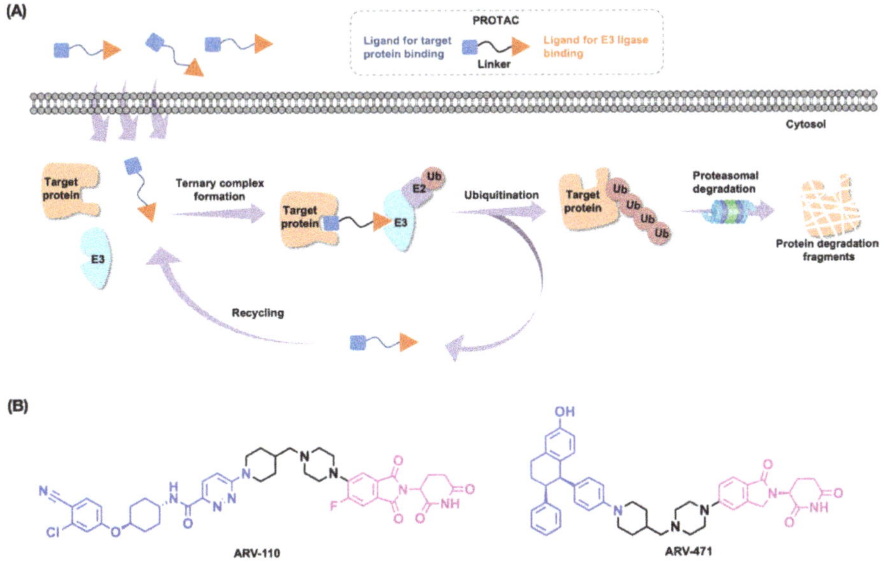

Figure 1. (**A**) Schematic illustrating the mechanism of action of PROTACs. (**B**) Chemical structures of PROTAC-based drug candidates, ARV-110 and ARV-471.

The specificity and efficiency of this process are critically dependent on the E3 ubiquitin ligase, the terminal enzyme in the ubiquitin-proteasome system (UPS) cascade. The human genome encodes for an estimated 600–700 E3 ligases, each possessing a distinct substrate scope, expression profile, and regulatory mechanism [10,11]. This diversity represents a vast, largely untapped reservoir of potential for directing protein degradation with precision. However, the historical development and current clinical landscape of PROTACs have been overwhelmingly reliant on a very small fraction of these enzymes [10,11]. The inception and validation of this field were built upon the recruitment of four canonical E3 ligases: Cereblon (CRBN) [12–14], von Hippel-Lindau (VHL) [15–18], Mouse double minute 2 homolog (MDM2) [19–22], and the inhibitor of apoptosis protein (IAP) family, which was first harnessed through the development of specific and nongenetic IAP-dependent protein erasers (SNIPERs) by Naito and colleagues [23–26]. The availability of well-characterized, high-affinity small molecule ligands for these E3 ligases, such as thalidomide and its analogs for CRBN [27,28], and hypoxia-inducible factor (HIF) mimetics for VHL [29], provided a crucial starting point that enabled the rapid validation of the PROTAC concept and the advancement of multiple candidates into clinical trials. This trajectory is reflected in the clinic. This foundation has propelled numerous candidates into the clinic, culminating in a landmark achievement for the entire field [30,31], notably including two orally administered degraders, ARV-110 (an androgen receptor degrader) and ARV-471 (an estrogen receptor degrader) (Figure 1B). The oral estrogen receptor PROTAC, vepdegestrant (ARV-471), recently met its primary endpoint in the pivotal Phase 3 VERITAC-2 trial for ER+/HER2- advanced breast cancer [32]. Vepdegestrant demonstrated a statistically significant and clinically meaningful improvement in progression-free survival (PFS) over the standard-of-care, fulvestrant, particularly in patients with resistance-driving ESR1 mutations. With a favorable safety profile, regulatory submissions have been filed, positioning it to potentially become the first approved PROTAC degrader [32]. This success provides definitive clinical validation for the PROTAC platform, confirming that this modality can yield best-in-class oral medicines capable of overcoming key mechanisms of drug resistance.

This foundational reliance, while instrumental to the field's success, has now become a significant bottleneck. The constraints imposed by this limited E3 ligase toolbox are manifold. Firstly, it restricts the degradable proteome; not all POIs can form stable and productive ternary complexes with CRBN or VHL, leaving many important disease-driving proteins inaccessible to current PROTACs. Secondly, it presents a clear avenue for acquired resistance. Clinical and preclinical studies have shown that mutations in the E3 ligase or its associated pathway components can abrogate PROTAC efficacy [33–35], a challenge that could be circumvented by having alternative E3 ligases to recruit. Thirdly, the ubiquitous expression of CRBN and VHL, while enabling broad applicability, complicates the development of tissue-selective degraders and can lead to on-target, off-tissue toxicities [36–38]. Recruiting an E3 ligase with a restricted expression pattern offers a compelling strategy to enhance the therapeutic window. Finally, the choice of E3 ligase can profoundly influence the degradation profile (e.g., kinetics, maximal degradation (D_{max}), and potency (DC_{50})) [39–41], and the optimal E3 for a given POI may not be among the canonical four.

In response to these challenges, a concerted effort across academia and industry has been directed towards expanding the repertoire of E3 ligases available for PROTAC technology [42,43]. This endeavor has begun to bear significant fruit, with a recent surge in the discovery of novel E3 ligases and the development of cognate ligands capable of hijacking their degradative capacity. This expansion is poised to fundamentally reshape the future of TPD.

Therefore, this review will provide a focused and comprehensive overview of the progress made in this critical area. We will move beyond the well-trodden ground of CRBN, VHL, MDM2, and IAP to exclusively survey the emerging landscape of novel E3 ligases and their respective ligands that have been successfully deployed for PROTAC development (Figure 2). We will systematically detail the discovery strategies, chemical matter, and biological validation for these new systems. By charting this expanding frontier, we aim to illuminate the immense potential of these novel E3 ligases to overcome current limitations and realize the full therapeutic promise of TPD.

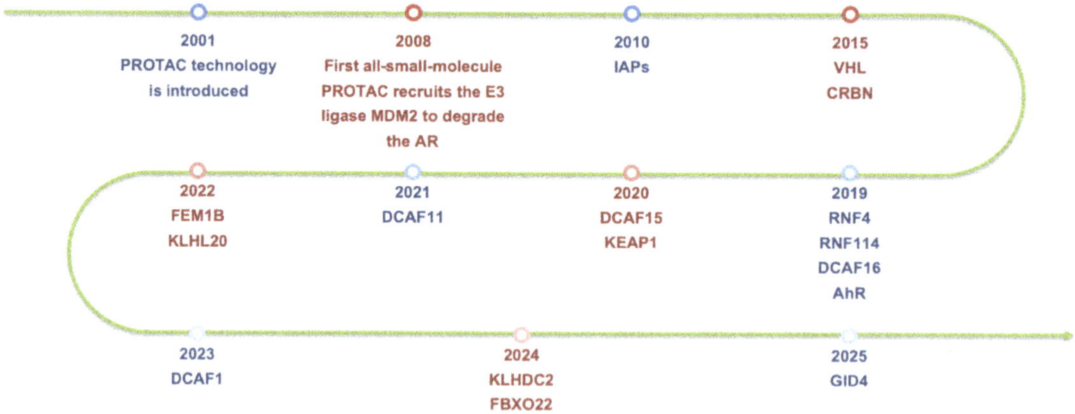

Figure 2. Timeline of the first use of E3 ligases in small-molecule PROTAC technology.

2. RNF Family

2.1. RNF4 E3 Ligase

In 2019, the Nomura group employed an activity-based protein profiling (ABPP) [44–47] covalent ligand screening platform to identify new chemical starting points for E3 ligase recruiters [48]. The screen identified TRH 1-23 as a hit that covalently binds to RING finger protein 4 (RNF4), an E3 ubiquitin ligase for which no chemical tools previously existed. Mass spectrometry analysis revealed that TRH 1-23 modifies the zinc-coordinating cysteines C132 and C135 within the RNF4 RING domain. Importantly, this binding was found to be non-functional, as it did not inhibit the autoubiquitination activity of RNF4, suggesting a non-functional binding mode suitable for ligase recruitment. Through structural optimization of this initial hit, they developed a more potent ligand, CCW 16, with an IC_{50} of 1.8 μM. To demonstrate its utility, CCW 16 was incorporated into a PROTAC degrader, CCW 28-3, by linking it to JQ1, a known inhibitor of the BET bromodomain protein family (Figure 3) [48]. CCW 28-3 demonstrated higher binding affinity for RNF4 than CCW 16, with an IC_{50} of 0.54 μM in competitive inhibition assays, successfully achieving degradation of bromodomain-containing protein 4 (BRD4) protein via both proteasome- and RNF4-dependent mechanisms. While acknowledging that the degradation efficiency and cellular target engagement of CCW 28-3 were modest compared to established degraders and that further optimization is needed, the study serves as a critical proof-of-concept. It validates the use of chemoproteomics-enabled covalent screening to successfully expand the arsenal of E3 ligase recruiters.

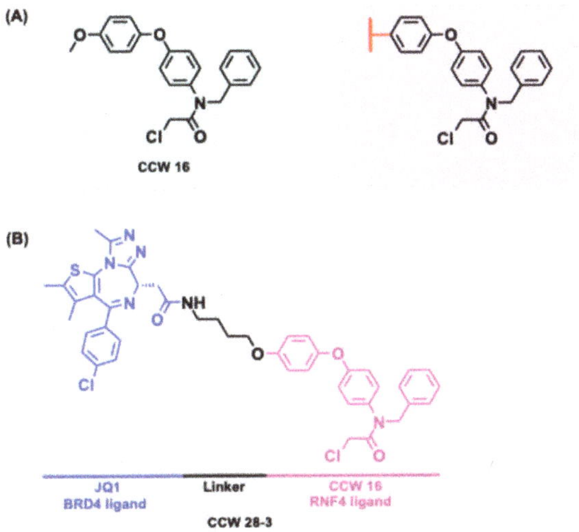

Figure 3. (**A**) Chemical structures of the RNF4 E3 ligase ligands. Red wavy lines indicate the typical attachment points for linker conjugation in PROTAC design. (**B**) Chemical structures of the RNF4-based PROTAC degraders.

2.2. RNF114 E3 Ligase

In 2019, the Nomura group positioned RING finger protein 114 (RNF114) [49] as a novel E3 ubiquitin ligase for small-molecule PROTACs, recruited via covalent engagement by the Neem-derived natural product nimbolide (Figure 4A) at cysteine-8 (C8) in its disordered N-terminal region, as revealed by ABPP in triple-negative breast cancer models [50]. This modification disrupts substrate recognition, inhibiting ubiquitination of tumor suppressors like p21 and p57, thereby stabilizing them and inducing cell cycle arrest and apoptosis. Harnessing this, nimbolide-JQ1 conjugates XH2 (Figure 4B) enable

RNF114-dependent, proteasome-mediated degradation of BRD4 with nanomolar potency and selectivity over other BET proteins, confirmed by proteomics and knockout studies, while also stabilizing additional substrates like PEG10 and CTGF. Cysteine-reactive screening identified simpler acrylamides that mimic nimbolide, offering tractable scaffolds to expand the limited E3 recruiter arsenal for targeted degradation.

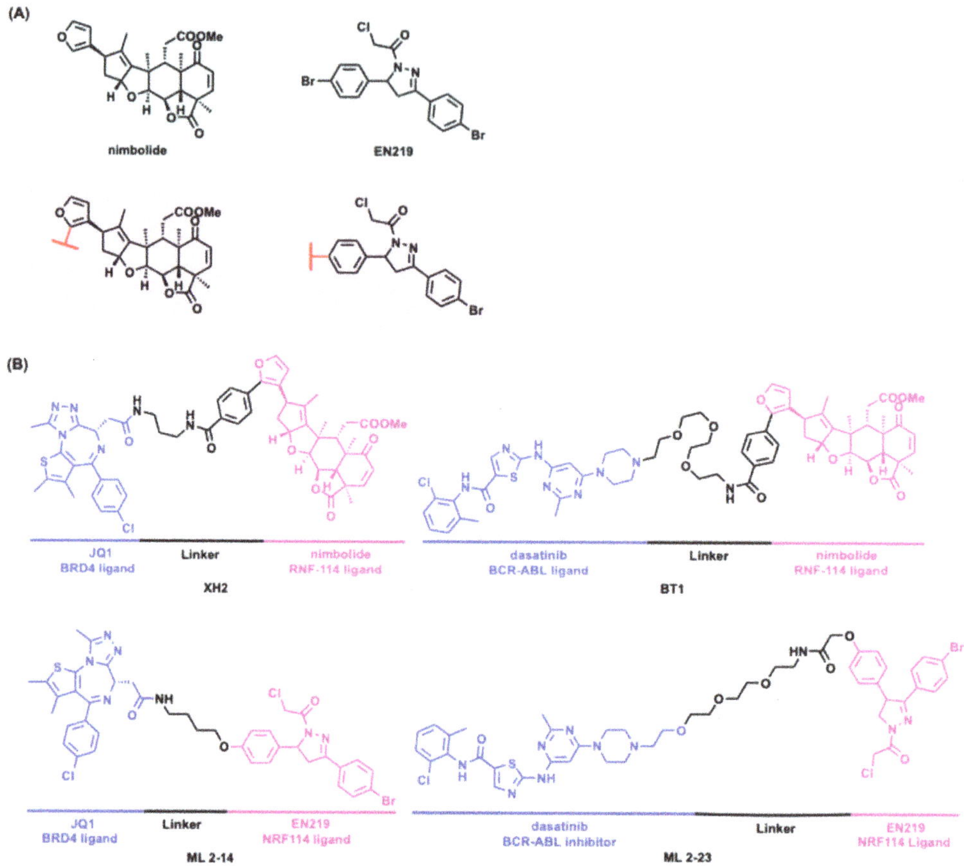

Figure 4. (**A**) Chemical structures of the RNF114 E3 ligase ligands. Red wavy lines indicate the typical attachment points for linker conjugation in PROTAC design. (**B**) Chemical structures of the RNF114-based PROTAC degraders.

In 2020, building on this concept, the Nomura group developed a series of novel PROTACs using nimbolide as a covalent recruiter for RNF114, designed to selectively degrade the oncogenic fusion protein BCR-ABL, which drives chronic myeloid leukemia (CML) [51]. By linking nimbolide with the BCR-ABL inhibitor dasatinib, they constructed a PROTAC BT1 (Figure 4B) capable of specifically degrading BCR-ABL. Experimental results demonstrated that, compared to PROTACs recruiting CRBN or VHL as E3 ligases, BT1 exhibited higher selectivity in degrading BCR-ABL and showed stronger anti-proliferative effects in leukemia cells. Additionally, BT1 treatment significantly increased levels of the tumor suppressor protein p21, suggesting a potential multi-target anticancer mechanism.

To overcome the inherent synthetic complexity of a natural product like nimbolide, the next logical step was to identify simpler, more tractable synthetic ligands. In 2021, the Nomura group screened 318 cysteine-reactive chloroacetic acid amine and acrylamide ligands using a gel-based competitive ABPP assay, and found that EN219 effectively inhibits the binding of the IA-rhodamine probe to the RNF114 protein, with an IC_{50} value of 470 nM (Figure 4A) [52]. EN219 inhibits the self-ubiquitination of RNF114 and the ubiquitination of its substrate p21 by covalently binding to the N-terminal cysteine (C8) of RNF114. Further studies linked EN219 with the BET inhibitor JQ1 or the BCR-ABL inhibitor dasatinib to construct PROTAC molecules (such as ML 2-14 and ML 2-23), achieving selective degradation of BRD4 and BCR-ABL, with the degradation process dependent on the proteasome and RNF114 (Figure 4B). This study not only provides new chemical tools for E3 ligase-based TPD applications but also demonstrates the potential of chemical biology methods to discover synthetic molecules that mimic the functions of natural products, offering new insights into targeting "undruggable" proteins.

3. DCAF Family

3.1. DCAF16 E3 Ligase

DDB1 and CUL4-associated factor 16 (DCAF16) is a largely uncharacterized substrate receptor for the Cullin–RING ubiquitin ligase 4 (CRL4) family [53]. By simultaneously docking to CUL4–DDB1 and a client protein, DCAF16 nucleates the formation of a ternary CRL4–DCAF16–substrate complex, thereby licensing poly-ubiquitination and proteasomal degradation of nuclear clients. Beyond housekeeping roles in DNA-damage responses and cell-cycle control [54,55], DCAF16 E3 ligase had not been exploited for ligand-directed protein degradation.

Using a chemical-proteomic platform, the Cravatt group screened heterobifunctional electrophilic PROTACs in which a broad-spectrum cysteine-reactive fragment (e.g., chloroacetamide KB02) was tethered to established ligands for FKBP12 (SLF) or BRD4 (JQ1) (Figure 5) [56]. Chemical-proteomic mapping, including isotopic tandem orthogonal proteolysis-activity-based protein profiling (isoTOP-ABPP) and affinity enrichment experiments, revealed that these PROTACs covalently modify a cluster of cysteines (likely C177 and/or C179) within DCAF16. Remarkably, occupancy of only 10–40% of the total DCAF16 pool was sufficient to drive robust, nuclear-restricted degradation of FKBP12 and BRD4, while the bulk of unmodified DCAF16 remained available for endogenous substrates. The covalent engagement also endowed the PROTACs with washout-resistant activity, a feature distinct from classic reversible degraders. This study positions DCAF16 as the first electrophile-engaged nuclear E3 ligase for PROTAC applications and demonstrates the potential of electro-affinity PROTACs in precisely regulating protein function.

Building on the identification of DCAF16 as a nuclear-restricted E3 ligase recruitable by covalent ligands, the Li group advanced the development of DCAF16-based PROTAC degraders targeting poly(ADP-ribose) polymerase 2 (PARP2) and cyclin-dependent kinase 4/6 (CDK4/6) for triple-negative breast cancer (TNBC) therapy [57,58]. They first tethered the PARP1/2 inhibitor olaparib to the covalent DCAF16 ligand KB02 to afford PROTAC C8 (Figure 5). Structural modeling predicted that C8's terminal α-chloroacetamide would self-react with Cys845 of PARP1, sterically preventing productive engagement of DCAF16 and thereby endowing selective PARP2 degradation. They next conjugated the CDK4/6 inhibitor palbociclib to KB02 to generate a series of DCAF16-based PROTACs with varied chains. Among these, the lead compound A4 induced concentration- and time-dependent CDK4/6 degradation in MDA-MB-231 cells via DCAF16 recruitment and proteasomal pathways.

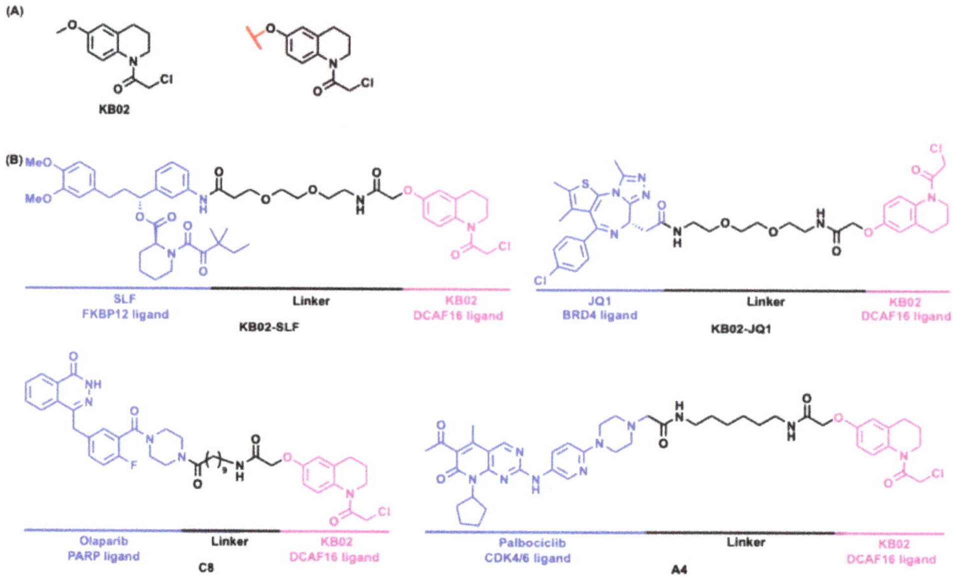

Figure 5. (**A**) Chemical structures of the DCAF16 E3 ligase ligand. Red wavy lines indicate the typical attachment points for linker conjugation in PROTAC design. (**B**) Chemical structures of the DCAF16-based PROTAC degraders.

3.2. DCAF15 E3 Ligase

In 2020, the Chen group reported a novel BRD4 degrader DP1 by conjugating the DDB1 and CUL4-associated factor 15 (DCAF15) ligand E7820 with the BRD4 inhibitor JQ1 (Figure 6), which can induce persistent degradation of the target protein and demonstrates therapeutic potential against hematological malignancies both in vitro and in vivo [59]. Although the current drug activity is not high, requiring μM levels to achieve better degradation, this study provides a series of evidence proving that DCAF15 can serve as a supplement to the limited E3 ligases in PROTAC design. In 2024, Lucas et al. optimized the SPLAM sulfonamide E7820 into the 53 nM DCAF15 ligand and generated a series PROTACs. Only one degraded BRD4, but competition and knockout assays showed the effect was DCAF15-independent [60]. Their study highlights that potent glue ligands do not guarantee functional PROTAC warheads and warns against using BRD4/JQ1 models without rigorous proof of mechanism.

3.3. DCAF11 E3 Ligase

In 2021, the Cravatt group identified DDB1 and CUL4-associated factor 11 (DCAF11) as a novel substrate adaptor capable of supporting ligand-induced protein degradation through covalent engagement [61]. Employing a cell-based screening strategy across diverse human cancer cell lines with a focused library of electrophilic PROTACs—bifunctional compounds featuring an FKBP12-binding ligand SLF linked to structurally varied α-chloroacetamide electrophiles—they discovered multiple hits that degraded cytosolic and nuclear FKBP12 in a proteasome- and Cullin-RING ligase-dependent manner, with compound 21-SLF exhibiting prostate cancer cell-specific activity (Figure 7). Chemical proteomics and CRISPR/Cas9 knockout experiments pinpointed DCAF11 as the key E3 component, revealing that 21-SLF covalently modifies conserved cysteines (notably C460, alongside C443 and C485) to form ternary complexes facilitating ubiquitination and degradation. Extending this to endogenous targets, they demonstrated DCAF11-dependent degradation of the androgen receptor using an analogous electrophilic PROTAC 21-ARL (Figure 7).

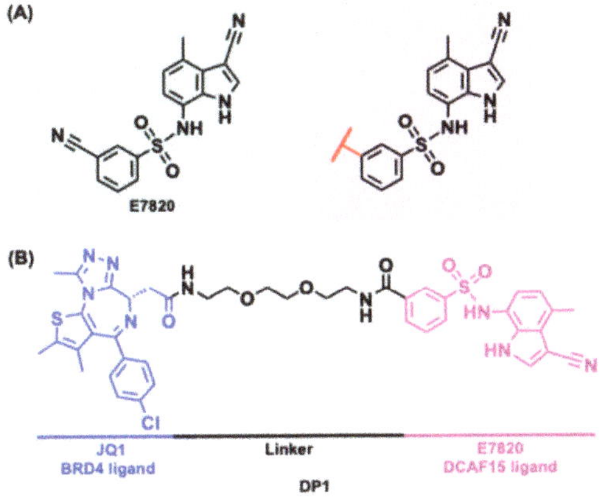

Figure 6. (**A**) Chemical structures of the DCAF15 E3 ligase ligand. Red wavy lines indicate the typical attachment points for linker conjugation in PROTAC design. (**B**) Chemical structures of the DCAF15-based PROTAC degraders.

In 2024, the Li group introduced alkenyl oxindoles as novel PROTAC moieties capable of recruiting the DCAF11 E3 ubiquitin ligase complex for efficient protein degradation, thereby expanding the limited repertoire of E3 ligases available for small-molecule degraders [62]. Alkenyl oxindoles previously was identified as autophagosome-tethering compounds (ATTECs) for mutant huntingtin degradation [63]. They synthesized a series of heterobifunctional conjugates linking these moieties to the BRD4 inhibitor JQ1 via diverse linkers, conducting structure-activity relationship analyses to optimize degradation potency. Contrary to expectations of autophagy-lysosomal mediation, mechanistic investigations revealed that these conjugates, including the lead compound HL435 (with DC_{50} values of 11.9 and 21.9 nM in MDA-MB-231 and MCF-7 cells, respectively), induced BRD4 ubiquitination and proteasomal degradation in a Cullin-RING ligase-dependent manner, unaffected by autophagy inhibitors but blocked by proteasome or neddylation inhibitors. Employing a pooled CRISPR interference (CRISPRi) screening, they identified $CRL4^{DCAF11}$ components (CUL4B, RBX1, DDB1, and DCAF11) as essential for this activity, with knockdown validations confirming DCAF11's role in ternary complex formation. This work not only repurposes alkenyl oxindoles as versatile E3 recruiters but also highlights the value of unbiased screening in uncovering underutilized ligases. Building on prior findings that alkenyl oxindoles covalently engage DCAF11, they further modified the aryl ring, linker types, and oxindole substituents to optimize structure-activity relationships [64]. Employing a novel high-throughput screening system based on high-content imaging, they identified L134 as a highly potent BRD4 degrader (DC_{50} = 7.36 nM, D_{max} > 98%), which operates via the ubiquitin-proteasome system in a DCAF11-dependent manner, as confirmed by mechanistic studies involving proteasome inhibitors and DCAF11 knockdown (Figure 7). L134 exhibited robust antitumor activity in cellular and in vivo models, underscoring its potential as a therapeutic candidate for BRD4-driven cancers.

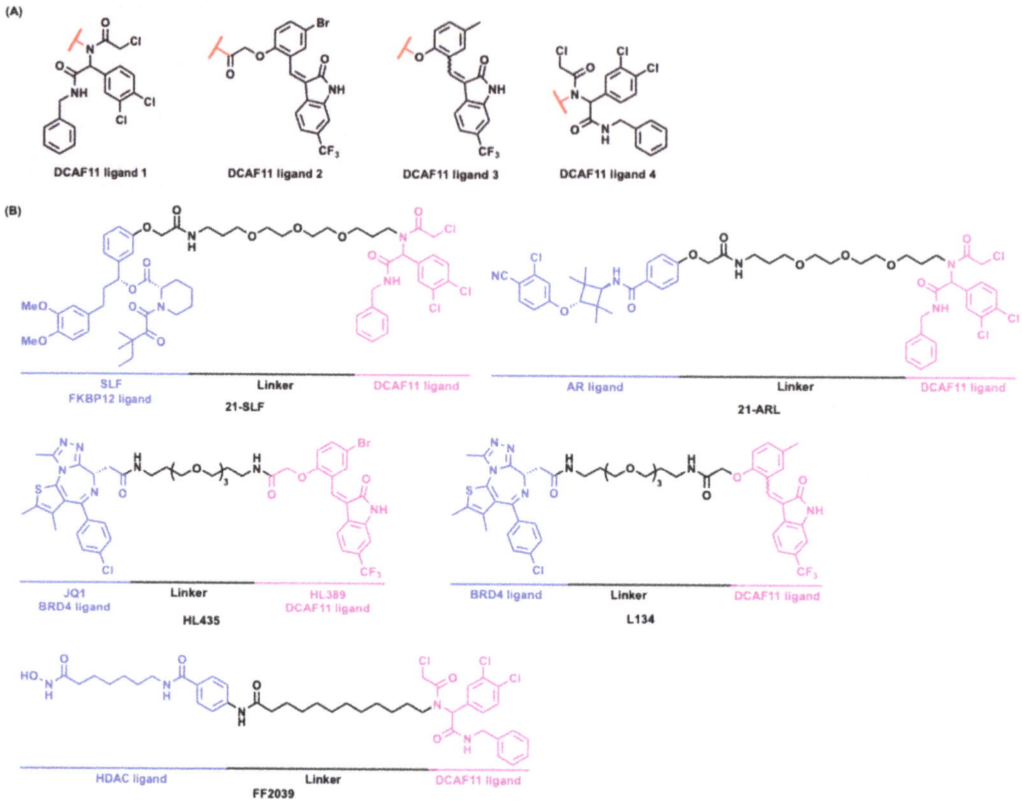

Figure 7. (**A**) Chemical structures of PROTAC moiety that recruits the DCAF11 E3 ligase. Red wavy lines indicate the typical attachment points for linker conjugation in PROTAC design. (**B**) Chemical structures of the DCAF11-based PROTAC degraders.

In 2025, the Hansen group developed novel histone deacetylase (HDAC) degraders that recruit DCAF11, transforming a selective HDAC6 PROTAC into a pan-HDAC degrader [65]. They synthesized eleven PROTACs on solid-phase, incorporating an alkenyl oxindole-based DCAF11 ligand and a pan-HDAC inhibitor, with FF2039 identified as a potent degrader of HDAC1 and HDAC6, as well as other isoforms across classes I, IIa, and IIb (Figure 7). Mechanistic studies confirmed that FF2039 operates via the ubiquitin-proteasome system in a DCAF11-dependent manner, inducing significant antiproliferative effects, cell cycle arrest, and apoptosis in both hematological and solid cancer cell lines. Compared to their prior CRBN-recruiting HDAC6-selective PROTAC A6 [66], switching to DCAF11 altered the degradation profile, highlighting the impact of E3 ligase choice on selectivity and efficacy.

3.4. DCAF1 E3 Ligase

In 2024, the Thoma group leveraged a selective DDB1 and CUL4-associated factor 1 (DCAF1) binder to develop potent PROTACs, including DBr-1 (targeting BRD9) and DBt-10 (targeting BTK), which efficiently degrade their respective targets via the DCAF1 E3 ligase, as validated through chemical and genetic rescue experiments (Figure 8) [67]. Notably, DBt-10 effectively degraded BTK in cells with acquired resistance to CRBN-based PROTACs, while DBr-1 circumvented intrinsic resistance to VHL-based degraders. Additionally, a dasatinib-based DCAF1 PROTAC successfully degraded multiple tyrosine kinases, broadening the scope of targetable proteins. This work highlights DCAF1 as a promising E3 ligase for PROTAC design, offering a strategy to address both intrinsic and acquired resistance in clinical settings.

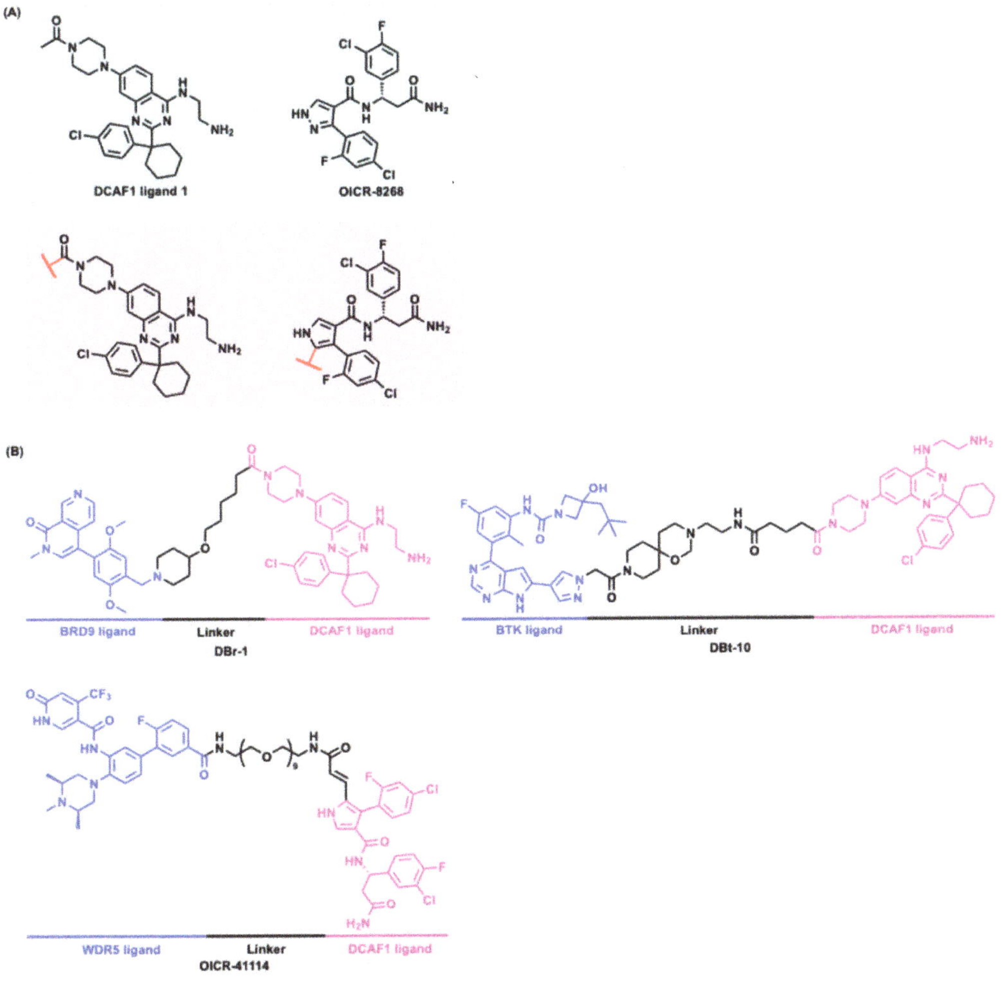

Figure 8. (**A**) Chemical structures of PROTAC moiety that recruits the DCAF1 E3 ligase. Red wavy lines indicate the typical attachment points for linker conjugation in PROTAC design. (**B**) Chemical structures of the DCAF1-based PROTAC degraders.

In a 2025 study, the Vedadi group advanced DCAF1-based PROTACs by targeting WDR5, a key cancer-associated protein [68], and provided structural insights into DCAF1's substrate specificity through high-resolution crystal structures of the DCAF1-PROTAC-WDR5 ternary complex (PDB ID: 9B9H, 9B9T, 9B9W and 9DLW) [69]. Using potent DCAF1 ligand OICR-8268, they designed four PROTACs that effectively degrade both endogenous and exogenous WDR5 via the ubiquitin-proteasome system, mediated by the CUL4-DCAF1 E3 ligase complex. Among the four DCAF1-based PROTACs, PROTAC OICR-41114 achieves the most robust WDR5 degradation (Figure 8). The crystal structures reveal that DCAF1's flexible loops confer surface plasticity, enabling diverse substrate interactions, which explains its broad substrate specificity.

4. AhR E3 Ligase

The Naito group broadens the E3 ligase repertoire by incorporating β-naphthoflavone (β-NF) as a ligand to recruit the aryl hydrocarbon receptor (AhR) E3 ligase complex, thereby creating a new class of degraders (Figure 9) [70]. Specifically, β-NF-ATRA, targeting cellular retinoic acid binding proteins (CRABPs), induced AhR-dependent degradation of CRABP-1 and CRABP-2 through the ubiquitin-proteasome pathway. Similarly, ITE-ATRA, employing an alternative AhR ligand ITE, also effectively degraded CRABPs. Extending this strategy, β-NF-JQ1 was developed to target BRD proteins, promoting AhR-BRD interactions and exhibiting potent anticancer activity that aligned with its protein knockdown efficacy.

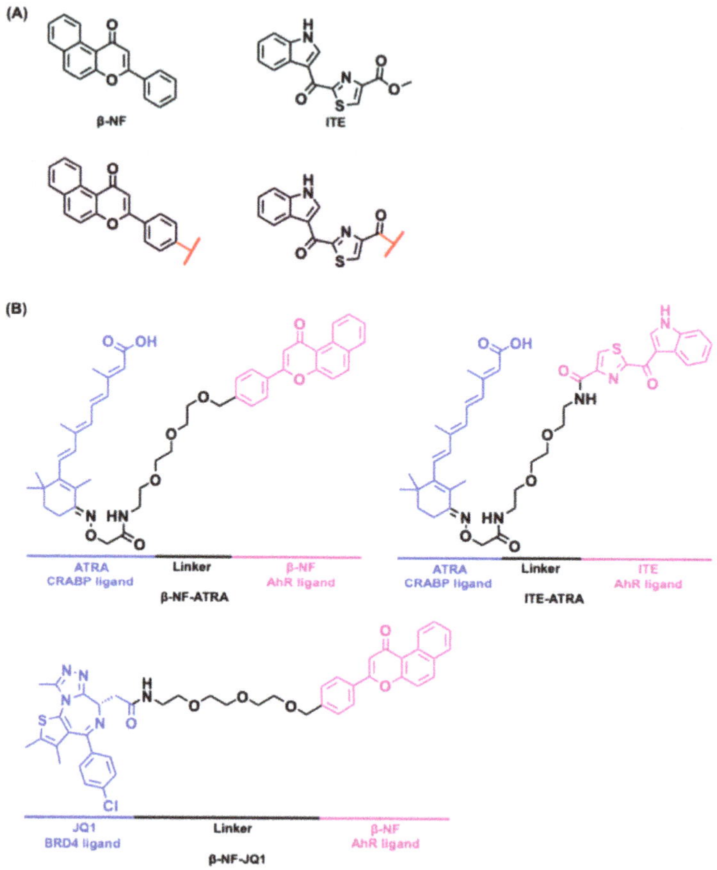

Figure 9. (**A**) Chemical structures of the AhR E3 ligase ligands. Red wavy lines indicate the typical attachment points for linker conjugation in PROTAC design. (**B**) Chemical structures of the AhR-based PROTAC degraders.

5. KEAP1 E3 Ligase

Kelch-like ECH-associated protein 1 (KEAP1), a critical substrate-recognition subunit of the CUL3-based E3 ubiquitin ligase complex, plays an essential role in orchestrating protein ubiquitination and proteasomal degradation [71,72]. By regulating the degradation of nuclear factor erythroid 2-related factor 2 (Nrf2), KEAP1 maintains cellular redox homeostasis, making it a key regulator of oxidative stress responses [73,74]. In 2020, the Nomura group pioneered the application of KEAP1 as an E3 ubiquitin ligase in TPD [75].

They engineered a novel PROTAC (CDDO-JQ1) by conjugating the KEAP1-Nrf2 activator bardoxolone with the BRD4 inhibitor JQ1, achieving potent BRD4 degradation (Figure 10). In the 231MFP human breast cancer cell line, CDDO-JQ1 demonstrated dose-dependent BRD4 degradation, which was abolished by pretreatment with proteasome inhibitor bortezomib and E1-activating enzyme inhibitor MLN7243, confirming its proteasome-dependent degradation mechanism. Additionally, the NEDD8-activating enzyme inhibitor MLN4924 attenuated degradation, underscoring the involvement of CUL3-family E3 ligases, which require NEDDylation for activity. Despite these advances, mechanistic questions remain, including whether CDDO-JQ1 targets KEAP1's BTB domain or other cysteine residues and whether additional E3 ligases contribute to the observed degradation. These findings lay a robust foundation for future chemoproteomic and genetic studies to elucidate the full scope of CDDO-JQ1's mechanism and for the development of next-generation KEAP1-based PROTACs.

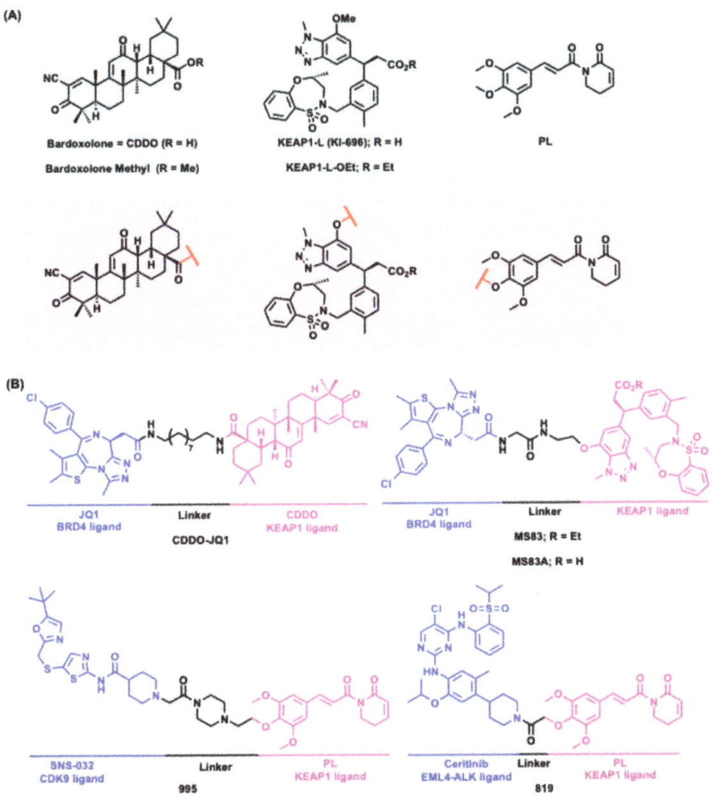

Figure 10. (**A**) Chemical structures of the KEAP1 E3 ligase ligands. Red wavy lines indicate the typical attachment points for linker conjugation in PROTAC design. (**B**) Chemical structures of the KEAP1-based PROTAC degraders.

In 2021, the Jin group further explored the potential of KEAP1 in PROTAC applications by developing MS83, a novel PROTAC linking a highly potent, selective, non-covalent KEAP1 ligand KEAP1-L-OEt to the BET bromodomain inhibitor JQ1 (Figure 10) [76]. In MDA-MB-468 cells, MS83 effectively induced degradation of BRD4 (both long and short isoforms) and BRD3 but showed minimal effects on BRD2. Notably, MS83 exhibited more sustained degradation kinetics compared to CRBN-recruiting PROTAC dBET1. MS83 was specifically designed and synthesized as a prodrug of MS83A using the ethyl ester derivative KEAP1-L-OEt to enhance cell permeability. This design allows MS83 to effectively degrade BRD4 at lower concentrations. However, this process requires a longer time, as it depends on the in-cell hydrolysis of its ethyl ester group to release the active carboxylic acid group. Interestingly, in MDA-MB-231 cells, MS83 selectively degraded the oncogenic BRD4 short isoform while sparing the potentially tumor-suppressive long isoform.

The Lv group expanded KEAP1 applications into novel territory by identifying piperlongumine (PL), a natural product, as a novel covalent E3 ligase ligand [77]. By conjugating PL with the CDK9 inhibitor SNS-032, they developed PROTAC 955 capable of degrading CDK9 (Figure 10). Through systematic synthesis of PL-SNS-032 conjugate and evaluation in MOLT4 cells, 955 demonstrated potent CDK9 degradation via both proteasomal and CRL E3 ligase-dependent mechanisms. TurboID-based proteomics confirmed KEAP1 as the recruited E3 ligase. They further investigated PL as a novel covalent KEAP1 ligand by synthesizing PL-ceritinib conjugates designed to degrade the EML4-ALK fusion oncoprotein, a driver of ceritinib resistance in NSCLC. The successful degradation of EML4-ALK by conjugate 819 confirms the potential of PL as a versatile covalent ligand for building E3 ligase-based protein degraders.

6. FEM1B E3 Ligase

Fem-1 homolog B (FEM1B), a substrate-recognition component of the CUL2 E3 ubiquitin protein ligase, plays a pivotal role in cellular reductive stress response by recognizing the reduced cysteine residue C186 on its substrate FNIP1, thereby mediating FNIP1 ubiquitination and degradation to restore mitochondrial activity and redox balance [78–80]. This unique mechanism positions FEM1B as a potential therapeutic target. In 2022, the Nomura group reported the discovery of EN106, the first covalent ligand targeting FEM1B [81]. EN106 forms a covalent bond with C186 of FEM1B, disrupting FEM1B-FNIP1 recognition and inhibiting FNIP1 degradation. Experimental results demonstrated that EN106 effectively stabilizes FNIP1 and exhibits high target specificity and degradation efficiency in cells. Building on this discovery, the researchers conjugated EN106 with the BET bromodomain inhibitor JQ1 to develop FEM1B-based BET degraders [81]. Among these, NJH-1-106 achieved the most potent degradation of BRD4, with a DC_{50} of 250 nM and a D_{max} of 94%. Additionally, the FEM1B-based kinase degrader NJH-2-142 successfully targeted the oncogenic BCR-ABL fusion protein, demonstrating the versatility of FEM1B as a recruiter in PROTAC applications (Figure 11).

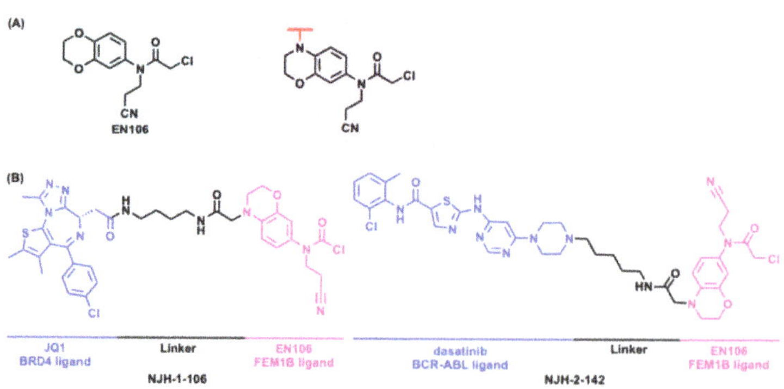

Figure 11. (**A**) Chemical structure of the covalent ligand EN106 targeting the FEM1B E3 ligase. Red wavy lines indicate the typical attachment points for linker conjugation in PROTAC design. (**B**) Chemical structures of the FEM1B-based PROTAC degraders.

7. KLHL20 E3 Ligase

In 2022, the Yeh group designed a synthetic macrocyclic ligand BTR2000 for KLHL20, a substrate adaptor of the CUL3 E3 ligase complex, and validated KLHL20 as an effective E3 ligase for PROTAC applications [82]. Through structure-guided design (PDB ID: 6GY5), the authors developed a bifunctional degrader BTR2003 that recruits KLHL20 to BET family proteins (Figure 12), achieving potent, dose-dependent degradation in cellular models via the ubiquitin-proteasome pathway, as confirmed by mechanistic studies involving ligase inhibitors and knockdown experiments. This macrocyclic PROTAC demonstrated improved efficiency over linear counterparts, highlighting the role of conformational constraints in enhancing ternary complex formation.

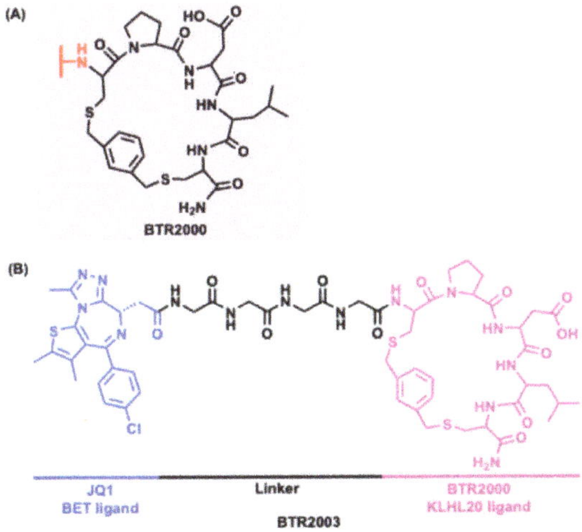

Figure 12. (**A**) Chemical structures of the KLHL20 E3 ligase ligand. Red wavy lines indicate the typical attachment points for linker conjugation in PROTAC design. (**B**) Chemical structures of the KLHL20-based PROTAC degraders.

8. KLHDC2 E3 Ligase

In 2024, the Békés group expanded the repertoire of recruitable E3 ligases for PROTACs by identifying and characterizing small-molecule ligands for KLHDC2, a CRL2 substrate adapter that recognizes glycine-ending C-end degrons [83]. Through biochemical assays, the authors developed potent KLHDC2 binders and functionalized them into PROTACs targeting BRD2/3/4 (K2-B4-3e and K2-B4-5e) and the AR (K2-AR-1), demonstrating KLHDC2-dependent ubiquitination and proteasomal degradation in cellular models (Figure 13). Structural insights from X-ray crystallography and cryo-EM revealed that KLHDC2 assembles into a dynamic tetramer stabilized by its C terminus (PDB ID: 8SGE and 8SGF), with ligand and substrate engagement modulating this oligomeric state to facilitate ternary complex formation. This work not only validates KLHDC2 as a versatile E3 ligase for PROTAC design, but also provides mechanistic understanding of its substrate specificity, potentially enabling broader applications in overcoming resistance and targeting undruggable proteins.

In the same year, through ligand screening and structure-based optimization, the Schulman group developed compounds engaging the U-shaped degron-binding pocket of KLHDC2, which were conjugated to the BET inhibitor JQ1 via tailored linkers to optimize exit vectors and induce cooperative ternary complexes [84]. The lead PROTAC, SJ46421, selectively recruited and ubiquitinated BRD3BD2 in vitro, while prodrug variants (e.g.,

SJ46420) addressed cell permeability challenges imposed by acidic degron-mimicking moieties, enabling robust BRD3 degradation in cells without affecting other BET paralogs like BRD2 or BRD4 (Figure 13). Mechanistic insights revealed KLHDC2's auto-inhibition overcome by neo-substrate engagement, ensuring ligase selectivity over related KLHDC1, KLHDC3, and KLHDC10. This work establishes a framework for designing selective PROTACs with C-degron E3 ligases, enhancing neo-substrate specificity and expanding the molecular toolkit to tackle challenges in drug discovery.

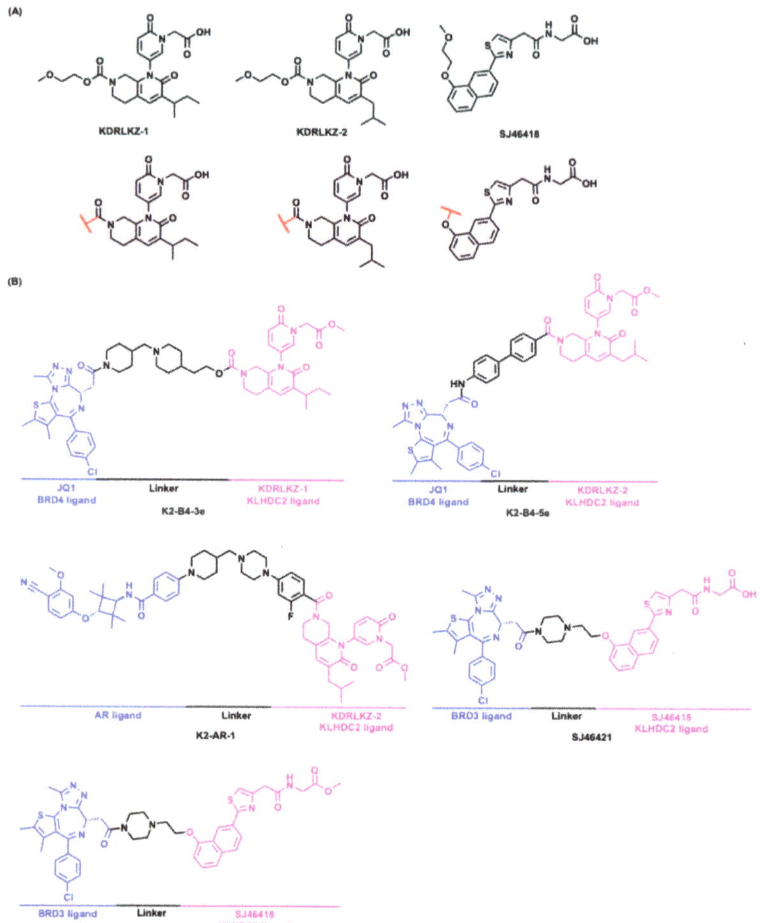

Figure 13. (**A**) Chemical structures of the KLHDC2 E3 ligase ligand. Red wavy lines indicate the typical attachment points for linker conjugation in PROTAC design. (**B**) Chemical structures of the KLHDC2-based PROTAC degraders.

9. FBXO22 E3 Ligase

In 2024, the Zhang group first identified FBXO22 as a novel recruitable ligase through a CRISPR-based transcriptional activation screen of human E3 ligases [85]. Using a heterobifunctional PROTAC, 22-SLF, which links an FKBP12-binding ligand to an electrophilic warhead, they demonstrated FBXO22-dependent degradation of FKBP12, mediated by covalent interactions with cysteines C227 and/or C228 in FBXO22, as confirmed by mechanistic studies. Further, FBXO22-based PROTACs effectively degraded additional endogenous proteins, including BRD4 (22-JQ1) and the oncogenic fusion protein EML4-ALK (22-TAE), in cellular models (Figure 14). This work validates FBXO22 as a versatile E3 ligase for PROTAC design, offering a new avenue to diversify the limited pool of ligases.

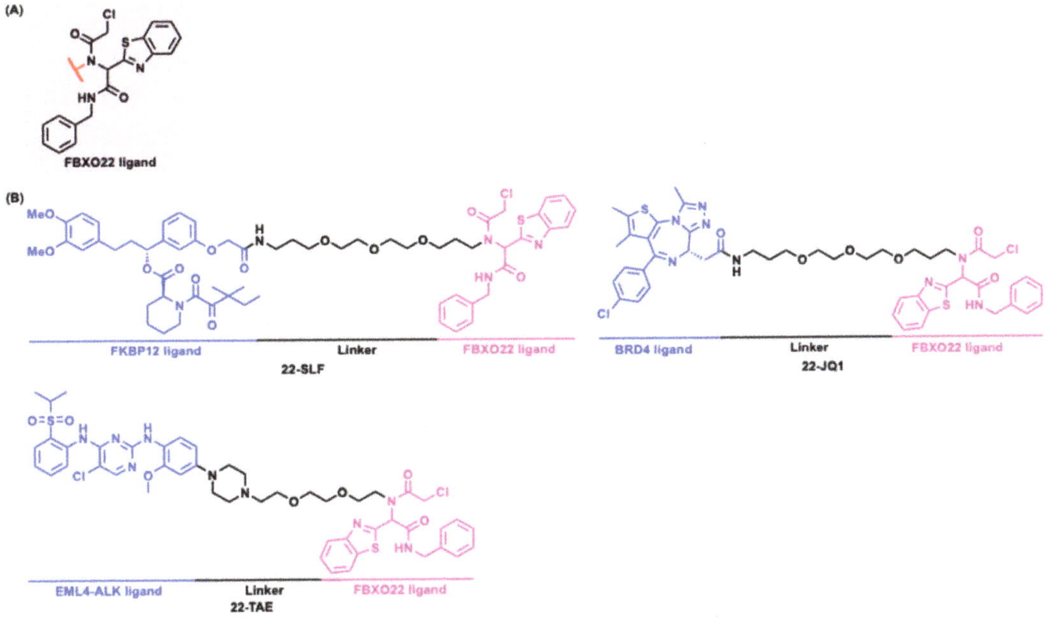

Figure 14. (**A**) Chemical structures of the FBXO22 E3 ligase ligand. Red wavy lines indicate the typical attachment points for linker conjugation in PROTAC design. (**B**) Chemical structures of the FBXO22-based PROTAC degraders.

10. GID4 E3 Ligase

The Dong group pioneered the use of glucose-induced degradation deficient complex 4 (GID4), a substrate receptor of the CTLH E3 ligase complex [86], for TPD using noncovalent small-molecule ligands [87]. By linking the E3 ligase GID4 antagonist PFI-7 [88,89] to warheads like JQ1, they developed PROTACs, NEP108 and the optimized NEP162 (Figure 15), which trigger ubiquitin-proteasome-dependent degradation of BRD4 and BRD2 (but not BRD3) in U2OS and SW480 cancer cell lines, demonstrating potent antiproliferative effects and tumor suppression in xenograft models with good tolerability. Crystal structures of GID4–PROTAC–BRD4 ternary complexes reveal plastic and dynamic interactions (PDB ID: 8X7G and 8X7H), providing crucial insights for rational design, while extending applicability to additional targets like ERα and SMARCA2 (Figure 15). GID4 is ubiquitously expressed in most tissue types, with notable enrichment in the testis, suggesting that future efforts can leverage GID4 recruiters to design additional PROTACs and develop tissue-specific degraders to enhance disease specificity or target selectivity.

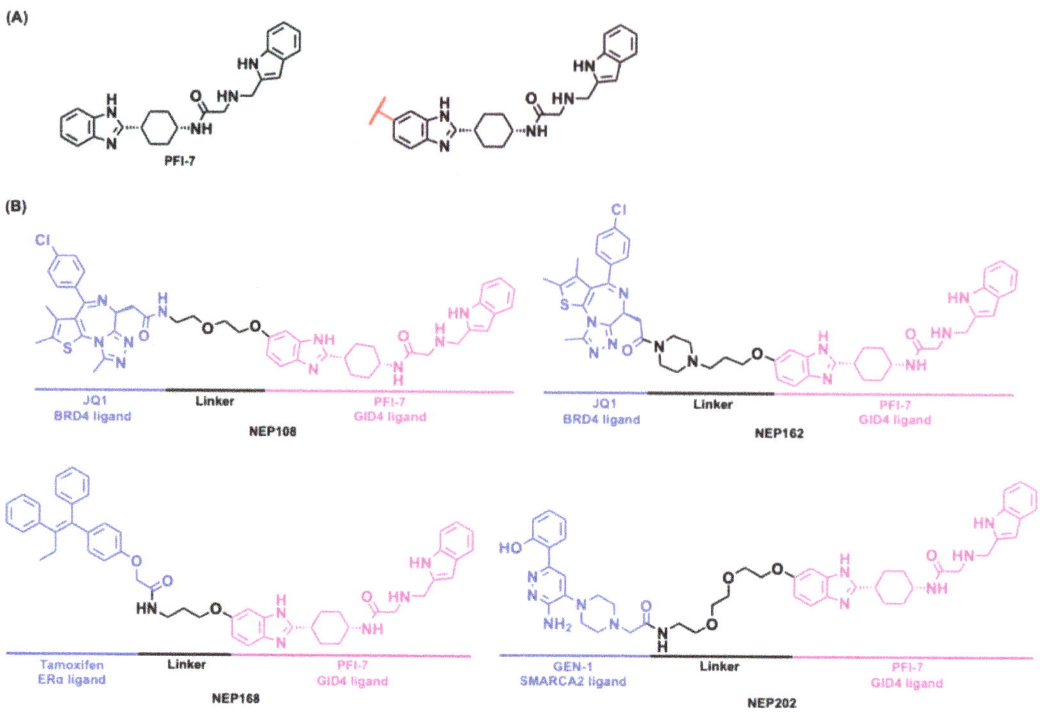

Figure 15. (**A**) Chemical structures of the GID4 E3 ligase ligand. Red wavy lines indicate the typical attachment points for linker conjugation in PROTAC design. (**B**) Chemical structures of the GID4-based PROTAC degraders.

11. Key Challenges and Future Perspectives

The successful recruitment of novel E3 ligases has broken new ground for PROTACs, yet it has also brought the complexity and scale of the remaining challenges into sharp relief. Moving from a handful of validated E3 Ligases to a truly versatile platform requires a concerted effort to bridge the gap between discovery and rational application.

11.1. The Ligand Discovery Bottleneck: From Binder to Functional Recruiter

The primary obstacle in TPD remains the discovery of high-quality ligands for E3 ligases. However, the core challenge extends beyond identifying mere binders; it involves the intricate and often unpredictable transformation of these binders into functional recruiters capable of inducing effective ubiquitination and degradation. Historical examples in the field abound where high-affinity ligands failed to translate into degraders due to issues like poor ternary complex formation or suboptimal exit vectors for linker attachment. To overcome this bottleneck, a multifaceted integration of technologies is imperative (Figure 16). High-throughput methods such as DNA-encoded libraries and phenotypic screening can generate initial hits efficiently, while chemoproteomics enables proteome-wide mapping of ligandable hotspots on the E3-ome. The pivotal step, however, is the optimization phase, where computational biology and machine learning emerge as transformative tools. These not only predict binding affinities but also simulate the dynamic conformational changes in ternary complexes that govern ubiquitination efficiency, thereby accelerating the design of potent, selective recruiters.

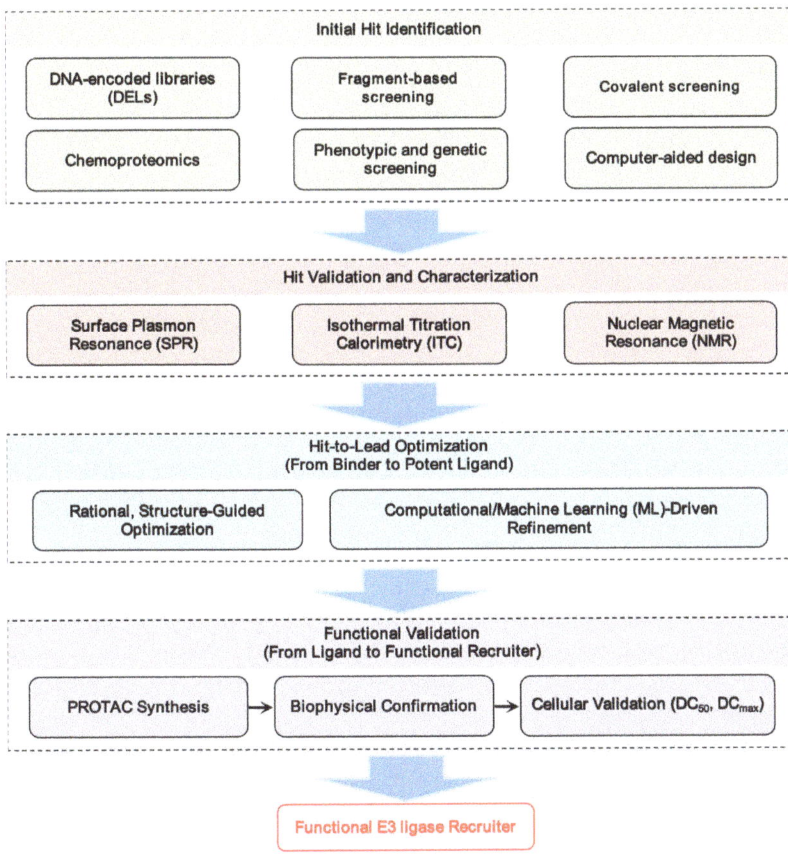

Figure 16. A schematic workflow for novel E3 ligase recruiter discovery and validation.

11.2. The Craft of Rational E3 Ligase Selection: A Paradigm Shift to Precision

With an expanding menu of E3 ligases, the paradigm of degrader design must shift from opportunistic discovery to rational, hypothesis-driven selection. This art of selection requires a multi-parameter calculus that weighs several critical factors:

(1) Tissue-Specific Expression: This remains the key goal for enhancing the therapeutic index. Systematically mapping the E3 ligase proteome across healthy and diseased tissues will provide the foundational atlas for designing tissue-restricted degraders.

(2) Subcellular Topography: A target protein's efficacy is intimately tied to its location. Matching the E3 ligase's subcellular address—be it nuclear, cytosolic, or membrane-associated—to that of the target is a non-negotiable prerequisite for success.

(3) The Principles of Ternary Complex Cooperativity: The rules governing productive ternary complex formation are exquisitely specific to each E3-target pair. Deciphering these complex principles of protein–protein interactions is essential for predicting which pairings will be synergistic and which will be futile.

(4) Exploiting Inherent Biology: Leveraging an E3's native function offers a compelling pathway to create multi-functional medicines. The recruitment of cIAP1, for instance, provides a dual-action strategy that couples target degradation with the induction of apoptosis. Similarly, co-opting MDM2 in a p53-wildtype cancer could not only eliminate a primary oncoprotein but also therapeutically stabilize the p53 tumor suppressor by sequestering its key negative regulator. This strategic approach transforms

the E3 ligase from a passive scaffold into an active, synergistic contributor to the overall therapeutic effect, adding a powerful dimension to rational degrader design.

To illustrate this calculus in action, consider the following case studies, which highlight how different criteria become the primary drivers of the decision depending on the therapeutic context:

Scenario 1: Targeting a Cytosolic Protein in a Neurodegenerative Disease

The primary challenge is to design a degrader for a pathogenic, cytosolic protein (e.g., a pathogenic kinase or a misfolded protein like Tau) expressed mainly in the central nervous system (CNS), while minimizing potential toxicities in peripheral organs. In this context, tissue expression becomes the paramount selection criterion. While the default E3 ligases, CRBN and VHL, are ubiquitously expressed and could be effective, they carry an inherent risk of on-target, off-tissue degradation in the periphery. A more rational, hypothesis-driven approach would prioritize the recruitment of a novel E3 ligase with enriched CNS expression, such as FEM1B or specific DCAF members, pending proteomic validation in relevant disease models. Although other criteria like subcellular localization are met by multiple candidates and serve as a necessary filter, the decision is ultimately driven by the goal of creating a brain-targeted degrader with a superior therapeutic index.

Scenario 2: Overcoming Acquired Resistance in Oncology

Consider a scenario where a potent, CRBN-based PROTAC against an oncogenic kinase shows initial clinical efficacy, but patients relapse with mutations in CRBN that preclude degrader binding. Here, the immediate problem is resistance risk, making the switch to a different E3 ligase family—such as VHL, cIAP1, or RNF114—the most logical step. However, the selection of the next E3 offers an opportunity for a more sophisticated strategy by exploiting inherent biology.

A simple switch to VHL might restore activity, but if the cancer type is known to be sensitive to apoptotic stimuli, recruiting cIAP1 becomes a highly compelling choice. A cIAP1-based degrader would provide a dual-action mechanism: it would not only degrade the primary oncogenic driver but also co-opt the intrinsic pro-apoptotic function of cIAP1 recruitment. This synergistic approach, which combines targeted degradation with apoptosis induction, could lead to deeper, more durable responses. Thus, while multiple E3 switches are possible, a rational selection informed by the tumor's biological vulnerabilities represents a superior strategy that actively leverages the E3 ligase's inherent function.

Mastering this art of selection is the gateway to unlocking true precision medicine in the TPD space, enabling the degradation machinery to be tailored to the specific context of the disease.

11.3. The Computational and Data-Driven Revolution in Degrader Design

While early degrader discovery relied heavily on empirical screening, the field is undergoing a computational revolution. In silico and data-driven tools are becoming indispensable for navigating the immense combinatorial complexity of E3-ligand-target space and are accelerating the entire design-build-test cycle.

The true challenge in degrader design lies in predicting the structure and dynamics of the ternary complex, which ultimately dictates degradation efficiency. Here, AI-based structure prediction tools like AlphaFold-Multimer and DiffDock are revolutionizing the field [90,91]. These models can now predict the three-dimensional structure of the entire POI-PROTAC-E3 complex with remarkable accuracy, enabling the in silico screening of thousands of potential E3-target pairings before any wet lab synthesis. This predictive power is further enhanced by molecular dynamics simulations, which model the flexibility of the linker and the subtle conformational rearrangements crucial for productive ubiquitination.

Looking ahead, the next frontier is the integration of these predictive AI/ML models with large-scale experimental data. For example, by combining chemoproteomics data that maps the "ligandable E3-ome" with tissue-specific proteomics and structural information, AI models could be trained to generate a predictive "E3-substrate compatibility matrix". Such a tool would guide the rational selection of the optimal E3 for any given target in a specific disease context. Ultimately, the fusion of these powerful computational and data-driven approaches is transitioning degrader development from an empirical art to a more predictable engineering science.

11.4. The Next Frontier: Expanding into Untapped E3 Ligase Families

While this review has focused on the successful recruitment of novel RING-type E3 ligases, a vast and largely uncharted territory of the "E3-ome" remains to be explored. To truly unlock the full potential of TPD, the field must venture beyond the currently accessible E3 ligases. Major and biologically important E3 ligase families—including the HECT family, the TRIM family, the membrane-embedded MARCH ligases, and the majority of the ~70 F-box protein substrate receptors—remain largely unexploited due to unique challenges in their catalytic mechanisms, structural complexity, or subcellular localization. Overcoming the hurdles to drugging these untapped families represents a key long-term goal and a major future direction for the field.

11.5. Conditionally Activated and Inducible Degraders for Spatiotemporal Control

The current generation of PROTACs are "always-on", which can lead to on-target, off-tissue toxicities. The next generation of degraders will be "smart", activated only at the desired time and place. This can be achieved through several exciting strategies for conditional activation, including light-activated (photodegraders) that can be switched on with spatiotemporal precision, and degraders that are activated by the unique conditions of a tumor microenvironment, such as hypoxia or low pH. These inducible systems promise to unlock an unprecedented level of control, significantly widening the therapeutic window.

11.6. Scope and Limitations of the Current Landscape

Finally, it is crucial to frame the exciting progress detailed in this review within the context of the field's current limitations. This review, by necessity, reflects the state of a rapidly evolving but still nascent area. Several key caveats should be acknowledged. First, while many novel E3 ligases have been successfully recruited, most of the cognate ligands discussed are tool compounds and have not yet been optimized for drug-like properties such as oral bioavailability or metabolic stability. Second, a significant portion of the literature on these emerging E3 ligases represents critical proof-of-concept work in cellular models. Rigorous and extensive in vivo validation of efficacy and safety for most of these new systems is a necessary next step that is still largely pending. Lastly, the scope of this review reflects the current focus of the TPD field, which has been predominantly concentrated on oncology. While immensely promising, the application of these novel E3 ligases in other key therapeutic areas, such as neurodegenerative and immune diseases, is still in its early stages and represents a major frontier for future research. Acknowledging these limitations highlights the immense opportunities that still lie ahead in translating this expanded E3 toolbox into broadly applicable, next-generation medicines.

12. Conclusions

The field of TPD stands at a pivotal juncture. The recent clinical success of vepdegestrant (ARV-471), a degrader recruiting a canonical E3 ligase, has provided definitive validation for the entire modality, proving that orally bioavailable, event-driven pharmacology can yield best-in-class medicines. This triumph, however, does not mark an endpoint but rather energizes the next great challenge: moving beyond the foundational reliance on CRBN and VHL. As this review has delineated, the E3 ligase toolbox is undergoing a dramatic and accelerating expansion, from the early explorations with MDM2 and cIAP1 to the recent breakthroughs in recruiting novel ligases such as RNF114, KEAP1, and members of the DCAF and CRL2 families (Table 1).

Table 1. Summary of Novel E3 Ligases and the Corresponding PROTACs.

Emerging E3 Ligase	Ligand	Ligand Type	PROTAC	Target	Degradation Efficiency (DC_{50}/D_{max})	Cell Lines	Ternary Complex (PDB ID)	Refs.
RNF4	CCW 16	Covalent	CCW 28-3	BRD4		231MFP, HeLa		[48]
RNF114	Nimbolide	Covalent	XH2	BRD4		231MFP		[50]
			BT1	BCR-ABL		K562		[51]
	EN219	Covalent	ML 2-14	BRD4	Long isoform: DC_{50} = 36 nM Short isoform: DC_{50} = 14 nM	231MFP		[52]
			ML 2-23	BCR-ABL		K562		[52]
DCAF16	KB02	Covalent	KB02-SLF	FKBP12		HEK293T, MDA-MB-231		[56]
			KB02-JQ1	BRD4		HEK293T		[56]
			C8	PARP2	DC_{50} = 2 μM, D_{max} > 92%	MDA-MB-231		[57]
			A4	CDK4/6	CDK4: DC_{50} = 6.5 μM, D_{max} > 89% CDK6: DC_{50} = 8 μM, D_{max} > 83%	MDA-MB-231		[58]
DCAF15	E7820	Noncovalent	DP1	BRD4	DC_{50} = 10.84 ± 0.92 μM, D_{max} = 98%	SU-DHL-4		[59]
DCAF11	DCAF11 ligand 1	Covalent	21-SLF	FKBP12		22Rv1		[61]
			21-ARL	AR		22Rv1		[61]
	DCAF11 ligand 2	Covalent	HL435	BRD4	DC_{50} = 11.9 nM (MDA-MB-231), DC_{50} = 21.9 nM (MCF-7), D_{max} > 99%	MDA-MB-231, MCF-7		[62]
	DCAF11 ligand 3	Covalent	L134	BRD4	DC_{50} = 7.36 nM, D_{max} > 98%	MDA-MB-231		[64]
	DCAF11 ligand 4	Covalent	FF2039	HDAC1, HDAC2, HDAC4, HDAC6	HDAC1: D_{max} = 90%, HDAC2, 4, and 6: D_{max} = 71–76%	MM.1S		[65]

Table 1. *Cont.*

Emerging E3 Ligase	Ligand	Ligand Type	PROTAC	Target	Degradation Efficiency (DC_{50}/D_{max})	Cell Lines	Ternary Complex (PDB ID)	Refs.
DCAF1	DCAF1 ligand 1	Noncovalent	DBr-1	BRD9		HEK293		[67]
			DBt-10	BTK		TMD8		[67]
	OICR-8268	Noncovalent	OICR-41114	WDR5	DC_{50} = 40 ± 24 nM, D_{max} = 49 ± 1.9%	MV4-11	9DLW	[69]
AhR	β-NF	Noncovalent	β-NF-ATRA	CRABP-1, CRABP-2		MCF-7, IMR-32		[70]
			β-NF-JQ1	BRD2, BRD3, BRD4		MCF-7		[70]
	ITE	Noncovalent	ITE-ATRA	CRABP-1, CRABP-2		MCF-7, IMR-32		[70]
KEAP1	CDDO	Covalent	CDDO-JQ1	BRD4		231MFP		[75]
	KEAP1-L-OEt	Noncovalent	MS83	BRD4, BRD3		MDA-MB-468, MDA-MB-231		[76]
	PL	Covalent	995	CDK9	DC_{50} = 9 nM	MOLT4		[77]
			819	EML4-ALK		NCI-H2228		[77]
FEM1B	EN106	Covalent	NJH-1-106	BRD4	DC_{50} = 250 nM, D_{max} = 94%	HEK293T		[81]
			NJH-2-142	BCR-ABL		K562		[81]
KLHL20	BTR2000	Noncovalent	BTR2003	BRD2, BRD3, BRD4	BRD2: DC_{50} = 46 nM, BRD3: DC_{50} = 87 nM, BRD4: DC_{50} = 777 nM	PC3		[82]
KLHDC2	KDRLKZ-1	Noncovalent	K2-B4-3e	BRD4	DC_{50} = 66 nM, D_{max} = 62%	PC3		[83]
	KDRLKZ-2	Noncovalent	K2-B4-5e	BRD4	DC_{50} = 6.2 nM, D_{max} = 93%	PC3		[83]
			K2-AR-1	AR		VCaP		[83]
	SJ46418	Noncovalent	SJ46421	BRD3		U2OS		[84]
			SJ46420	BRD3		U2OS		[84]
FBXO22	FBXO22 ligand		22-SLF	FKBP12	DC_{50} = 0.5 μM, D_{max} = 89%	HEK293T		[85]
			22-JQ1	BRD4		A549		[85]
			22-TAE	EML4-ALK		H2228		[85]
GID4	PFI-7	Noncovalent	NEP108	BRD4	DC_{50} = 3.8 μM	U2OS	8X7G	[87]
			NEP162	BRD4	DC_{50} = 1.2 μM (SW480), DC_{50} = 1.6 μM (U2OS)	SW480, U2OS	8X7H	[87]
			NEP168	ERα		MCF-7		[87]
			NEP202	SMARCA2		U2OS		[87]

This strategic diversification is not merely a numerical addition of E3 ligases; it represents a fundamental leap towards precision medicine. Each new E3 ligase, discovered through innovative engines like chemoproteomics, phenotypic screening, and structure-based design, brings a unique set of biological properties to the table. This expanding repertoire offers tangible solutions to the field's most pressing challenges: it provides the tools to achieve tissue-selective degradation, new avenues to overcome acquired resistance, and the potential to expand the degradable proteome to targets previously inaccessible to the canonical recruiters. The art of rational E3 selection is therefore becoming a central pillar of modern degrader design, enabling a hypothesis-driven approach to tailor the degradation machinery to the specific context of a disease.

Looking ahead, the true promise of this expanded toolbox will be realized through its convergence with the powerful computational and AI-driven platforms that are reshaping drug discovery. The ability to accurately model and predict ternary complex formation, combined with a growing arsenal of E3 recruiters, is transitioning degrader development from an empirical art to a more predictable engineering science. By continuing to explore this vast E3 ligase landscape and integrating our discoveries with these powerful new design tools, the field is poised to deliver a new generation of precision-engineered degraders, capable of selectively erasing the root causes of disease with unprecedented control.

Author Contributions: Z.L.: literature review, manuscript preparation, writing—original draft. X.H. and X.Z.: literature review, manuscript preparation. Y.Z.: review and edit. P.L.: review and edit. All authors have read and agreed to the published version of the manuscript.

Funding: This work was supported by the National Natural Science Foundation of China (22407110), Gansu Provincial Science and Technology Plan Project (25RCKA027), Gansu Provincial Youth Science and Technology Fund Project (24JRRA143), Lanzhou Youth Science and Technology Talents Innovation Project (2024-QN-34), and Gansu Provincial Higher Education Innovation Fund project (2024B-044).

Institutional Review Board Statement: Not applicable.

Informed Consent Statement: Not applicable.

Data Availability Statement: Not applicable.

Conflicts of Interest: The authors declare no conflicts of interest.

Abbreviations

The following abbreviations are used in this review:

TPD	Targeted protein degradation
PROTAC	Proteolysis-targeting chimera
POI	Protein of interest
UPS	Ubiquitin-proteasome system
CRBN	Cereblon
VHL	von Hippel-Lindau
MDM2	Mouse double minute 2
IAP	Inhibitor of apoptosis protein
SNIPERs	Specific and nongenetic IAP-dependent protein erasers
HIF	Hypoxia-inducible factor
ABPP	Activity-based protein profiling
RNF4	RING finger protein 4
BRD4	Bromodomain-containing protein 4
RNF114	RING finger protein 114

CML	Chronic myeloid leukemia
DCAF16	DDB1 and CUL4-associated factor 16
CRL4	Cullin–RING ubiquitin ligase 4
PARP2	Poly(ADP-ribose) polymerase 2
CDK4/6	Cyclin-dependent kinase 4/6
TNBC	Triple-negative breast cancer
DCAF15	DDB1 and CUL4-associated factor 15
DCAF11	DDB1 and CUL4-associated factor 11
ATTECs	Autophagosome-tethering compounds
CRISPRi	CRISPR interference
HDAC	Histone deacetylase
DCAF1	DDB1 and CUL4-associated factor 1
β-NF	β-naphthoflavone
AhR	Aryl hydrocarbon receptor
CRABPs	Cellular retinoic acid binding proteins
KEAP1	Kelch-like ECH-associated protein 1
Nrf2	Nuclear factor erythroid 2-related factor 2
FEM1B	Fem-1 homolog B
GID4	Glucose-induced degradation deficient complex 4

References

1. Mullard, A. First targeted protein degrader hits the clinic. *Nat. Rev. Drug Discov.* **2019**, *18*, 237–239. [CrossRef] [PubMed]
2. Lai, A.C.; Crews, C.M. Induced protein degradation: An emerging drug discovery paradigm. *Nat. Rev. Drug Discov.* **2017**, *16*, 101–114. [CrossRef] [PubMed]
3. Nalawansha, D.A.; Crews, C.M. PROTACs: An Emerging Therapeutic Modality in Precision Medicine. *Cell Chem. Biol.* **2020**, *27*, 998–1014. [CrossRef]
4. Békés, M.; Langley, D.R.; Crews, C.M. PROTAC targeted protein degraders: The past is prologue. *Nat. Rev. Drug Discov.* **2022**, *21*, 181–200. [CrossRef]
5. Dale, B.; Cheng, M.; Park, K.S.; Kaniskan, H.; Xiong, Y.; Jin, J. Advancing targeted protein degradation for cancer therapy. *Nat. Rev. Cancer* **2021**, *21*, 638–654. [CrossRef]
6. Bondeson, D.P.; Mares, A.; Smith, I.E.D.; Ko, E.; Campos, S.; Miah, A.H.; Mulholland, K.E.; Routly, N.; Buckley, D.L.; Gustafson, J.L.; et al. Catalytic in vivo protein knockdown by small-molecule PROTACs. *Nat. Chem. Biol.* **2015**, *11*, 611–617. [CrossRef]
7. Burslem, G.M.; Smith, B.E.; Lai, A.C.; Jaime-Figueroa, S.; McQuaid, D.C.; Bondeson, D.P.; Toure, M.; Dong, H.Q.; Qian, Y.M.; Wang, J.; et al. The Advantages of Targeted Protein Degradation Over Inhibition: An RTK Case Study. *Cell Chem. Biol.* **2018**, *25*, 67–77. [CrossRef]
8. Bai, L.C.; Zhou, H.B.; Xu, R.Q.; Zhao, Y.J.; Chinnaswamy, K.; McEachern, D.; Chen, J.Y.; Yang, C.Y.; Liu, Z.M.; Wang, M.; et al. A Potent and Selective Small-Molecule Degrader of STAT3 Achieves Complete Tumor Regression In Vivo. *Cancer Cell* **2019**, *36*, 498–511. [CrossRef]
9. Henley, M.J.; Koehler, A.N. Advances in targeting 'undruggable' transcription factors with small molecules. *Nat. Rev. Drug Discov.* **2021**, *20*, 669–688. [CrossRef]
10. Kannt, A.; Dikic, I. Expanding the arsenal of E3 ubiquitin ligases for proximity-induced protein degradation. *Cell Chem. Biol.* **2021**, *28*, 1014–1031. [CrossRef] [PubMed]
11. Belcher, B.P.; Ward, C.C.; Nomura, D.K. Ligandability of E3 Ligases for Targeted Protein Degradation Applications. *Biochemistry* **2023**, *62*, 588–600. [CrossRef]
12. Winter, G.E.; Buckley, D.L.; Paulk, J.; Roberts, J.M.; Souza, A.; Dhe-Paganon, S.; Bradner, J.E. Phthalimide conjugation as a strategy for in vivo target protein degradation. *Science* **2015**, *348*, 1376–1381. [CrossRef]
13. Lu, J.; Qian, Y.M.; Altieri, M.; Dong, H.Q.; Wang, J.; Raina, K.; Hines, J.; Winkler, J.D.; Crew, A.P.; Coleman, K.; et al. Hijacking the E3 Ubiquitin Ligase Cereblon to Efficiently Target BRD4. *Chem. Biol.* **2015**, *22*, 755–763. [CrossRef]
14. Min, J.; Mayasundari, A.; Keramatnia, F.; Jonchere, B.; Yang, S.W.; Jarusiewicz, J.; Actis, M.; Das, S.; Young, B.; Slavish, J.; et al. Phenyl-Glutarimides: Alternative Cereblon Binders for the Design of PROTACs. *Angew. Chem.-Int. Ed.* **2021**, *60*, 26663–26670. [CrossRef] [PubMed]
15. Zengerle, M.; Chan, K.H.; Ciulli, A. Selective Small Molecule Induced Degradation of the BET Bromodomain Protein BRD4. *ACS Chem. Biol.* **2015**, *10*, 1770–1777. [CrossRef]

16. Raina, K.; Lu, J.; Qian, Y.M.; Altieri, M.; Gordon, D.; Rossi, A.M.K.; Wang, J.; Chen, X.; Dong, H.Q.; Siu, K.; et al. PROTAC-induced BET protein degradation as a therapy for castration-resistant prostate cancer. *Proc. Natl. Acad. Sci. USA* **2016**, *113*, 7124–7129. [CrossRef]
17. Han, X.; Wang, C.; Qin, C.; Xiang, W.G.; Fernandez-Salas, E.; Yang, C.Y.; Wang, M.; Zhao, L.J.; Xu, T.F.; Chinnaswamy, K.; et al. Discovery of ARD-69 as a Highly Potent Proteolysis Targeting Chimera (PROTAC) Degrader of Androgen Receptor (AR) for the Treatment of Prostate Cancer. *J. Med. Chem.* **2019**, *62*, 941–964. [CrossRef] [PubMed]
18. Smalley, J.P.; Baker, I.M.; Pytel, W.A.; Lin, L.Y.; Bowman, K.J.; Schwabe, J.W.R.; Cowley, S.M.; Hodgkinson, J.T. Optimization of Class I Histone Deacetylase PROTACs Reveals thatHDAC1/2 Degradation is Critical to Induce Apoptosis and Cell Arrest in Cancer Cells. *J. Med. Chem.* **2022**, *65*, 5642–5659. [CrossRef]
19. Schneekloth, A.R.; Pucheault, M.; Tae, H.S.; Crews, C.M. Targeted intracellular protein degradation induced by a small molecule: En route to chemical proteomics. *Bioorg. Med. Chem. Lett.* **2008**, *18*, 5904–5908. [CrossRef] [PubMed]
20. He, S.P.; Ma, J.H.; Fang, Y.X.; Liu, Y.; Wu, S.C.; Dong, G.Q.; Wang, W.; Sheng, C.Q. Homo-PROTAC mediated suicide of MDM2 to treat non-small cell lung cancer. *Acta Pharm. Sin. B* **2021**, *11*, 1617–1628. [CrossRef] [PubMed]
21. Hines, J.; Lartigue, S.; Dong, H.Q.; Qian, Y.M.; Crews, C.M. MDM2-Recruiting PROTAC Offers Superior, Synergistic Antiproliferative Activity via Simultaneous Degradation of BRD4 and Stabilization of p53. *Cancer Res.* **2019**, *79*, 251–262. [CrossRef]
22. De Dominici, M.; Porazzi, P.; Xiao, Y.C.; Chao, A.; Tang, H.Y.; Kumar, G.; Fortina, P.; Spinelli, O.; Rambaldi, A.; Peterson, L.F.; et al. Selective inhibition of Ph-positive ALL cell growth through kinase-dependent and -independent effects by CDK6-specific PROTACs. *Blood* **2020**, *135*, 1560–1573. [CrossRef] [PubMed]
23. Itoh, Y.; Ishikawa, M.; Naito, M.; Hashimoto, Y. Protein Knockdown Using Methyl Bestatin−Ligand Hybrid Molecules: Design and Synthesis of Inducers of Ubiquitination-Mediated Degradation of Cellular Retinoic Acid-Binding Proteins. *J. Am. Chem. Soc.* **2010**, *132*, 5820–5826. [CrossRef]
24. Ohoka, N.; Okuhira, K.; Ito, M.; Nagai, K.; Shibata, N.; Hattori, T.; Ujikawa, O.; Shimokawa, K.; Sano, O.; Koyama, R.; et al. In Vivo Knockdown of Pathogenic Proteins via Specific and Nongenetic Inhibitor of Apoptosis Protein (IAP)-dependent Protein Erasers (SNIPERs). *J. Biol. Chem.* **2017**, *292*, 4556–4570. [CrossRef] [PubMed]
25. Ohoka, N.; Morita, Y.; Nagai, K.; Shimokawa, K.; Ujikawa, O.; Fujimori, I.; Ito, M.; Hayase, Y.; Okuhira, K.; Shibata, N.; et al. Derivatization of inhibitor of apoptosis protein (IAP) ligands yields improved inducers of estrogen receptor degradation. *J. Biol. Chem.* **2018**, *293*, 6776–6790. [CrossRef] [PubMed]
26. Miah, A.H.; Smith, I.E.D.; Rackham, M.; Mares, A.; Thawani, A.R.; Nagilla, R.; Haile, P.A.; Votta, B.J.; Gordon, L.J.; Watt, G.; et al. Optimization of a Series of RIPK2 PROTACs. *J. Med. Chem.* **2021**, *64*, 12978–13003. [CrossRef]
27. Ito, T.; Ando, H.; Suzuki, T.; Ogura, T.; Hotta, K.; Imamura, Y.; Yamaguchi, Y.; Handa, H. Identification of a Primary Target of Thalidomide Teratogenicity. *Science* **2010**, *327*, 1345–1350. [CrossRef]
28. Zhu, Y.X.; Braggio, E.; Shi, C.-X.; Bruins, L.A.; Schmidt, J.E.; Van Wier, S.; Chang, X.-B.; Bjorklund, C.C.; Fonseca, R.; Bergsagel, P.L.; et al. Cereblon expression is required for the antimyeloma activity of lenalidomide and pomalidomide. *Blood* **2011**, *118*, 4771–4779. [CrossRef]
29. Buckley, D.L.; Van Molle, I.; Gareiss, P.C.; Tae, H.S.; Michel, J.; Noblin, D.J.; Jorgensen, W.L.; Ciulli, A.; Crews, C.M. Targeting the von Hippel-Lindau E3 Ubiquitin Ligase Using Small Molecules To Disrupt the VHL/HIF-1α Interaction. *J. Am. Chem. Soc.* **2012**, *134*, 4465–4468. [CrossRef]
30. He, M.; Cao, C.G.; Ni, Z.H.; Liu, Y.B.; Song, P.L.; Hao, S.; He, Y.N.; Sun, X.Y.; Rao, Y. PROTACs: Great opportunities for academia and industry (an update from 2020 to 2021). *Signal Transduct. Target. Ther.* **2022**, *7*, 181. [CrossRef]
31. Chirnomas, D.; Hornberger, K.R.; Crews, C.M. Protein degraders enter the clinic—A new approach to cancer therapy. *Nat. Rev. Clin. Oncol.* **2023**, *20*, 265–278. [CrossRef]
32. Ma, Z.H.; Zhou, J. NDA Submission of Vepdegestrant (ARV-471) to US FDA: The Beginning of a New Era of PROTAC Degraders. *J. Med. Chem.* **2025**, *68*, 14129–14136. [CrossRef]
33. Zhang, L.; Riley-Gillis, B.; Vijay, P.; Shen, Y. Acquired Resistance to BET-PROTACs (Proteolysis-Targeting Chimeras) Caused by Genomic Alterations in Core Components of E3 Ligase Complexes. *Mol. Cancer Ther.* **2019**, *18*, 1302–1311. [CrossRef]
34. Shirasaki, R.; Matthews, G.M.; Gandolfi, S.; Simoes, R.d.M.; Buckley, D.L.; Vora, J.R.; Sievers, Q.L.; Bruggenthies, J.B.; Dashevsky, O.; Poarch, H.; et al. Functional Genomics Identify Distinct and Overlapping Genes Mediating Resistance to Different Classes of Heterobifunctional Degraders of Oncoproteins. *Cell Rep.* **2021**, *34*, 108532. [CrossRef] [PubMed]
35. Hanzl, A.; Casement, R.; Imrichova, H.; Hughes, S.J.; Barone, E.; Testa, A.; Bauer, S.; Wright, J.; Brand, M.; Ciulli, A.; et al. Functional E3 ligase hotspots and resistance mechanisms to small-molecule degraders. *Nat. Chem. Biol.* **2023**, *19*, 323–333. [CrossRef]
36. Bondeson, D.P.; Smith, B.E.; Burslem, G.M.; Buhimschi, A.D.; Hines, J.; Jaime-Figueroa, S.; Wang, J.; Hamman, B.D.; Ishchenko, A.; Crews, C.M. Lessons in PROTAC Design from Selective Degradation with a Promiscuous Warhead. *Cell Chem. Biol.* **2018**, *25*, 78–87. [CrossRef]

37. Donovan, K.A.; Ferguson, F.M.; Bushman, J.W.; Eleuteri, N.A.; Bhunia, D.; Ryu, S.; Tan, L.; Shi, K.; Yue, H.; Liu, X.; et al. Mapping the Degradable Kinome Provides a Resource for Expedited Degrader Development. *Cell* **2020**, *183*, 1714–1731. [CrossRef]
38. Luo, X.; Archibeque, I.; Dellamaggiore, K.; Smither, K.; Homann, O.; Lipford, J.R.; Mohl, D. Profiling of diverse tumor types establishes the broad utility of VHL-based ProTaCs and triages candidate ubiquitin ligases. *iScience* **2022**, *25*, 103985. [CrossRef]
39. Lai, A.C.; Toure, M.; Hellerschmied, D.; Salami, J.; Jaime-Figueroa, S.; Ko, E.; Hines, J.; Crews, C.M. Modular PROTAC Design for the Degradation of Oncogenic BCR-ABL. *Angew. Chem.-Int. Ed.* **2016**, *55*, 807–810. [CrossRef] [PubMed]
40. Girardini, M.; Maniaci, C.; Hughes, S.J.; Testa, A.; Ciulli, A. Cereblon versus VHL: Hijacking E3 ligases against each other using PROTACs. *Bioorg. Med. Chem.* **2019**, *27*, 2466–2479. [CrossRef] [PubMed]
41. Steinebach, C.; Kehm, H.; Lindner, S.; Lan Phuong, V.; Koepff, S.; Marmol, A.L.; Weiler, C.; Wagner, K.G.; Reichenzeller, M.; Kroenke, J.; et al. PROTAC-mediated crosstalk between E3 ligases. *Chem. Commun.* **2019**, *55*, 1821–1824. [CrossRef] [PubMed]
42. Ishida, T.; Ciulli, A. E3 Ligase Ligands for PROTACs: How They Were Found and How to Discover New Ones. *Slas Discov.* **2021**, *26*, 484–502. [CrossRef]
43. Lee, J.; Lee, Y.; Jung, Y.M.; Park, J.H.; Yoo, H.S.; Park, J. Discovery of E3 Ligase Ligands for Target Protein Degradation. *Molecules* **2022**, *27*, 6515. [CrossRef]
44. Wang, C.; Weerapana, E.; Blewett, M.M.; Cravatt, B.F. A chemoproteomic platform to quantitatively map targets of lipid-derived electrophiles. *Nat. Methods* **2014**, *11*, 79–85. [CrossRef]
45. Backus, K.M.; Correia, B.E.; Lum, K.M.; Forli, S.; Horning, B.D.; González-Páez, G.E.; Chatterjee, S.; Lanning, B.R.; Teijaro, J.R.; Olson, A.J.; et al. Proteome-wide covalent ligand discovery in native biological systems. *Nature* **2016**, *534*, 570–574. [CrossRef]
46. Hacker, S.M.; Backus, K.M.; Lazear, M.R.; Forli, S.; Correia, B.E.; Cravatt, B.F. Global profiling of lysine reactivity and ligandability in the human proteome. *Nat. Chem.* **2017**, *9*, 1181–1190. [CrossRef]
47. Grossman, E.A.; Ward, C.C.; Spradlin, J.N.; Bateman, L.A.; Huffman, T.R.; Miyamoto, D.K.; Kleinman, J.I.; Nomura, D.K. Covalent Ligand Discovery against Druggable Hotspots Targeted by Anti-cancer Natural Products. *Cell Chem. Biol.* **2017**, *24*, 1368–1376. [CrossRef]
48. Ward, C.C.; Kleinman, J.I.; Brittain, S.M.; Lee, P.S.; Chung, C.Y.S.; Kim, K.; Petri, Y.; Thomas, J.R.; Tallarico, J.A.; McKenna, J.M.; et al. Covalent Ligand Screening Uncovers a RNF4 E3 Ligase Recruiter for Targeted Protein Degradation Applications. *ACS Chem. Biol.* **2019**, *14*, 2430–2440. [CrossRef] [PubMed]
49. Yang, H.; Ping, X.Y.; Zhou, J.Y.; Ailifeire, H.; Wu, J.; Nadal-Nicolás, F.M.; Miyagishima, K.J.; Bao, J.; Huang, Y.X.; Cui, Y.L.; et al. Reversible cold-induced lens opacity in a hibernator reveals a molecular target for treating cataracts. *J. Clin. Investig.* **2024**, *134*, e169666. [CrossRef]
50. Spradlin, J.N.; Hu, X.R.; Ward, C.C.; Brittain, S.M.; Jones, M.D.; Ou, L.S.; To, M.; Proudfoot, A.; Ornelas, E.; Woldegiorgis, M.; et al. Harnessing the anti-cancer natural product nimbolide for targeted protein degradation. *Nat. Chem. Biol.* **2019**, *15*, 747–755. [CrossRef] [PubMed]
51. Tong, B.Q.; Spradlin, J.N.; Novaes, L.F.T.; Zhang, E.; Hu, X.R.; Moeller, M.; Brittain, S.M.; McGregor, L.M.; McKenna, J.M.; Tallarico, J.A.; et al. A Nimbolide-Based Kinase Degrader Preferentially Degrades Oncogenic BCR-ABL. *ACS Chem. Biol.* **2020**, *15*, 1788–1794. [CrossRef] [PubMed]
52. Luo, M.; Spradlin, J.N.; Boike, L.; Tong, B.Q.; Brittain, S.M.; McKenna, J.M.; Tallarico, J.A.; Schirle, M.; Maimone, T.J.; Nomura, D.K. Chemoproteomics-enabled discovery of covalent RNF114-based degraders that mimic natural product function. *Cell Chem. Biol.* **2021**, *28*, 559–566. [CrossRef]
53. Lee, J.; Zhou, P. DCAFs, the Missing Link of the CUL4-DDB1 Ubiquitin Ligase. *Mol. Cell* **2007**, *26*, 775–780. [CrossRef]
54. Sutani, A.; Shima, H.; Hijikata, A.; Hosokawa, S.; Katoh-Fukui, Y.; Takasawa, K.; Suzuki, E.; Doi, S.; Shirai, T.; Morio, T.; et al. WDR11 is another causative gene for coloboma, cardiac anomaly and growth retardation in 10q26 deletion syndrome. *Eur. J. Med. Genet.* **2020**, *63*, 103626. [CrossRef] [PubMed]
55. Hall, J.R.; Bereman, M.S.; Nepomuceno, A.I.; Thompson, E.A.; Muddiman, D.C.; Smart, R.C. C/EBPα regulates CRL4Cdt2-mediated degradation of p21 in response to UVB-induced DNA damage to control the G1/S checkpoint. *Cell Cycle* **2014**, *13*, 3602–3610. [CrossRef]
56. Zhang, X.Y.; Crowley, V.M.; Wucherpfennig, T.G.; Dix, M.M.; Cravatt, B.F. Electrophilic PROTACs that degrade nuclear proteins by engaging DCAF16. *Nat. Chem. Biol.* **2019**, *15*, 737–746. [CrossRef] [PubMed]
57. Pu, C.L.; Tong, Y.; Liu, Y.Y.; Lan, S.K.; Wang, S.R.; Yan, G.Y.; Zhang, H.J.; Luo, D.; Ma, X.Y.; Yu, S.; et al. Selective degradation of PARP2 by PROTACs via recruiting DCAF16 for triple-negative breast cancer. *Eur. J. Med. Chem.* **2022**, *236*, 114321. [CrossRef]
58. Pu, C.L.; Liu, Y.Y.; Deng, R.; Xu, Q.J.; Wang, S.R.; Zhang, H.J.; Luo, D.; Ma, X.Y.; Tong, Y.; Li, R. Development of PROTAC degrader probe of CDK4/6 based on DCAF16. *Bioorg. Chem.* **2023**, *138*, 106637. [CrossRef]
59. Li, L.; Mi, D.Z.; Pei, H.X.; Duan, Q.H.; Wang, X.Y.; Zhou, W.B.; Jin, J.P.; Li, D.L.; Liu, M.Y.; Chen, Y.H. In vivo target protein degradation induced by PROTACs based on E3 ligase DCAF15. *Signal Transduct. Target. Ther.* **2020**, *5*, 129. [CrossRef]

60. Lucas, S.C.C.; Ahmed, A.; Ashraf, S.N.; Argyrou, A.; Bauer, M.R.; De Donatis, G.M.; Demanze, S.; Eisele, F.; Fusani, L.; Hock, A.; et al. Optimization of Potent Ligands for the E3 Ligase DCAF15 and Evaluation of Their Use in Heterobifunctional Degraders. *J. Med. Chem.* **2024**, *67*, 5538–5566. [CrossRef]
61. Zhang, X.Y.; Luukkonen, L.M.; Eissler, C.L.; Crowley, V.M.; Yamashita, Y.; Schafroth, M.A.; Kikuchi, S.; Weinstein, D.S.; Symons, K.T.; Nordin, B.E.; et al. DCAF11 Supports Targeted Protein Degradation by Electrophilic Proteolysis-Targeting Chimeras. *J. Am. Chem. Soc.* **2021**, *143*, 5141–5149. [CrossRef]
62. Wang, Y.; Wei, T.Z.; Zhao, M.; Huang, A.M.; Sun, F.; Chen, L.; Lin, R.S.; Xie, Y.B.; Zhang, M.; Xu, S.Y.; et al. Alkenyl oxindole is a novel PROTAC moiety that recruits the CRL4DCAF11 E3 ubiquitin ligase complex for targeted protein degradation. *PLoS Biol.* **2024**, *22*, e3002550. [CrossRef] [PubMed]
63. Li, Z.Y.; Wang, C.; Wang, Z.Y.; Zhu, C.G.; Li, J.; Sha, T.; Ma, L.X.; Gao, C.; Yang, Y.; Sun, Y.M.; et al. Allele-selective lowering of mutant HTT protein by HTT-LC3 linker compounds. *Nature* **2019**, *575*, 203–209. [CrossRef] [PubMed]
64. Zhao, M.; Ma, W.J.; Liang, J.Y.; Xie, Y.B.; Wei, T.Z.; Zhang, M.; Qin, J.J.; Lao, L.Y.; Tian, R.L.; Wu, H.Q.; et al. Design, Synthesis, and Activity Evaluation of BRD4 PROTAC Based on Alkenyl Oxindole-DCAF11 Pair. *J. Med. Chem.* **2024**, *67*, 19428–19447. [CrossRef]
65. Feller, F.; Weber, H.; Miranda, M.; Honin, I.; Hanl, M.; Hansen, F.K. Replacing a Cereblon Ligand by a DDB1 and CUL4 Associated Factor 11 (DCAF11) Recruiter Converts a Selective Histone Deacetylase 6 PROTAC into a Pan-Degrader. *ChemMedChem* **2025**, *20*, e202500035. [CrossRef]
66. Sinatra, L.; Yang, J.; Schliehe-Diecks, J.; Dienstbier, N.; Vogt, M.; Gebing, P.; Bachmann, L.M.; Sönnichsen, M.; Lenz, T.; Stühler, K.; et al. Solid-Phase Synthesis of Cereblon-Recruiting Selective Histone Deacetylase 6 Degraders (HDAC6 PROTACs) with Antileukemic Activity. *J. Med. Chem.* **2022**, *65*, 16860–16878. [CrossRef]
67. Schröder, M.; Renatus, M.; Liang, X.Y.; Meili, F.; Zoller, T.; Ferrand, S.; Gauter, F.; Li, X.Y.; Sigoillot, F.; Gleim, S.; et al. DCAF1-based PROTACs with activity against clinically validated targets overcoming intrinsic- and acquired-degrader resistance. *Nat. Commun.* **2024**, *15*, 275. [CrossRef] [PubMed]
68. Guarnaccia, A.D.; Tansey, W.P. Moonlighting with WDR5: A Cellular Multitasker. *J. Clin. Med.* **2018**, *7*, 21. [CrossRef]
69. Mabanglo, M.F.; Wilson, B.; Noureldin, M.; Kimani, S.W.; Mamai, A.; Krausser, C.; González-Alvarez, H.; Srivastava, S.; Mohammed, M.; Hoffer, L.; et al. Crystal structures of DCAF1-PROTAC-WDR5 ternary complexes provide insight into DCAF1 substrate specificity. *Nat. Commun.* **2024**, *15*, 10165. [CrossRef]
70. Ohoka, N.; Tsuji, G.; Shoda, T.; Fujisato, T.; Kurihara, M.; Demizu, Y.; Naito, M. Development of Small Molecule Chimeras That Recruit AhR E3 Ligase to Target Proteins. *ACS Chem. Biol.* **2019**, *14*, 2822–2832. [CrossRef]
71. Suzuki, T.; Yamamoto, M. Stress-sensing mechanisms and the physiological roles of the Keap1-Nrf2 system during cellular stress. *J. Biol. Chem.* **2017**, *292*, 16817–16824. [CrossRef]
72. Oskomic, M.; Tomic, A.; Barbaric, L.; Matic, A.; Kindl, D.C.; Matovina, M. KEAP1-NRF2 Interaction in Cancer: Competitive Interactors and Their Role in Carcinogenesis. *Cancers* **2025**, *17*, 447. [CrossRef] [PubMed]
73. Lee, S.M.; Hu, L.Q. Nrf2 activation through the inhibition of Keap1-Nrf2 protein-protein interaction. *Med. Chem. Res.* **2020**, *29*, 846–867. [CrossRef]
74. Saha, S.; Sachivkina, N.; Karamyan, A.; Novikova, E.; Chubenko, T. Advances in Nrf2 Signaling Pathway by Targeted Nanostructured-Based Drug Delivery Systems. *Biomedicines* **2024**, *12*, 403. [CrossRef]
75. Tong, B.Q.; Luo, M.; Xie, Y.; Spradlin, J.N.; Tallarico, J.A.; McKenna, J.M.; Schirle, M.; Maimone, T.J.; Nomura, D.K. Bardoxolone conjugation enables targeted protein degradation of BRD4. *Sci. Rep.* **2020**, *10*, 15543. [CrossRef] [PubMed]
76. Wei, J.L.; Meng, F.Y.; Park, K.S.; Yim, H.; Velez, J.; Kumar, P.; Wang, L.; Xie, L.; Chen, H.; Shen, Y.D.; et al. Harnessing the E3 Ligase KEAP1 for Targeted Protein Degradation. *J. Am. Chem. Soc.* **2021**, *143*, 15073–15083. [CrossRef] [PubMed]
77. Pei, J.; Xiao, Y.F.; Liu, X.G.; Hu, W.Y.; Sobh, A.; Yuan, Y.X.; Zhou, S.; Hua, N.; Mackintosh, S.G.; Zhang, X.; et al. Piperlongumine conjugates induce targeted protein degradation. *Cell Chem. Biol.* **2023**, *30*, 203–213. [CrossRef]
78. Akhtar, M.N.; Singh, A.; Manjunath, L.E.; Dey, D.; Kumar, S.D.; Vasu, K.; Das, A.; Eswarappa, S.M. Hominini-specific regulation of the cell cycle by stop codon readthrough of FEM1B. *J. Cell Sci.* **2024**, *137*, jcs261921. [CrossRef]
79. Lecoquierre, F.; Punt, A.M.; Ebstein, F.; Wallaard, I.; Verhagen, R.; Studencka-Turski, M.; Duffourd, Y.; Moutton, S.B.; Mau-Them, F.T.; Philippe, C.; et al. A recurrent missense variant in the E3 ubiquitin ligase substrate recognition subunit FEM1B causes a rare disorder. *Genet. Med.* **2024**, *26*, 101119. [CrossRef]
80. Manford, A.G.; Rodríguez-Pérez, F.; Shih, K.Y.; Shi, Z.; Berdan, C.A.; Choe, M.Y.; Titov, D.V.; Nomura, D.K.; Rape, M. A Cellular Mechanism to Detect and Alleviate Reductive Stress. *Cell* **2020**, *183*, 46–61. [CrossRef]
81. Henning, N.J.; Manford, A.G.; Spradlin, J.N.; Brittain, S.M.; Zhang, E.; McKenna, J.M.; Tallarico, J.A.; Schirle, M.; Rape, M.; Nomura, D.K. Discovery of a Covalent FEM1B Recruiter for Targeted Protein Degradation Applications. *J. Am. Chem. Soc.* **2022**, *144*, 701–708. [CrossRef]
82. Farrell, B.M.; Gerth, F.; Yang, C.H.R.; Yeh, J.T.H. A synthetic KLHL20 ligand to validate CUL3KLHL20 as a potent E3 ligase for targeted protein degradation. *Genes Dev.* **2022**, *36*, 1031–1042. [CrossRef]

83. Hickey, C.M.; Digianantonio, K.M.; Zimmermann, K.; Harbin, A.; Quinn, C.; Patel, A.; Gareiss, P.; Chapman, A.; Tiberi, B.; Dobrodziej, J.; et al. Co-opting the E3 ligase KLHDC2 for targeted protein degradation by small molecules. *Nat. Struct. Mol. Biol.* **2024**, *31*, 311–322. [CrossRef]
84. Scott, D.C.; Dharuman, S.; Griffith, E.; Chai, S.C.; Ronnebaum, J.; King, M.T.; Tangallapally, R.; Lee, C.; Gee, C.T.; Yang, L.; et al. Principles of paralog-specific targeted protein degradation engaging the C-degron E3 KLHDC2. *Nat. Commun.* **2024**, *15*, 8829. [CrossRef]
85. Basu, A.A.; Zhang, C.L.; Riha, I.A.; Magassa, A.; Campos, M.A.; Caldwell, A.G.; Ko, F.; Zhang, X.Y. A CRISPR activation screen identifies FBXO22 supporting targeted protein degradation. *Nat. Chem. Biol.* **2024**, *20*, 1608–1616. [CrossRef] [PubMed]
86. Dong, C.; Zhang, H.; Li, L.; Tempel, W.; Loppnau, P.; Min, J.R. Molecular basis of GID4-mediated recognition of degrons for the Pro/N-end rule pathway. *Nat. Chem. Biol.* **2018**, *14*, 466–473. [CrossRef]
87. Li, Y.R.; Bao, K.W.; Sun, J.Y.; Ge, R.X.; Zhang, Q.Q.; Zhang, B.; Yan, X.J.; Li, J.L.; Shi, F.Y.; Zhang, M.L.; et al. Design of PROTACs utilizing the E3 ligase GID4 for targeted protein degradation. *Nat. Struct. Mol. Biol.* **2025**, *32*, 1825–1837. [CrossRef] [PubMed]
88. Yazdi, A.K.; Perveen, S.; Dong, C.; Song, X.; Dong, A.P.; Szewczyk, M.M.; Calabrese, M.F.; Casimiro-Garcia, A.; Chakrapani, S.; Dowling, M.S.; et al. Chemical tools for the Gid4 subunit of the human E3 ligase C-terminal to LisH (CTLH) degradation complex. *RSC Med. Chem.* **2024**, *15*, 1066–1071. [CrossRef]
89. Owens, D.D.G.; Maitland, M.E.R.; Yazdi, A.K.; Song, X.S.; Reber, V.; Schwalm, M.P.; Machado, R.A.C.; Bauer, N.; Wang, X.; Szewczyk, M.M.; et al. A chemical probe to modulate human GID4 Pro/N-degron interactions. *Nat. Chem. Biol.* **2024**, *20*, 1164–1175. [CrossRef] [PubMed]
90. Spitz, M.L.; Kashkush, A.; Benhamou, R.I. Advancing target validation with PROTAC technology. *Expert Opin. Drug Discov.* **2025**, *20*, 551–563. [CrossRef]
91. Xue, F.L.; Zhang, M.H.; Li, S.Q.; Gao, X.Y.; Wohlschlegel, J.A.; Huang, W.B.; Yang, Y.; Deng, W.X. SE(3)-equivariant ternary complex prediction towards target protein degradation. *Nat. Commun.* **2025**, *16*, 5514. [CrossRef] [PubMed]

targets

MDPI

Review

Potential of MMP-2 and MMP-9 Gelatinase Blockade as a Therapeutic Strategy in Fibrosarcoma Treatment: A Decadal Review

Alireza Shoari

Department of Cancer Biology, Mayo Clinic, Jacksonville, FL 32224, USA; shoari.alireza@mayo.edu

Abstract: Fibrosarcoma represents a significant challenge in oncology, characterized by high invasiveness and a poor prognosis. Gelatinases, particularly matrix metalloproteinases MMP-2 and MMP-9, play a pivotal role in the degradation of the extracellular matrix, facilitating tumor invasion and metastasis. Inhibiting these enzymes has emerged as a promising therapeutic strategy. This review evaluates the progress in the development and therapeutic potential of gelatinase inhibitors as treatments for fibrosarcoma over the last decade, highlighting molecular mechanisms and future directions. A comprehensive literature review was conducted, focusing on studies published from 2013 to 2023. Research articles and review papers relevant to gelatinase inhibition and fibrosarcoma were examined to assess the efficacy and mechanisms of gelatinase inhibitors. Gelatinase inhibitors have shown the potential to reduce tumor progression, invasion, and metastasis in fibrosarcoma. Clinical trials, although limited, have indicated that these inhibitors can be effectively integrated into existing therapeutic regimens, offering a reduction in metastatic spread and potentially improving patient survival rates. Mechanistic studies suggest that the inhibition of MMP-2 and MMP-9 disrupts critical pathways involved in tumor growth and cell invasion. Gelatinase inhibition represents a viable and promising approach to fibrosarcoma treatment. Future research should focus on developing more specific inhibitors, understanding long-term outcomes, and integrating gelatinase inhibition into multimodal treatment strategies to enhance efficacy.

Keywords: fibrosarcoma; gelatinase inhibition; MMP-2; MMP-9; matrix metalloproteinases; cancer therapy

1. Fibrosarcoma

Fibrosarcoma is a rare type of cancer that arises from fibrous connective tissue, primarily affecting the fibroblasts, which are the cells that form the collagen-rich fibrous tissue throughout the body [1]. This type of cancer belongs to a larger group of cancers known as soft-tissue sarcomas. It primarily develops in the fibroblasts and is characterized by the formation of malignant fibrous tumors [2]. Fibrosarcoma can occur anywhere in the body but is most commonly found in the legs, arms, and trunk. It can also appear in the bones and is occasionally found in the lungs or other internal organs [3]. While it can affect individuals of any age, there are variations such as infantile fibrosarcoma that occurs in children, typically presenting a different, often less aggressive, behavior compared to adult forms [4].

The symptoms of fibrosarcoma depend on the tumor's location but typically include a noticeable lump or swelling that may be painful, pain or soreness caused by compressed nerves or muscles, and limited movement in the nearby joints if the tumor is located close to joints [5]. Diagnosing fibrosarcoma involves a combination of medical imaging and a tissue biopsy. Imaging techniques such as MRI, CT scans, and X-rays are used to visualize the tumor's size, location, and impact on the surrounding structures. A biopsy, where a sample of the tumor is removed and examined microscopically, is essential for a definitive diagnosis [6].

A complete surgical excision of the tumor is the primary treatment approach. Efforts are made to remove the tumor with some surrounding tissue to minimize the risk of recurrence [7]. Radiation therapy can be used post-surgery to eliminate residual cancer cells or in cases where surgery is not fully possible [8]. The role of chemotherapy is more limited but can be considered depending on the case, particularly for more advanced or metastatic disease [9]. The prognosis for fibrosarcoma varies widely based on factors such as the tumor's size, its location, the extent of surgical removal, the presence of metastasis, and the patient's overall health. Early detection and complete surgical removal typically offer the best outcomes [10]. Recurrence is possible, which requires ongoing monitoring. Understanding fibrosarcoma and its treatment options is crucial for effective management and improving patient outcomes. Regular follow-ups and imaging are often recommended as part of the post-treatment care plan to monitor for any signs of recurrence [11].

Fibrosarcoma is considered a rare type of cancer, with its prevalence being relatively low compared to other types of cancer. Fibrosarcoma accounts for approximately 1% of all soft-tissue sarcomas. The incidence rate varies, but, typically, it is estimated to affect around 0.2 to 0.3 people per 100,000 in the population annually [12]. Although fibrosarcoma can occur at any age, it generally has two peak incidences. In infants (as infantile fibrosarcoma, a less aggressive form) and in adults aged 30–55 years. The adult form tends to be more aggressive [13]. There is a slight male predominance in the incidence of fibrosarcoma [14]. Data on racial disparities are less clear, but like many soft-tissue sarcomas, there is no significant evidence of a racial predilection. There is no significant geographic variation noted with fibrosarcoma; it appears to affect individuals worldwide without major differences in incidence rates across different regions [15].

There are no well-established risk factors for fibrosarcoma except for a history of previous radiation therapy, which has been linked to a slightly increased risk of developing this and other types of sarcomas [16]. The survival rates for fibrosarcoma vary significantly depending on factors like tumor size, location, grade, and the success of surgical removal. Generally, the 5-year survival rate for localized fibrosarcoma can be quite high (approximately 50–70%), but this decreases significantly if the cancer is high-grade or has metastasized [16]. Understanding the prevalence and epidemiology of fibrosarcoma helps guide research and treatment strategies. Despite its rarity, the impact on affected individuals can be significant, highlighting the need for ongoing research and tailored treatment approaches.

2. Matrix Metalloproteinases (MMPs)

MMPs are a group of enzymes crucial to the regulation of extracellular matrix (ECM) components such as collagen, elastin, and gelatin, which play a fundamental role in tissue remodeling, wound healing, and other physiological and pathological processes [17]. Due to their significant involvement in these biological processes, MMPs are studied extensively in the context of diseases such as cancer, where they can facilitate tumor invasion and metastasis [18,19]. MMPs are characterized by several structural domains that are common across the different enzymes in this family, including the signal peptide; the propeptide domain, which is approximately 80 amino acids long and contains a conserved cysteine switch motif, typically PRCGXPD; the catalytic domain; the hinge region (also known as the linker peptide); and the hemopexin-like domain, which is involved in substrate recognition and specificity (Figure 1) [20].

MMPs can be classified based on their substrate specificity, domain organization, and sequence homology into multiple major groups, including the following: collagenases (e.g., MMP-1, MMP-8, and MMP-13), which primarily degrade fibrillar collagens, which are critical components of the extracellular matrix; gelatinases (e.g., MMP-2 and MMP-9), which can degrade gelatin and denatured collagens, playing significant roles in cancer metastasis; stromelysins, which can degrade a variety of ECM proteins but have broad specificity (e.g., MMP-3 and MMP-10); and membrane-type MMPs (MT-MMPs), which are membrane-bound and participate in pericellular proteolysis, crucial for cell migration and other functions [21]. MMPs are typically secreted as inactive proenzymes and need to be activated extracellularly. This activation can occur through various mechanisms, such as autoactivation, activation by other MMPs, or disruption of the cysteine switch by chemical agents or reactive oxygen species [22].

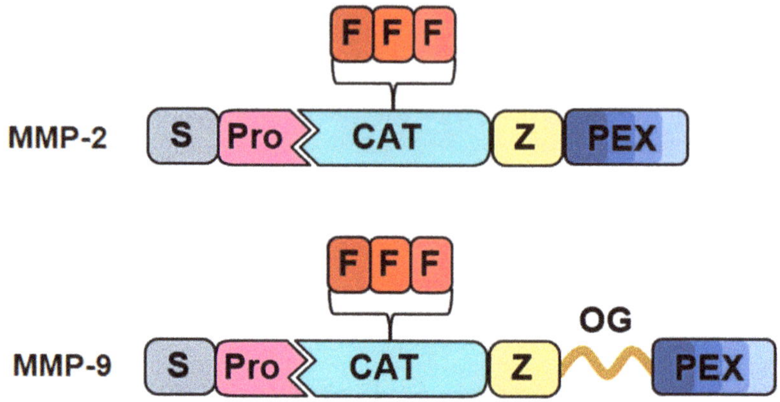

Figure 1. The multidomain structure of MMP-2 and MMP-9. The following domains are shown: the signal peptide (silver), the propeptide (pink), the catalytic domain (cyan), the three fibronectin repeats (red), the metal binding site (beige), the OG domain (brown), and the PEX domain (blue).

MMP activity is tightly regulated at multiple levels, such as gene transcription, zymogen activation, and inhibition by tissue inhibitors of metalloproteinases (TIMPs), which bind to the active sites of MMPs and block their enzymatic activity [23].

Understanding the structure, classification, and regulation of MMPs is critical in the context of physiological processes and in the development of therapeutic agents, especially for diseases characterized by excessive ECM degradation, such as arthritis, fibrosis, and cancer. MMPs play a significant and complex role in cancer progression. These enzymes are capable of degrading various components of the ECM, which not only influences tumor growth and development but also impacts metastasis, the process by which cancer spreads to new areas of the body [24]. Understanding the function of MMPs in cancer can provide insights into potential therapeutic targets and strategies for inhibiting tumor progression and metastasis.

3. MMP Roles in Fibrosarcoma

MMPs play a critical role in the progression and behavior of fibrosarcoma [25]. Understanding the interaction between MMPs and the tumor microenvironment in fibrosarcoma provides insights into the aggressive nature of this cancer and potential therapeutic targets. Fibrosarcoma cells produce MMPs that degrade the ECM, facilitating not only the local invasion of the tumor into surrounding tissues but also supporting tumor expansion. This ECM breakdown is pivotal in tumor progression because it liberates space for growing tumor cells and modifies the biomechanical properties of the tissue matrix [26]. The degradation of ECM components by MMPs releases bound growth factors such as transforming growth factor-beta (TGF-β) and fibroblast growth factors (FGFs), which are involved in promoting cell proliferation and angiogenesis and are essential for tumor growth and survival [27]. MMPs clear a path through the ECM, enabling fibrosarcoma cells to invade adjacent tissues and, potentially, enter the bloodstream or lymphatic system, leading to metastasis [28]. Additionally, they can induce epithelial-to-mesenchymal transition (EMT)-like processes in tumor cells, increasing their motility and invasive capabilities [20]. This is particularly relevant in fibrosarcoma, given its mesenchymal origin. By altering chemokine and cytokine profiles through ECM remodeling, MMPs can modulate the immune microenvironment around fibrosarcomas. This can lead to reduced effectiveness of the immune response against the tumor, aiding in tumor survival and progression [29].

In fibrosarcoma, certain MMPs are often upregulated. For example, the gelatinases MMP-2 (gelatinase A) and MMP-9 (gelatinase B) are frequently found to be elevated and are associated with poor prognostic outcomes. Their expression levels can correlate with tumor grade, size, metastatic potential, and the overall aggressiveness of the disease. MMP-2 and MMP-9 are instrumental in angiogenesis within fibrosarcoma tumors [30]. They degrade the basement membrane and interstitial collagens, facilitating endothelial cell migration and the formation of new blood vessels. This neovascularization is crucial for providing the tumor with nutrients and oxygen and for metabolic waste removal. The ability of these enzymes to break down the structural barriers in the ECM and basement membranes is particularly significant in the context of tumor invasion, metastasis, and angiogenesis [31].

Gelatinases degrade type IV collagen, which is a major component of basement membranes. This degradation is crucial for the disruption of these membranes and may allow fibrosarcoma cells to escape from their original site and invade surrounding tissues [30] (Figure 2). By breaking down the physical barriers in the ECM and basement membranes, gelatinases clear a path for tumor cells to migrate. This enhanced migratory capability is essential for local invasion and distant metastasis, the latter being a leading cause of cancer-related mortality [32]. The activity of gelatinases, particularly MMP-9, is associated with the mobilization of VEGF (vascular endothelial growth factor) from the ECM. VEGF is a potent pro-angiogenic factor that stimulates the formation of new blood vessels within the tumor, facilitating tumor growth and providing a means for cancer cells to access the systemic circulation [33]. The degradation products of basement membranes, generated by gelatinase activity, have been shown to promote the migration and organization of endothelial cells, which are critical steps in the formation of new blood vessels [34]. Beyond their roles in structural ECM degradation, gelatinases can process various bioactive molecules, including cytokines and growth factors. This processing can activate these molecules or modulate their activity, thus influencing inflammatory responses, tumor growth, and immune surveillance within the tumor microenvironment [35]. By modifying the ECM and releasing sequestered growth factors, gelatinases may also directly or indirectly promote tumor cell proliferation and survival. This effect supports not only primary tumor growth but also the establishment and expansion of metastases [36].

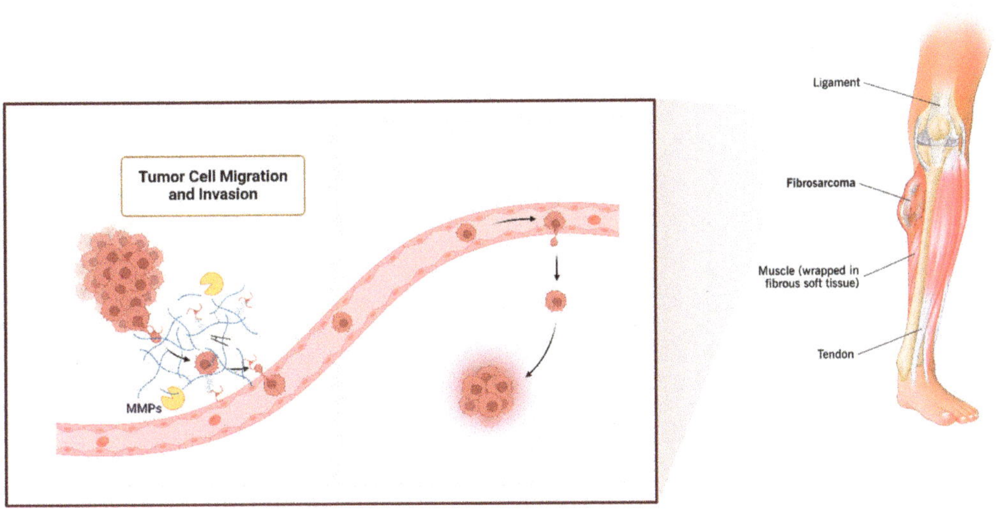

Figure 2. Gelatinases facilitate the migration and invasion of tumor cells by degrading the ECM barrier and promoting the remodeling of the surrounding tissue. This allows cancer cells to escape

from the primary tumor site and invade adjacent tissues, contributing to the aggressiveness and metastatic potential of fibrosarcoma. (Illustration created with BioRender.com) (https://app.biorender.com accessed on 4 June 2024).

4. Therapeutic Implications of Gelatinase Inhibition in Fibrosarcoma

Given their crucial roles in fibrosarcoma progression, gelatinases represent valuable targets for therapeutic intervention. Several inhibitors that specifically target MMP-2 and MMP-9 have been investigated for their potential to halt tumor progression by preventing ECM degradation, invasion, and angiogenesis [37]. These include both small-molecule inhibitors and biological agents like monoclonal antibodies. Integrating gelatinase inhibitors with conventional therapies such as chemotherapy and radiation might enhance the overall treatment efficacy, particularly by preventing tumor spread and overcoming resistance mechanisms [38].

However, the therapeutic targeting of gelatinases comes with challenges. Specific inhibitors that can selectively target MMP-2 or MMP-9 without affecting other MMPs are needed to minimize potential side effects, as broad-spectrum MMP inhibitors have been associated with adverse clinical outcomes [39]. The redundancy within the MMP family and the compensatory pathways involving other proteases can undermine the effectiveness of gelatinase-specific inhibitors [40].

HT-1080 cells are widely used in cancer research, particularly for studying the biological mechanisms and therapeutic strategies related to fibrosarcoma and other similar cancers. HT-1080 is a cell line derived from a human fibrosarcoma that was established from a 35-year-old male patient [41]. These cells are known for their ability to invade and migrate, making them mainly useful for studies on cancer metastasis, tumor progression, and drug testing, particularly in fibrosarcoma-related studies [42].

Ongoing research on fibrosarcoma is focused on improving the specificity of gelatinase inhibitors and understanding their role in the tumor microenvironment. This includes studying the interplay between gelatinases and other cell types within the tumor, such as immune cells and fibroblasts, to develop more comprehensive therapeutic strategies. Gelatinases are pivotal in the progression of fibrosarcoma through their involvement in ECM remodeling, invasion, angiogenesis, and the modulation of the tumor microenvironment. The effective targeting of these enzymes holds promise for improving clinical outcomes in fibrosarcoma treatment. In the following paragraphs, I will concentrate on the potential of gelatinase inhibitors as a therapeutic approach for treating fibrosarcoma.

5. Natural-Based Inhibitors

Natural-based anticancer compounds are substances derived from natural sources—such as plants, fungi, bacteria, and marine organisms—that exhibit potential anticancer properties. These compounds often play a pivotal role in the development of new cancer therapies, offering alternatives to synthetic drugs. The exploration of natural compounds is driven by their diverse chemical structures and mechanisms of action, which can target various cellular pathways involved in cancer progression [43]. Natural anticancer compounds can exert their effects through various mechanisms, including inducing apoptosis, inhibiting angiogenesis, suppressing metastasis, and modulating the immune system to enhance the body's ability to fight cancer [44]. Research into natural-based anticancer compounds is significant due to their potential to offer more targeted, less toxic treatment options and their ability to overcome resistance to current therapies [45]. Ongoing research focuses on isolating new compounds, understanding their mechanisms, and developing ways to synthesize them efficiently and test them in clinical settings. This field holds a promising avenue for the discovery of novel therapies that can contribute to cancer treatment [46]. Natural-based anticancer compounds have garnered interest for their potential in treating various types of cancer, including fibrosarcoma (Table 1). In the following paragraphs, there are some notable natural compounds that have shown potential in research for treating fibrosarcoma in the last decade.

Table 1. List of natural-based compounds explored as gelatinase inhibitors in fibrosarcoma cells.

Name	Source	Structure	Function	Ref.
Eugenol	Cloves	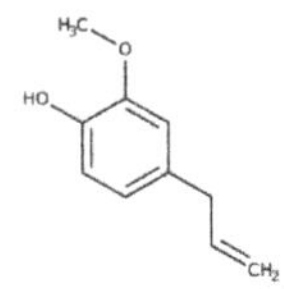	Inhibited MMP-9 activities related to metastasis in PMA-stimulated HT-1080 cells	[47]
Fisetin	*Dalbergia odorifera*	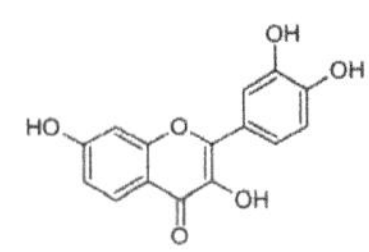	Inhibited MMP-9 more efficiently than a naturally occurring MMP inhibitor, tetracycline; dose-dependently inhibits proliferation of fibrosarcoma HT-1080 cells	[48]
Quercetin	Polyphenol found abundantly in a variety of foods	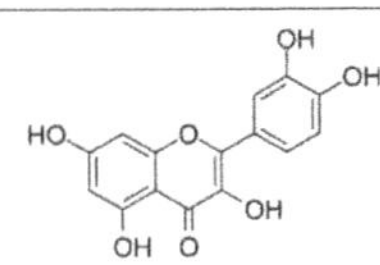	Inhibited PMS-induced increases in cell motility in HT-1080 cells; attenuated PMS-induced MMP-2 activation	[49]
Gallic acid, delphinidin-3-glucoside	Polyphenols found in a variety of foods	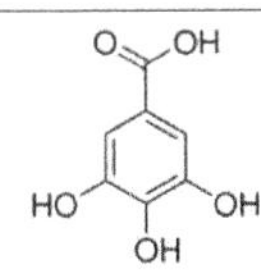	Anti-invasive activity; inhibitory activity on secreted and activated gelatinases; inhibited MMP-2 and MMP-9 proteolytic activity	[50]
Scutellarein	*Scutellaria lateriflora*	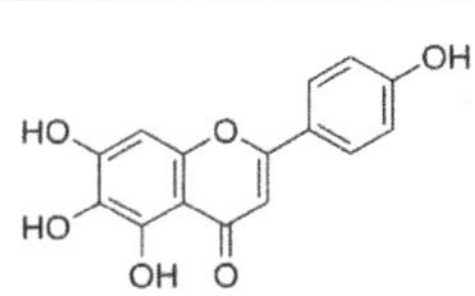	Proliferation rate of cells was significantly suppressed; volume and weight of the tumors were markedly reduced; potently inhibited cell migration, invasion, and the expression and activity of MMP-2 and MMP-9	[51]
Galangin, kaempferol	Flavonoids	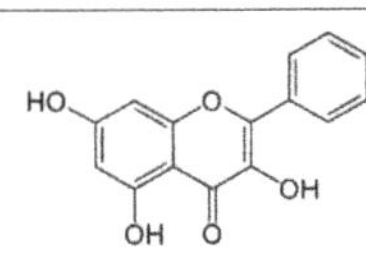	Efficiently decreased MMP-9 secretion; decreased transcription of MMP-9 mRNA	[52]
p- Cymene	Plants	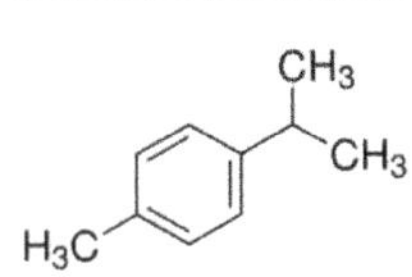	Dose-dependently inhibited the TPA-augmented production and gene expression of MMP-9; enhanced the TPA-augmented production and gene expression of TIMP-1; TPA-augmented invasiveness was inhibited	[53]
SE1	*Hippocampus kuda* Bleeler	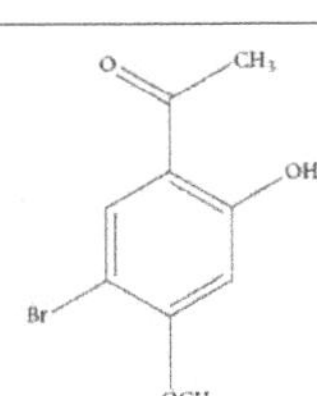	Potently inhibited gelatin digestion by MMP-9 induced by PMA; inhibited migration of HT-1080 cells in a dose-dependent manner	[54]

Table 1. *Cont.*

Name	Source	Structure	Function	Ref.
Chlorophyllin	Plants and some algae	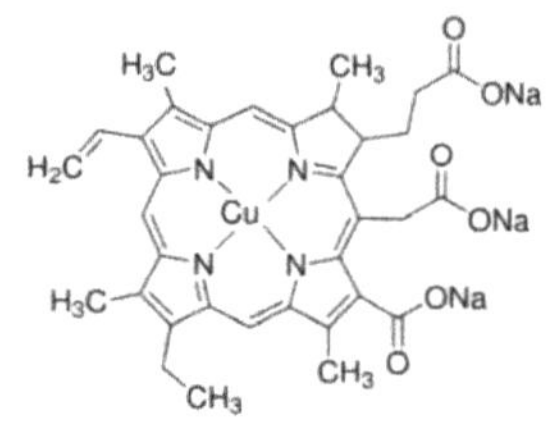	Cell proliferation was significantly decreased; expression of MMP-2 and MMP-9 decreased; cell invasion through Matrigel and cell migration were also reduced	[55]
Kahweol	Coffee	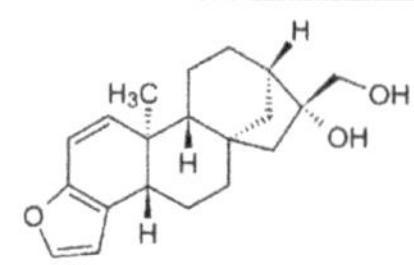	Inhibited cell proliferation enhanced by PMA; attenuated PMA-induced cell migration; suppressed PMA-enhanced activation of MMP-9	[56]
Lapachol	*Bignoniaceae*	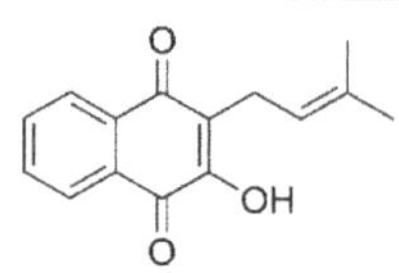	Inhibited the activation of MMP-2 and MMP-9 stimulated by PMA; protein and gene expression levels of MMP-2 remarkably decreased; increased the expression level of TIMP-1	[57]
Salidroside, 8(E)-nuezhenide, ligustroside	*Ligustrum japonicum*	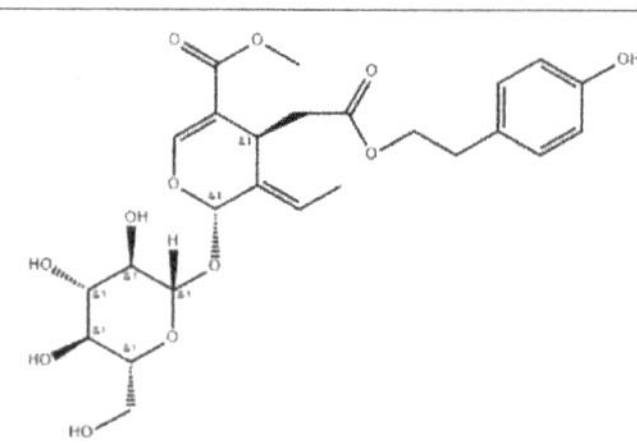	Significantly lowered the amount of MMP-2 and MMP-9 released; RNA and protein expression levels of MMP-2 and MMP-9 were notably suppressed	[58]
β-Caryophyllene oxide	*Piper nigrum*	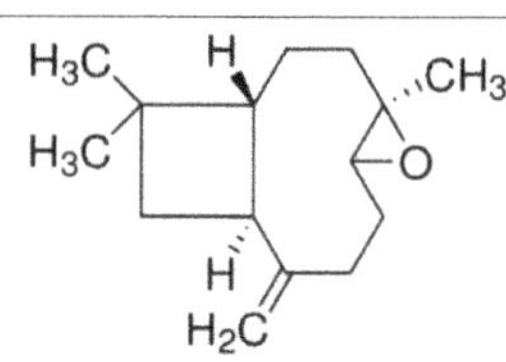	Reduced MMP-2 and MMP-9 activity; expression level of MMP-2 was reduced; reduced cell invasion	[59]
Silibinin	*Silybum marianum*	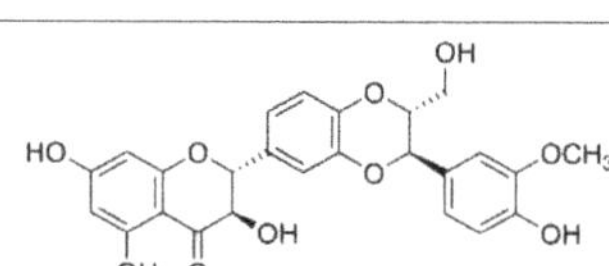	Showed growth inhibitory effects on fibrosarcoma cells; remarkably inhibited the levels of MMP-2 and MMP-9 activation; inhibited cell invasion on HT-1080 cells	[60]

Eugenol is a naturally occurring compound commonly found in clove oil, though it is also present in other essential oils like cinnamon, nutmeg, and bay leaf. It is a phenylpropene, which is a type of aromatic compound that gives cloves their distinctive smell and flavor [61]. Eugenol has a variety of uses, both in traditional medicine and in commercial applications. Eugenol has been studied for its potential anticancer properties, and some research suggests it might have beneficial effects in fighting various types of cancer [62]. In 2013, Nam and Kim investigated the potential of eugenol to suppress the expression and activity of MMPs, as well as its antioxidant effects. They specifically examined the impact of eugenol on MMP-9, which is associated with metastasis, using gelatin zymography and Western blot analysis. Their findings revealed that eugenol effectively inhibited MMP-9 activity in HT-1080 cells, a human fibrosarcoma cell line, stimulated by PMA (phorbol 12-myristate 13-acetate). Additionally, the study indicated that the inhibitory effects of eugenol on MMP-9 are mediated through the inactivation of the ERK (Extracellular signal-Regulated Kinase) pathway [47]. In the same year, Park et al. evaluated fisetin as an anticancer compound. Fisetin is a plant flavonol, a type of flavonoid, which is found in many fruits and vegetables. Fisetin inhibited MMP-9 more effectively than tetracycline, a naturally occurring MMP inhibitor. It also suppressed the proliferation of fibrosarcoma HT-1080 cells and human umbilical vascular endothelial cells (HUVECs) in a dose-dependent manner. Additionally, fisetin reduced the invasiveness of HT-1080 cells and the ability of HUVECs to form tubes in vitro [48].

In a further study in 2013 by Roomi and colleagues, they evaluated a nutrient mixture (NM) on different adult human sarcoma cell lines. The NM, prepared by VitaTech (Hayward, CA, USA), was composed of the following ingredients in the relative amounts indicated: vitamin C (as ascorbic acid and as Mg, Ca, and palmitate ascorbate) 700 mg; L-lysine 1000 mg; standardized green tea extract (80% polyphenol) 1000 mg; L-proline 750 mg; L-arginine 500 mg; N-acetyl cysteine 200 mg; selenium 30 μg; copper 2 mg; and manganese 1 mg. In gelatinase zymography, HT-1080 cells showed strong expression of inactive MMP-2 and weak expression of active MMP-2, both more pronounced than MMP-9. These were inhibited by the NM in a dose-dependent manner, achieving almost complete inhibition of MMP-9 at 100 μg/mL and MMP-2 at 250 μg/mL. Treatment with PMA (100 ng/mL) increased the expression of active MMP-2 and MMP-9 in HT-1080 cells; the NM effectively blocked both MMP-2 and MMP-9 in a dose-dependent manner, completely inhibiting active MMP-2 at 500 μg/mL and nearly completely inhibiting MMP-9 at 1000 μg/mL. The activity of TIMPs was upregulated by the NM in the HT-1080 cell line in a dose-dependent manner [63]. In 2018, this group also explored the effects of 10, 25, and 50 μM of chlorophyllin on HT-1080 cell proliferation and migration. Cell proliferation in HT-1080 cells was significantly reduced at a 50 μM dose of chlorophyllin. The expression of MMP-2 and MMP-9 also decreased in a dose-dependent manner, with significant inhibition observed at 25 μM and levels becoming virtually undetectable at 50 μM. Additionally, cell invasion through Matrigel and cell migration were reduced as the concentrations of chlorophyllin increased, with complete inhibition of invasion at 50 μM. H&E staining at 10 μM of chlorophyllin revealed some cellular changes indicative of apoptosis, while more significant apoptotic morphological changes were observed at higher doses of chlorophyllin [55].

Quercetin is a type of flavonoid antioxidant that is found in a variety of fruits, vegetables, and grains. Some research has indicated that quercetin may help slow cancer cell growth and induce apoptosis in various types of cancer cells [64]. Lee et al. conducted research to determine if quercetin could suppress the activities of MMP-2 and MMP-9 by reducing the formation of reactive oxygen species (ROS), a process anticipated to diminish cell motility. To trigger prolonged formation of ROS, cells were exposed to phenazine methosulfate (PMS; 1 μM). Quercetin, at noncytotoxic levels, was found to inhibit the increase in cell motility induced by PMS in HT-1080 cells. After 24 h of PMS treatment, nearly all cells demonstrated migration; however, quercetin significantly ($p < 0.01$) reduced this effect. Additionally, quercetin at concentrations up to 10 μg/mL was observed to decrease PMS-induced MMP-2 activation. Quercetin treatments significantly reduced the formation of ROS induced by PMS ($p < 0.01$) and led to decreased cell motility, accompanied by a decrease in MMP-2 and -9 activity in HT-1080 cells, even without PMS treatment [49].

One year later, in 2014, Filipiak et al. explored whether the inhibitory effect of gallic acid on gelatinases correlated with its cytotoxic activity in HT-1080 cells and assessed whether delphinidin-3-glucoside exhibited similar properties. Gallic acid and delphinidin-3-glucoside, which are prevalent bioactive compounds in many fruits and vegetables, are particularly concentrated in berries. Gallic acid is noted for its cytotoxic effects across various cancer cell lines and has demonstrated the ability to inhibit carcinogenesis in animal studies. Gallic acid and delphinidin-3-glucoside have exhibited selective cytotoxicity toward HT-1080 cells. In migration and invasion assays, both gallic acid and delphinidin-3-

glucoside demonstrated anti-invasive properties. A zymographic analysis revealed that gallic acid inhibits both secreted and activated gelatinases. Additionally, gallic acid was shown to suppress the proteolytic activities of MMP-2 and MMP-9 with similar effectiveness. Nuclear magnetic resonance (NMR) and molecular modeling studies have confirmed that gallic acid interacts with MMP-2, likely at the catalytic center [50].

Scutellarein is a flavone, a type of flavonoid compound that naturally occurs in certain plants. It can be found in plants like *Scutellaria lateriflora*, and it has been the subject of research for its various biological activities. Preliminary studies have indicated that scutellarein may have anticancer effects, including inhibiting the proliferation of certain types of cancer cells and inducing apoptosis [65]. Shi et al.'s analysis showed that scutellarein significantly inhibited the growth of HT-1080 by inducing apoptosis. Additionally, an in vivo experiment using Balb/c nude mice demonstrated that both the volume and weight of the tumors were significantly decreased after treatment with scutellarein. The further outcomes revealed that scutellarein potently blocked cell migration, invasion, and the expression and activity of MMP-2 and MMP-9 [51].

Marine plants, including algae, seaweeds, and other aquatic vegetation, have garnered significant interest in the field of cancer research due to their unique bioactive compounds [66]. In an interesting study, eight different species of brown algae were collected, and their extracts were analyzed for their anti-MMP effects. Based on the gelatin zymography results, *Ecklonia cava*, *Ecklonia bicyclis*, and *Ishige okamurae* demonstrated stronger inhibitory effects on MMP-2 and MMP-9 activities at concentrations of 10, 50, and 100 μg/mL, compared to the other samples tested. All samples effectively prevented cell migration in a dose-dependent manner after 24 h of incubation. Among all the samples tested, *Ecklonia cava* and *Ishige okamurae* were the most effective at inhibiting the invasion of HT-1080 cells [67]. In a similar study, Barletta et al. evaluated the anticancer effect of a hydrophilic extract from *Posidonia oceanica*. *Posidonia oceanica*, commonly known as Neptune grass or Mediterranean tapeweed, is a seagrass species that forms extensive meadows in the Mediterranean Sea. Their results showed for the first time in the highly invasive HT-1080 human fibrosarcoma cell line that hydrophilic extract from *P. oceanica* significantly reduced both gene and protein expressions of the gelatinases MMP-2 and MMP-9. Additionally, it directly inhibited gelatinolytic activity in a dose-dependent manner in vitro. Furthermore, their findings exhibited that this extract strongly suppressed the migration and invasion of HT-1080 cells [68]. *Limonium tetragonum*, commonly referred to as sea lavender, is a species of flowering plant in the *Plumbaginaceae* family. In traditional medicine, particularly in Korea, *Limonium tetragonum* has been used to treat various ailments due to its supposed anti-inflammatory and hepatoprotective properties. In an interesting study, *Limonium tetragonum* was investigated for its capability to inhibit MMP-2 and MMP-9 in HT-1080 fibrosarcoma cells. The study found that extracts from *L. tetragonum* effectively reduced the enzymatic activity and mRNA expression of MMP-2 and MMP-9, as evidenced by gelatin zymography and RT-PCR assays, respectively [69].

In a study in 2017, Lee et al. investigated the inhibitory effect of *Hizikia fusiformis* solvent-partitioned fractions on the invasion and MMP activity of HT-1080 cells. Hizikia fusiformis, a type of brown algae, has been found to exhibit various bioactivities such as antiviral, antimicrobial, and anti-inflammatory effects. These properties are attributed, in part, to its content of bioactive polysaccharides. *H. fusiformis* crude extract was fractionated with organic solvents, H_2O, n-BuOH, 85% aqueous MeOH, and n-hexane (n-Hex). All fractions successfully inhibited the enzymatic activities of MMP-2 and MMP-9, as evidenced by the gelatin zymography assay. The n-Hex fraction also significantly curtailed cell migration. These fractions reduced the mRNA and protein levels of MMP-2 and MMP-9 while increasing those of TIMP-1 and TIMP-2. Among the fractions, the H_2O fraction was the least effective, whereas the n-Hex fraction was the most effective [70]. In a similar study one year later, the effect of *Sargassum horneri*, a species of brown algae, was examined by researchers on HT-1080 cells. All fractions inhibited the enzymatic activities of MMP-2 and MMP-9, as demonstrated by gelatin zymography. Apart from the H_2O fraction, all fractions significantly hindered cell migration. Additionally, every tested fraction suppressed the mRNA and protein levels of MMP-2, MMP-9, TIMP-1, and TIMP-2 [71]. In a further study by this group, *Ishige okamurae*, a brown alga, blocked the enzymatic activities of MMP-2 and MMP-9. Additionally, cell migration was significantly reduced by the 85% aqueous MeOH fraction. The fractions decreased the mRNA and protein levels of MMP-2 and MMP-9 while increasing the levels of TIMP-1 and TIMP-2. Among these, the H_2O fraction was the least effective, while the 85% aq. MeOH fraction was the most effective [72].

Galangin and kaempferol are both naturally occurring flavonoids widely recognized for their potent antioxidant and anti-inflammatory properties. They are found in a variety of plants and are commonly consumed through dietary sources [73,74]. In a study in 2015, Choi and colleagues showed that galangin and kaempferol effectively reduced MMP-9 secretion in HT-1080 cells, while fisetin had only a slight effect on decreasing its secretion. Galangin and kaempferol did not impact cell viability at concentrations up to 30 μM. Luciferase reporter assays indicated that galangin and kaempferol reduced the transcription of MMP-9 mRNA [52].

p-Cymene, also known as 4-isopropyltoluene, is a naturally occurring aromatic organic compound. It is classified as a monoterpene and is a constituent of a variety of essential oils derived from plants, particularly cumin and thyme [75]. A study by Li et al. showed that *p*-cymene inhibited the production and gene expression of MMP-9 in HT-1080 cells in a dose-dependent manner when augmented by 12-O-tetradecanoylphorbol 13-acetate (TPA). Conversely, *p*-cymene increased the TPA-augmented production and gene expression of TIMP-1 in these cells. However, the baseline levels of MMP-9 and TIMP-1 mRNAs, as well as TIMP-1 protein, remained unchanged in the cells treated with *p*-cymene. Additionally, the study found that *p*-cymene dose-dependently reduced the TPA-enhanced invasiveness of HT-1080 cells in vitro [53].

The hippocampus, used in traditional Chinese medicine, is known for treating conditions such as tumors, aging, thrombosis, inflammation, hypertension, and prostatic hyperplasia [76]. 1-(5-Bromo-2-hydroxy-4-methoxyphenyl) ethanone (SE1), derived from the seahorse (*Hippocampus kuda Bleeler*), has demonstrated efficacy in suppressing proinflammatory responses. In a study by Gong et al., SE1 significantly inhibited gelatin digestion by MMP-9, which was induced by PMA, and reduced the migration of HT-1080 cells in a dose-dependent manner. The molecular docking results revealed that SE1 interacts with Tyr^{245} and His^{226} of MMP-9 through hydrogen and Pi-Pi bonds, leading to the inhibition of MMP-9 activity [54].

Kahweol acetate (KA), a coffee-specific diterpene, has shown antitumor properties in human tumor cells [77]. In a study by Choi and colleagues, KA significantly reduced cell proliferation induced by PMA in human fibrosarcoma cells. It also decreased PMA-induced cell migration and invasion in a concentration-dependent manner. KA inhibited the activation of MMP-9, primarily through the suppression of nuclear factor kappa B (NF-κB) activation. Additionally, KA hindered the PMA-induced phosphorylation of p38 MAP kinase, c-Jun N-terminal kinase (JNK) 1/2, and Akt, which are key signaling molecules upstream of MMP-9 expression [56].

Lapachol is a natural chemical compound with the molecular formula $C_{15}H_{14}O_3$. It is classified as a naphthoquinone and is primarily found in the bark of lapacho trees, which are native to South America. Lapachol has been studied for its potential medicinal properties, including its antibacterial, antifungal, and anti-inflammatory effects [78]. In HT-1080 cells, lapachol was found to inhibit the activation of MMP-2 and MMP-9 when stimulated by PMA at concentrations above 0.5 μM. Notably, both the protein and gene expression levels of MMP-2, when stimulated by PMA, were significantly reduced with 1 μM of lapachol compared to the group treated with PMA alone. Furthermore, lapachol also elevated the expression level of TIMP-1 compared to the PMA-only treatment group [57].

In a study by Kim et al., a phenyl ethanoid, salidroside (SAL), and two secoiridoids, 8(E)-nuezhenide (NZD) and ligustroside (LIG), were isolated from the fruits of *Ligustrum japonicum*. In fibrosarcoma cells, all the compounds substantially reduced the release of MMP-2 and MMP-9, as confirmed by gelatin zymography and ELISA tests. Moreover, the mRNA and protein expression levels of MMP-2 and MMP-9 were significantly decreased, as evidenced by RT-PCR and Western blot analyses. The Western blot assays revealed that SAL and LIG effectively diminished the expression of MMP-2 in a dose-dependent fashion. Similarly, NZD decreased the expression of MMP-9 [58].

β-Caryophyllene oxide (CPO) is a natural organic compound classified as a sesquiterpene oxide. It is a derivative of β-caryophyllene, a major component in many essential oils found in spices and aromatic herbs. In terms of biological activity, β-caryophyllene oxide has attracted interest for its potential therapeutic properties. It has been studied for its anti-inflammatory, anticancer, and analgesic properties [79]. The study by Jo et al. found that β-caryophyllene oxide (CPO) at a concentration of 32 μM decreased MMP-9 activity by 28% and MMP-2 activity by 60%. In the presence of 32 μM CPO, MMP-2 expression was reduced by 45%. Additionally, CPO lowered the expression levels of MMP-2 and MMP-9 in the immunofluorescence staining assay. An invasion assay on PMA-treated fibrosarcoma cells demonstrated that CPO reduced cell invasion in a dose-dependent manner, starting at a concentration of 2 μM [59].

Recently, in 2023, the antitumor effect of silibinin was explored on fibrosarcoma cells. Silibinin, also known as silybin, is a bioactive compound extracted from the seeds of the milk thistle plant (*Silybum marianum*). Recent studies have explored its effectiveness in inhibiting the growth of cancer cells in the liver, prostate, skin, and breast [80]. Silibinin, at concentrations above 20 μM, significantly reduced the activation levels of MMP-2 and MMP-9 when stimulated by PMA. Additionally, at a concentration of 25 μM, silibinin decreased the expression levels of MMP-2, IL-1β, ERK-1/2, and phosphorylated p38. Furthermore, silibinin at concentrations exceeding 10 μM blocked cell invasion in HT-1080 cells [60].

When considering natural-based medications for treating conditions like fibrosarcoma, the disadvantages can be particularly significant. For serious conditions like cancer, it is crucial to use treatments that have been proven effective through rigorous clinical trials. Natural-based treatments often lack sufficient scientific evidence supporting their efficacy in treating cancer, which can make them unreliable and potentially dangerous as a primary treatment option. Due to variations in the source and preparation of natural products, their potency and effectiveness can fluctuate significantly [81]. This inconsistency can be particularly problematic when treating something as aggressive and life-threatening as fibrosarcoma. Natural products can interfere with conventional cancer treatments, such as chemotherapy and radiation. They might either potentiate or inhibit the effects of these treatments, leading to increased toxicity or reduced efficacy [82]. While conventional cancer treatments undergo extensive safety testing, the safety profile of many natural medications is not as well established, especially in the context of cancer. This uncertainty can pose additional risks to patients. Relying on unproven natural treatments might cause delays in receiving more conventional therapies that have a proven track record of success. This delay can be critical, especially in fast-growing tumors like fibrosarcoma. For patients considering natural-based medications for fibrosarcoma, it is important to discuss these options thoroughly with oncologists and healthcare providers. Integrating such treatments into a comprehensive care plan should be carried out cautiously, ensuring that they do not undermine standard, evidence-based oncological therapies.

6. Synthetic Small-Molecule Inhibitors

Chemical and small-molecule medications are a cornerstone of cancer treatment. These drugs, designed to interact with specific molecular targets within cancer cells or related pathways, offer a broad range of therapeutic approaches [83]. Traditional chemotherapy drugs are small molecules that interfere with cell division. They target rapidly dividing cells, a hallmark of cancer cells, but can also affect normal cells, leading to side effects. These are designed to specifically target and inhibit cancer-specific molecules and signaling pathways, which are essential for tumor growth and survival [84]. Targeted therapies tend to have fewer side effects compared to traditional chemotherapy as they are designed to be more selective for cancer cells. Many targeted therapies inhibit specific enzymes that are involved in cancer cell signaling pathways, which are crucial for many cell functions, including division and survival [85]. Several chemical and small-molecule inhibitors have been developed, such as non-peptide small-molecule inhibitors. These newer inhibitors are designed to be more selective, aiming to reduce side effects and increase efficacy by specifically targeting MMP-2 and MMP-9. While the primary focus has been on the metastatic potential of cancers, MMP inhibitors also have the potential to improve outcomes by reducing angiogenesis and tumor growth [21]. In fibrosarcoma, where rapid and invasive growth is a major concern, inhibiting gelatinases can be particularly beneficial.

The treatment of HT-1080 cells with doxycycline resulted in a dose-dependent reduction in MMP-2 and MMP-9, with linear-trend R-squared values of 0.780 and 0.798, respectively. When HT-1080 cells were treated with PMA, there was a dose-dependent decrease in MMP-9 (R2 = 0.543) observed with the doxycycline treatment compared to the control, while MMP-2 levels remained unchanged. Actinomycin-D administration also led to a dose-dependent reduction in MMP-2, showing a strong linear trend (R2 = 0.978). Dexamethasone at a concentration of 50 μM had no impact on MMP-9 levels but reduced MMP-2 levels by 38% [86].

p-Dodecylaminophenol (*p*-DDAP) is a synthetic compound derived from fenretinide, which itself is a derivative of all-trans-retinoic acid used in cancer treatment. This compound has been developed to offer potent anticancer effects with fewer side effects than fenretinide, has exhibited strong antiproliferative properties against various cancers, including prostate cancer, and has shown effectiveness both in vitro and in vivo [87]. Takahashi et al. explored the anti-invasive effects of *p*-DDAP (1), which was produced based on N-(4-hydroxyphenyl)retinamide (2), a synthetic amide of all-trans-retinoic acid (3) (Figure 3). In HT-1080 cells, *p*-DDAP inhibited growth, induced apoptosis, and arrested the cell cycle in the S phase in a dose-dependent manner. Additionally, *p*-DDAP significantly suppressed cell invasion as well as the activity and mRNA expression of MMP-9. Moreover, the expression of the reversion-inducing cysteine-rich protein with Kazal motifs (RECK), a negative regulator of MMP-9, was increased following treatment. These results suggested that *p*-DDAP could be an effective anticancer agent, not only suppressing cell growth through apoptosis induction and cell cycle arrest but also inhibiting cell invasion by decreasing MMP-9 expression due to an increase in RECK expression [88].

In an interesting study, eleven biflavones, labeled 7a–b and 9a–i, were synthesized using a streamlined and effective method and evaluated for their inhibitory effects on MMP-2 and MMP-9 enzymes. Notably, the compound known as (I-3,II-3)-biacacetin (9h), resembling natural products, emerged as the most effective inhibitor (Figure 4). Unlike many conventional inhibitors, molecular docking studies revealed that 9h inhibits MMP-2 and MMP-9 in fibrosarcoma cells through interactions that do not involve zinc binding [89].

Aminoethyl-chitooligosaccharides (AE-COSs) are chemically modified derivatives of chitooligosaccharides, where hydroxyl groups are substituted with aminoethyl groups at specific positions, commonly the C-6 position. This modification enhances their biological activities, making AE-COSs particularly useful in various biomedical applications. Research has demonstrated the effectiveness of AE-COSs in several areas. Notably, AE-COSs have shown significant potential for inhibiting cancer cell proliferation and invasion [90]. The results from MTT assays, gelatin zymography, and Western blot analyses, which measure MMP-2 and MMP-9, indicate that aminoethyl-chitooligosaccharides (AE-COSs) (Figure 5) can inhibit the invasive and metastatic potential of HT-1080 cells by blocking MMP-2 and MMP-9 activity and expression [91].

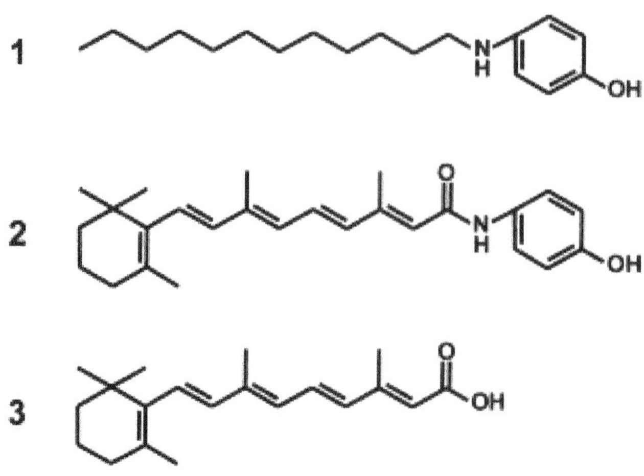

Figure 3. Chemical structures of *p*-DDAP (1), which was produced based on N-(4-hydroxyphenyl) retinamide (2), a synthetic amide of all-trans-retinoic acid (3). Adapted with permission from [88], Elsevier, 2013.

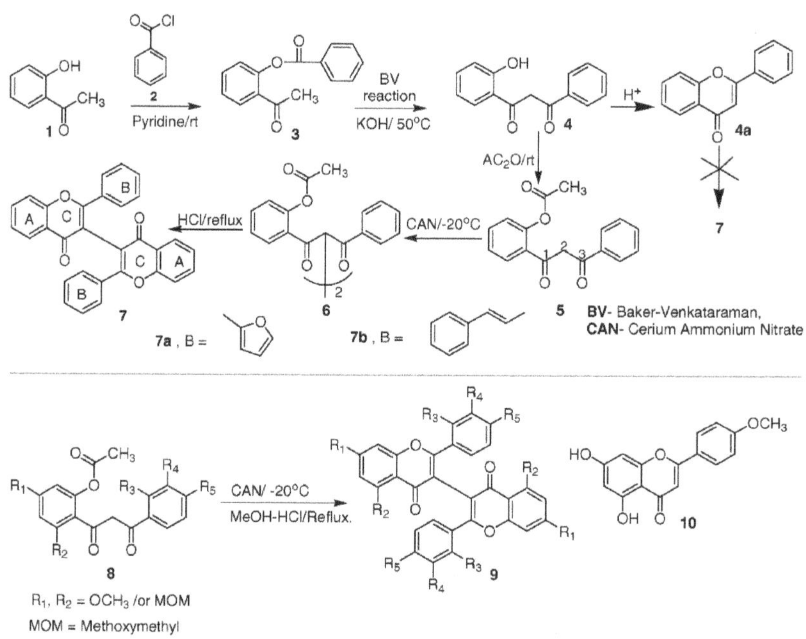

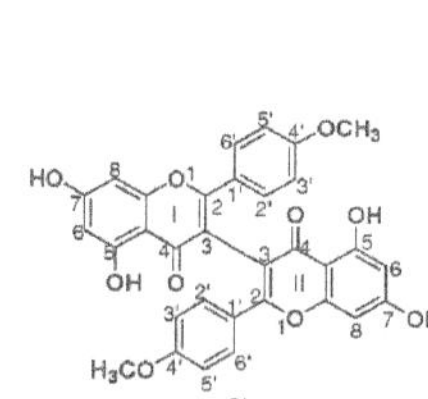

Entry	R_1	R_2	R_3	R_4	R_5	Yield (%)
9a	OCH_3	H	H	H	H	85
9b	H	H	OCH_3	H	H	88
9c	H	H	H	H	OCH_3	83
9d	OCH_3	OCH_3	H	H	H	82
9e	OCH_3	OCH_3	H	H	OCH_3	86
9f	OH	H	H	H	H	85
9g	OH	OH	H	H	H	83
9h	OH	OH	H	H	OCH_3	80
9i	OH	OH	H	OCH_3	OCH_3	78

Figure 4. Retro-synthetic analysis of (I-3,II-3)-biflavones. (compound in blue). Adapted with permission from [89], Elsevier, 2015.

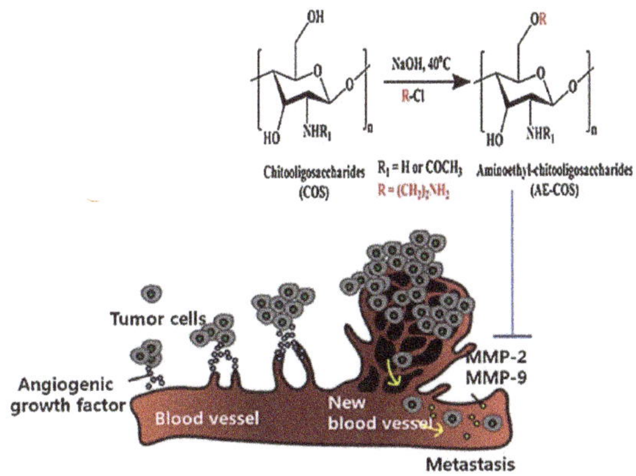

Figure 5. Diagram depicting the process by which AE-COS inhibits the invasion of human fibrosarcoma cells. AE-COS hinders the infiltration of human fibrosarcoma cells by obstructing the activity of MMP-2 and MMP-9. Adapted with permission from [91], Elsevier, 2016.

LY294002, PD98059, and SB203580 are all chemical inhibitors used in biochemical research to study signaling pathways. LY294002 is a specific inhibitor of phosphoinositide 3-kinases (PI3Ks) [92], and PD98059 is an inhibitor of mitogen-activated protein kinase (MEK), which is a part of the MAPK/ERK pathway [93]. SB203580 primarily inhibits p38 MAP kinase, which is involved in inflammatory responses and stress [94]. A study by Yu and Kim revealed that when these inhibitors were applied along with salinomycin (Figure 6), there was a suppression in metastatic potential through the reduction in MMP-2 expression as well as migration and invasion in HT-1080 cells. In a further study, they also found that 5-azacytidine (5-aza C), a methyltransferase inhibitor, can induce MMP-9 activity in HT-1080 cells and promote migration and invasion significantly, which can be inhibited by its inhibitors (LY294002 and PD98059) (Figure 6) [95].

Combination therapy that uses radiation along with small-molecule inhibitors is a promising approach for enhancing cancer treatment effectiveness. This strategy aims to target cancer cells more precisely and reduce the side effects typically associated with conventional therapies [96]. 4-Methylumbelliferone (4-MU) (Figure 6) is a derivative of coumarin, primarily known for its use as a hyaluronic acid synthesis inhibitor. A study showed that there was a higher reduction in the viability of HT-1080 cells cultured with a combination of 2 Gy IR and 100 μM 4-MU compared with untreated control cells. Moreover, this treatment led to a substantial decline in the invasion rate and MMP-2 and MMP-9 expression in the fibrosarcoma cell line [97].

In a study by Alford et al., they developed a series of compounds derived from the lead molecule N-[4-(difluoromethoxy)phenyl]-2-[(4-oxo-6-propyl-1H-pyrimidin-2-yl)sulfanyl]-acetamide, aimed at targeting the hemopexin-like domain of MMP-9. This group identified N-(4-fluorophenyl)-4-(4-oxo-3,4,5,6,7,8-hexahydroquinazolin-2-ylthio)butanamide, or 3c, as a potent lead compound with a dissociation constant (Kd) of 320 nM, exhibiting specificity for the proMMP-9 hemopexin-like domain. They showed that 3c inhibits MMP-9 homodimerization, and 500 nM of 3c effectively blocks fibrosarcoma cell invasion through the basement membrane and decreases angiogenesis [98].

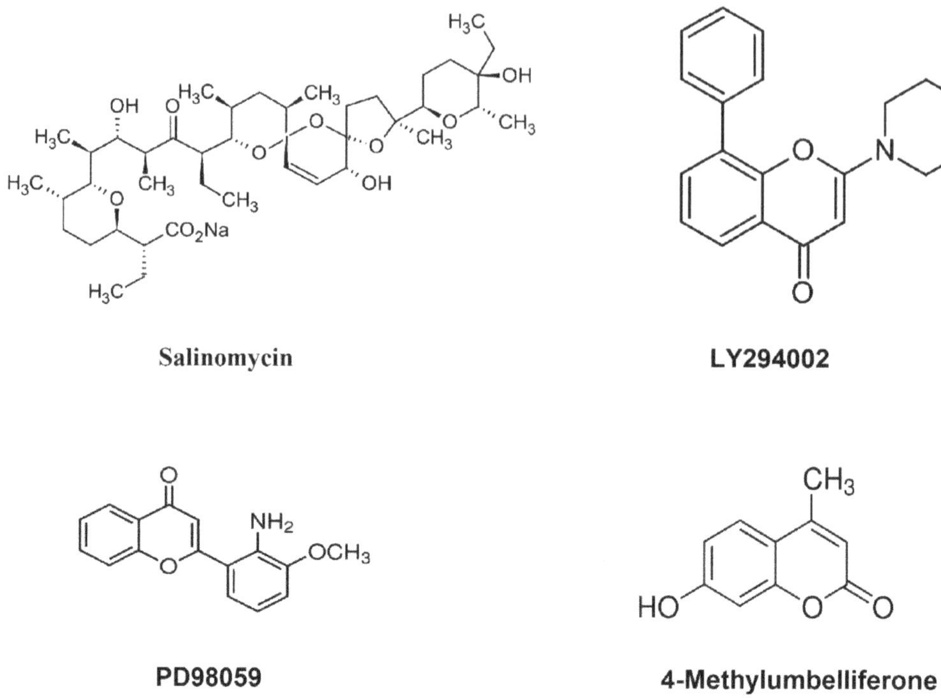

Figure 6. Chemical structure of salinomycin, LY294002, PD98059, and 4-Methylumbelliferone.

The synthetic chemical small molecules used in cancer therapy often offer significant advantages, such as the ability to target specific molecular pathways involved in cancer progression. However, they also come with several disadvantages. Many synthetic small molecules can be highly toxic, not only to cancer cells but also to normal cells, leading to various side effects. For example, they might cause nausea, vomiting, hair loss, and a decrease in blood cell counts, which can lead to increased susceptibility to infections [99]. Cancer cells can develop resistance to these treatments over time. This resistance can occur through various mechanisms, such as mutations in the target molecule, increased drug efflux, or compensatory activation of alternative pathways. This makes the treatment less effective over time and necessitates the use of additional or alternative therapies [100]. While designed to be specific, some synthetic small molecules may still affect other cellular targets, leading to off-target effects and unintended consequences. This lack of absolute specificity can compromise their effectiveness and safety [101]. The development, production, and formulation of synthetic small molecules can be expensive, which may make these drugs less accessible to patients, especially in low-income regions [102]. Issues related to the absorption, distribution, metabolism, and excretion of these drugs can affect their efficacy and safety. For instance, a drug might not reach the tumor in sufficient concentrations to be effective or could be metabolized too quickly to have a therapeutic effect [103]. Some tumors, such as fibrosarcoma, have a microenvironment that hinders the effective penetration and distribution of these drugs, which can limit their efficacy in killing cancer cells throughout the tumor. These challenges highlight the need for continuous research to improve the specificity, effectiveness, and safety of synthetic chemical small molecules in cancer therapy.

7. Protein- and Peptide-Based Inhibitors

Protein- and peptide-based therapies offer a promising approach to cancer treatment, focusing on using peptides and proteins to target specific biological processes involved in cancer growth and progression. Proteins and peptides can be designed to target specific receptors, enzymes, or other proteins associated with cancer cells with high specificity, potentially reducing the risk of harming normal cells [104]. These molecules can be engineered to carry out a variety of functions, such as directly inhibiting tumor growth, blocking signals that promote cancer cell survival, or delivering cytotoxic agents specifically to cancer cells [18,105]. Compared to traditional chemotherapy, peptide and protein therapies often have lower toxicity, leading to fewer and less severe side effects [106]. Being natural biological molecules, proteins and peptides are biodegradable and less likely to accumulate in the body, reducing long-term side effects [107]. The development of protein- and peptide-based therapies for fibrosarcoma is still in the experimental stages, with many potential treatments undergoing preclinical trials or early clinical testing. The complexity of fibrosarcoma's biological mechanisms often necessitates combination therapies, where protein- or peptide-based therapies might be used in conjunction with other treatments like chemotherapy or radiation.

In 2014, Kim et al. developed a carnosine–gallic acid peptide (CGP) to explore a more effective MMP inhibitor than carnosine. They examined the inhibitory impacts of CGP on MMP-2 and MMP-9 within the human fibrosarcoma cell line. CGP significantly reduced the expression levels of MMP-2 and MMP-9 without causing cytotoxic effects. Furthermore, CGP potentially hinders migration and invasion in HT-1080 cells by targeting the urokinase plasminogen activator (uPA)–uPA receptor signaling pathways, thereby suppressing MMP-2 and MMP-9 activity [108].

CD97 is a tumor-associated adhesion-class G-protein-coupled receptor that plays a role in regulating cell migration. Hsiao et al.'s findings indicated that CD97 expression in HT-1080 fibrosarcoma cells increases the secretion of TIMP-2, which subsequently reduces the activity of MMP-MT1 and MMP-2. This reduction in enzyme activity hampers cell migration and invasion in vitro [109].

In a study by Ko and colleagues in 2018, they isolated a heptameric peptide (Leu-Leu-Ala-Pro-Pro-Glu-Arg) from the marine microalga *Pavlova lutheri* and studied its inhibitory influence and mechanism of action on MMP-9 in HT-1080 cells. The outcomes exhibited that MMP-9's mRNA and protein expression levels were decreased in the presence of peptides. The activity of MMP-9 when stimulated by PMA was similar to that of the control group; however, treatment with the peptide considerably reduced MMP-9 activity at concentrations of 250 and 500 $\mu g/mL^{-1}$ ($p < 0.05$) [110].

Phage display is a versatile tool for the development of protease inhibitors. This technology involves the expression of peptide or protein libraries on the surface of phages. The displayed peptides or proteins can be engineered to bind to specific targets, including proteases [111]. In our two related studies in 2019 and 2022, we found that a phage display-isolated linear and cyclic type of peptide (MHPNAGHGSLMR) can significantly block the gelatinolytic activity of MMP-2 (pro- and active form) and proMMP-9 in a dose-dependent manner in HT-1080 cells. The results from cell invasion also showed inhibition of cell invasion at a concentration of 50 μM of peptide [112,113].

Currently, there are no antibodies specifically approved for the treatment of fibrosarcoma. However, research is ongoing to explore various antibodies that might target molecules associated with fibrosarcoma cells. Antibodies that inhibit growth factor receptors, which are often overexpressed in various cancers, including sarcomas, could potentially be effective. Research may focus on developing monoclonal antibodies that target specific antigens expressed in fibrosarcoma cells. These could include surface proteins unique to fibrosarcoma or associated with aggressive tumor behavior. Patients with fibrosarcoma might be eligible to participate in clinical trials testing new antibody therapies or combinations of existing treatments. These trials are crucial for discovering effective therapies for rare cancers like fibrosarcoma.

Proteins and peptides offer promising therapeutic potential for cancer treatment due to their specificity and ability to target key molecular pathways. However, their use as therapeutic agents also presents several disadvantages and challenges. Proteins and peptides are generally less stable in the body compared to small-molecule drugs. They are prone to degradation by proteases and other enzymes in the bloodstream and digestive tract, which can significantly reduce their effective lifespan and activity. Effective delivery of proteins and peptides to the target site within the body is challenging. Their large size and complex structure can hinder their ability to cross cellular membranes, and they may be poorly absorbed when administered orally. This often necessitates intravenous or subcutaneous administration, which can be less convenient and more painful for patients. Proteins and peptides can potentially be recognized as foreign substances by the patient's immune system, leading to immune responses that can neutralize the therapeutic agents or cause adverse reactions. Managing immunogenicity is a major concern, particularly for repeated administrations. Due to issues with bioavailability and degradation, high doses of protein and peptide drugs may be required to achieve therapeutic effects, which can increase the risk of side effects and further escalate treatment costs. The effective penetration of proteins and peptides into solid tumors is often limited, which can reduce their efficacy against certain types of cancer.

8. Conclusions

Over the past decade, the study of the gelatinases MMP-2 and MMP-9 as therapeutic targets in fibrosarcoma treatment has revealed their significant potential in inhibiting tumor progression and metastasis. Blockade of these enzymes can disrupt the extracellular matrix remodeling that fibrosarcomas exploit for growth and dissemination [33]. Clinical and experimental research into MMP inhibitors has uncovered promising approaches to curbing these processes, with several inhibitors showing efficacy in reducing tumor size and improving survival rates in preclinical models [114]. However, challenges remain in translating these findings into clinical successes due to issues such as side effects, resistance, and the specificity of inhibitors. The therapeutic potential of gelatinase blockade in fibrosarcoma treatment remains substantial but requires further refinement and investigation to optimize dosing, minimize adverse effects, and enhance delivery mechanisms.

The potential of targeting gelatinases as a therapeutic strategy in the treatment of fibrosarcoma is promising, as highlighted in this decadal review of inhibitors categorized into natural, chemical, and protein- and peptide-based groups. Natural inhibitors have shown significant potential due to their biocompatibility and fewer side effects, making them an appealing option for long-term therapy. Chemical inhibitors, on the other hand, are often more potent and can be designed to selectively target MMP-2 and MMP-9, minimizing off-target effects and enhancing therapeutic efficacy. Protein- and peptide-based inhibitors provide a unique advantage due to their specificity and ability to disrupt enzyme–substrate interactions directly at the molecular level. While each category of inhibitors has its advantages and limitations, the integration of these strategies could lead to more effective and personalized therapeutic approaches in fibrosarcoma treatment. Future research should focus on overcoming the challenges related to the resistance, delivery, and stability of these inhibitors to fully harness their potential in clinical settings.

Future research should also explore combination therapies that may enhance the effectiveness of gelatinase inhibitors and reduce the likelihood of tumor resistance. Continued exploration in this field holds promise for more effective fibrosarcoma treatments, potentially improving outcomes for patients suffering from this aggressive cancer.

Funding: This research received no external funding.

Data Availability Statement: Not applicable.

Conflicts of Interest: The author declares no conflicts of interest.

References

1. Augsburger, D.; Nelson, P.J.; Kalinski, T.; Udelnow, A.; Knosel, T.; Hofstetter, M.; Qin, J.W.; Wang, Y.; Gupta, A.S.; Bonifatius, S.; et al. Current diagnostics and treatment of fibrosarcoma -perspectives for future therapeutic targets and strategies. *Oncotarget* **2017**, *8*, 104638–104653. [CrossRef] [PubMed]
2. Genadry, K.C.; Pietrobono, S.; Rota, R.; Linardic, C.M. Soft Tissue Sarcoma Cancer Stem Cells: An Overview. *Front. Oncol.* **2018**, *8*, 475. [CrossRef] [PubMed]
3. Huvos, A.G.; Higinbotham, N.L. Primary fibrosarcoma of bone. A clinicopathologic study of 130 patients. *Cancer* **1975**, *35*, 837–847. [CrossRef] [PubMed]
4. Atalay, I.B.; Togral, G. Unusual localization and aggressive progression of large infantile fibrosarcoma. *Acta Orthop. Traumatol. Turc.* **2019**, *53*, 507–511. [CrossRef]
5. Scott, S.M.; Reiman, H.M.; Pritchard, D.J.; Ilstrup, D.M. Soft tissue fibrosarcoma. A clinicopathologic study of 132 cases. *Cancer* **1989**, *64*, 925–931. [CrossRef] [PubMed]
6. Ferreira, C.R.; da Fonseca, L.G.; Piotto, G.H.M.; Geyer, F.C.; de Alcantara, P.S.M. Fibrosarcoma: A challenging diagnosis. *Autops. Case Rep.* **2013**, *3*, 21–29. [CrossRef] [PubMed]
7. Dernell, W.S.; Withrow, S.J.; Kuntz, C.A.; Powers, B.E. Principles of treatment for soft tissue sarcoma. *Clin. Tech. Small Anim. Pract.* **1998**, *13*, 59–64. [CrossRef] [PubMed]
8. Windeyer, B.; Dische, S.; Mansfield, C.M. The place of radiotherapy in the management of fibrosarcoma of the soft tissues. *Clin. Radiol.* **1966**, *17*, 32–40. [CrossRef]
9. Chew, W.; Benson, C.; Thway, K.; Hayes, A.; Miah, A.; Zaidi, S.; Lee, A.T.J.; Messiou, C.; Fisher, C.; van der Graaf, W.T.; et al. Clinical Characteristics and efficacy of chemotherapy in sclerosing epithelioid fibrosarcoma. *Med. Oncol.* **2018**, *35*, 138. [CrossRef]
10. Zhang, P.; Zhao, L.; Zhu, Y.J.; Qiu, B.; Guo, S.P.; Li, Y.; Liu, Q.; Liu, M.Z.; Xi, M. Prognosis of Fibrosarcoma in Patients With and Without a History of Radiation for Nasopharyngeal Carcinoma. *Ann. Surg. Oncol.* **2017**, *24*, 434–440. [CrossRef]
11. Cantin, J.; McNeer, G.P.; Chu, F.C.; Booher, R.J. The problem of local recurrence after treatment of soft tissue sarcoma. *Ann. Surg.* **1968**, *168*, 47–53. [CrossRef]
12. Wibmer, C.; Leithner, A.; Zielonke, N.; Sperl, M.; Windhager, R. Increasing incidence rates of soft tissue sarcomas? A population-based epidemiologic study and literature review. *Ann. Oncol.* **2010**, *21*, 1106–1111. [CrossRef] [PubMed]
13. Coffin, C.M.; Dehner, L.P. Fibroblastic-myofibroblastic tumors in children and adolescents: A clinicopathologic study of 108 examples in 103 patients. *Pediatr. Pathol.* **1991**, *11*, 569–588. [CrossRef] [PubMed]
14. Bahrami, A.; Folpe, A.L. Adult-type fibrosarcoma: A reevaluation of 163 putative cases diagnosed at a single institution over a 48-year period. *Am. J. Surg. Pathol.* **2010**, *34*, 1504–1513. [CrossRef] [PubMed]
15. Burningham, Z.; Hashibe, M.; Spector, L.; Schiffman, J.D. The epidemiology of sarcoma. *Clin. Sarcoma Res.* **2012**, *2*, 14. [CrossRef] [PubMed]
16. Folpe, A.L. Fibrosarcoma: A review and update. *Histopathology* **2014**, *64*, 12–25. [CrossRef]
17. Wang, X.; Khalil, R.A. Matrix Metalloproteinases, Vascular Remodeling, and Vascular Disease. *Adv. Pharmacol.* **2018**, *81*, 241–330. [CrossRef] [PubMed]
18. Shoari, A.; Khalili-Tanha, G.; Coban, M.A.; Radisky, E.S. Structure and computation-guided yeast surface display for the evolution of TIMP-based matrix metalloproteinase inhibitors. *Front. Mol. Biosci.* **2023**, *10*, 1321956. [CrossRef]
19. Radisky, E.S. Extracellular proteolysis in cancer: Proteases, substrates, and mechanisms in tumor progression and metastasis. *J. Biol. Chem.* **2024**, *300*, 107347. [CrossRef]
20. Radisky, E.S.; Radisky, D.C. Matrix metalloproteinase-induced epithelial-mesenchymal transition in breast cancer. *J. Mammary Gland. Biol. Neoplasia* **2010**, *15*, 201–212. [CrossRef]
21. Almutairi, S.; Kalloush, H.M.; Manoon, N.A.; Bardaweel, S.K. Matrix Metalloproteinases Inhibitors in Cancer Treatment: An Updated Review (2013–2023). *Molecules* **2023**, *28*, 5567. [CrossRef]
22. Klein, T.; Bischoff, R. Physiology and pathophysiology of matrix metalloproteases. *Amino Acids* **2011**, *41*, 271–290. [CrossRef]
23. Cabral-Pacheco, G.A.; Garza-Veloz, I.; Castruita-De la Rosa, C.; Ramirez-Acuna, J.M.; Perez-Romero, B.A.; Guerrero-Rodriguez, J.F.; Martinez-Avila, N.; Martinez-Fierro, M.L. The Roles of Matrix Metalloproteinases and Their Inhibitors in Human Diseases. *Int. J. Mol. Sci.* **2020**, *21*, 9739. [CrossRef] [PubMed]
24. Walker, C.; Mojares, E.; Del Rio Hernandez, A. Role of Extracellular Matrix in Development and Cancer Progression. *Int. J. Mol. Sci.* **2018**, *19*, 3028. [CrossRef] [PubMed]
25. Xu, J.; Du, Y.; Liu, W.J.; Li, L.; Li, Y.; Wang, X.F.; Yi, H.F.; Shan, C.K.; Xia, G.M.; Liu, X.J.; et al. Intensive fibrosarcoma-binding capability of the reconstituted analog and its antitumor activity. *Drug Deliv.* **2018**, *25*, 102–111. [CrossRef]
26. Benassi, M.S.; Gamberi, G.; Magagnoli, G.; Molendini, L.; Ragazzini, P.; Merli, M.; Chiesa, F.; Balladelli, A.; Manfrini, M.; Bertoni, F.; et al. Metalloproteinase expression and prognosis in soft tissue sarcomas. *Ann. Oncol.* **2001**, *12*, 75–80. [CrossRef]
27. Lu, P.; Takai, K.; Weaver, V.M.; Werb, Z. Extracellular matrix degradation and remodeling in development and disease. *Cold Spring Harb. Perspect. Biol.* **2011**, *3*, a005058. [CrossRef]
28. Elgundi, Z.; Papanicolaou, M.; Major, G.; Cox, T.R.; Melrose, J.; Whitelock, J.M.; Farrugia, B.L. Cancer Metastasis: The Role of the Extracellular Matrix and the Heparan Sulfate Proteoglycan Perlecan. *Front. Oncol.* **2019**, *9*, 1482. [CrossRef] [PubMed]
29. Niland, S.; Riscanevo, A.X.; Eble, J.A. Matrix Metalloproteinases Shape the Tumor Microenvironment in Cancer Progression. *Int. J. Mol. Sci.* **2021**, *23*, 146. [CrossRef]

30. Yang, H.K.; Jeong, K.C.; Kim, Y.K.; Jung, S.T. Role of matrix metalloproteinase (MMP) 2 and MMP-9 in soft tissue sarcoma. *Clin. Orthop. Surg.* **2014**, *6*, 443–454. [CrossRef]
31. Roebuck, M.M.; Helliwell, T.R.; Chaudhry, I.H.; Kalogrianitis, S.; Carter, S.; Kemp, G.J.; Ritchie, D.A.; Jane, M.J.; Frostick, S.P. Matrix metalloproteinase expression is related to angiogenesis and histologic grade in spindle cell soft tissue neoplasms of the extremities. *Am. J. Clin. Pathol.* **2005**, *123*, 405–414. [CrossRef] [PubMed]
32. Di Martino, J.S.; Akhter, T.; Bravo-Cordero, J.J. Remodeling the ECM: Implications for Metastasis and Tumor Dormancy. *Cancers* **2021**, *13*, 4916. [CrossRef] [PubMed]
33. Partridge, J.J.; Madsen, M.A.; Ardi, V.C.; Papagiannakopoulos, T.; Kupriyanova, T.A.; Quigley, J.P.; Deryugina, E.I. Functional analysis of matrix metalloproteinases and tissue inhibitors of metalloproteinases differentially expressed by variants of human HT-1080 fibrosarcoma exhibiting high and low levels of intravasation and metastasis. *J. Biol. Chem.* **2007**, *282*, 35964–35977. [CrossRef]
34. Nguyen, M.; Arkell, J.; Jackson, C.J. Human endothelial gelatinases and angiogenesis. *Int. J. Biochem. Cell Biol.* **2001**, *33*, 960–970. [CrossRef] [PubMed]
35. Farina, A.R.; Mackay, A.R. Gelatinase B/MMP-9 in Tumour Pathogenesis and Progression. *Cancers* **2014**, *6*, 240–296. [CrossRef] [PubMed]
36. Bjorklund, M.; Koivunen, E. Gelatinase-mediated migration and invasion of cancer cells. *Biochim. Biophys. Acta* **2005**, *1755*, 37–69. [CrossRef] [PubMed]
37. Augoff, K.; Hryniewicz-Jankowska, A.; Tabola, R.; Stach, K. MMP9: A Tough Target for Targeted Therapy for Cancer. *Cancers* **2022**, *14*, 1847. [CrossRef] [PubMed]
38. Hodge, J.W.; Ardiani, A.; Farsaci, B.; Kwilas, A.R.; Gameiro, S.R. The tipping point for combination therapy: Cancer vaccines with radiation, chemotherapy, or targeted small molecule inhibitors. *Semin. Oncol.* **2012**, *39*, 323–339. [CrossRef] [PubMed]
39. Vandenbroucke, R.E.; Libert, C. Is there new hope for therapeutic matrix metalloproteinase inhibition? *Nat. Rev. Drug Discov.* **2014**, *13*, 904–927. [CrossRef]
40. Moliere, S.; Jaulin, A.; Tomasetto, C.L.; Dali-Youcef, N. Roles of Matrix Metalloproteinases and Their Natural Inhibitors in Metabolism: Insights into Health and Disease. *Int. J. Mol. Sci.* **2023**, *24*, 10649. [CrossRef]
41. Rasheed, S.; Nelson-Rees, W.A.; Toth, E.M.; Arnstein, P.; Gardner, M.B. Characterization of a newly derived human sarcoma cell line (HT-1080). *Cancer* **1974**, *33*, 1027–1033. [CrossRef] [PubMed]
42. Zhang, L.; Hill, R.P. Hypoxia enhances metastatic efficiency in HT1080 fibrosarcoma cells by increasing cell survival in lungs, not cell adhesion and invasion. *Cancer Res.* **2007**, *67*, 7789–7797. [CrossRef] [PubMed]
43. Wang, J.; Jiang, Y.F. Natural compounds as anticancer agents: Experimental evidence. *World J. Exp. Med.* **2012**, *2*, 45–57. [CrossRef] [PubMed]
44. Nan, Y.; Su, H.; Zhou, B.; Liu, S. The function of natural compounds in important anticancer mechanisms. *Front. Oncol.* **2022**, *12*, 1049888. [CrossRef] [PubMed]
45. Lin, S.R.; Chang, C.H.; Hsu, C.F.; Tsai, M.J.; Cheng, H.; Leong, M.K.; Sung, P.J.; Chen, J.C.; Weng, C.F. Natural compounds as potential adjuvants to cancer therapy: Preclinical evidence. *Br. J. Pharmacol.* **2020**, *177*, 1409–1423. [CrossRef] [PubMed]
46. Chunarkar-Patil, P.; Kaleem, M.; Mishra, R.; Ray, S.; Ahmad, A.; Verma, D.; Bhayye, S.; Dubey, R.; Singh, H.N.; Kumar, S. Anticancer Drug Discovery Based on Natural Products: From Computational Approaches to Clinical Studies. *Biomedicines* **2024**, *12*, 201. [CrossRef] [PubMed]
47. Nam, H.; Kim, M.M. Eugenol with antioxidant activity inhibits MMP-9 related to metastasis in human fibrosarcoma cells. *Food Chem. Toxicol.* **2013**, *55*, 106–112. [CrossRef] [PubMed]
48. Park, J.H.; Jang, Y.J.; Choi, Y.J.; Jang, J.W.; Kim, J.H.; Rho, Y.K.; Kim, I.J.; Kim, H.J.; Leem, M.J.; Lee, S.T. Fisetin inhibits matrix metalloproteinases and reduces tumor cell invasiveness and endothelial cell tube formation. *Nutr. Cancer* **2013**, *65*, 1192–1199. [CrossRef] [PubMed]
49. Lee, D.E.; Chung, M.Y.; Lim, T.G.; Huh, W.B.; Lee, H.J.; Lee, K.W. Quercetin suppresses intracellular ROS formation, MMP activation, and cell motility in human fibrosarcoma cells. *J. Food Sci.* **2013**, *78*, H1464–H1469. [CrossRef]
50. Filipiak, K.; Hidalgo, M.; Silvan, J.M.; Fabre, B.; Carbajo, R.J.; Pineda-Lucena, A.; Ramos, A.; de Pascual-Teresa, B.; de Pascual-Teresa, S. Dietary gallic acid and anthocyanin cytotoxicity on human fibrosarcoma HT1080 cells. A study on the mode of action. *Food Funct.* **2014**, *5*, 381–389. [CrossRef]
51. Shi, X.; Chen, G.; Liu, X.; Qiu, Y.; Yang, S.; Zhang, Y.; Fang, X.; Zhang, C.; Liu, X. Scutellarein inhibits cancer cell metastasis in vitro and attenuates the development of fibrosarcoma in vivo. *Int. J. Mol. Med.* **2015**, *35*, 31–38. [CrossRef] [PubMed]
52. Choi, Y.J.; Lee, Y.H.; Lee, S.T. Galangin and kaempferol suppress phorbol-12-myristate-13-acetate-induced matrix metalloproteinase-9 expression in human fibrosarcoma HT-1080 cells. *Mol. Cells* **2015**, *38*, 151–155. [CrossRef] [PubMed]
53. Li, J.; Liu, C.; Sato, T. Novel Antitumor Invasive Actions of p-Cymene by Decreasing MMP-9/TIMP-1 Expression Ratio in Human Fibrosarcoma HT-1080 Cells. *Biol. Pharm. Bull.* **2016**, *39*, 1247–1253. [CrossRef]
54. Gong, F.; Zhang, Y.; Lin, J.; Li, C.; Zhou, C.; Hong, P.; Ryu, B.; Qian, Z.J. 1-(5-Bromo-2-hydroxy-4-methoxyphenyl)ethanone [SE1] Inhibits MMP-9 Expression by Regulating NF-kappaB and MAPKs Signaling Pathways in HT1080 Human Fibrosarcoma Cells. *Evid. Based Complement. Alternat Med.* **2018**, *2018*, 5639486. [CrossRef] [PubMed]
55. Roomi, M.W.; Bhanap, B.; Niedzwiecki, A.; Rath, M. Chlorophyllin suppresses growth, mmp secretion, invasion and cell migration of fibrosarcoma cell line ht-1080. *Med. Res. Arch.* **2018**, *6*. [CrossRef]

56. Choi, J.H.; Hwang, Y.P.; Jin, S.W.; Lee, G.H.; Kim, H.G.; Han, E.H.; Kim, S.K.; Kang, K.W.; Chung, Y.C.; Jeong, H.G. Suppression of PMA-induced human fibrosarcoma HT-1080 invasion and metastasis by kahweol via inhibiting Akt/JNK1/2/p38 MAPK signal pathway and NF-kappaB dependent transcriptional activities. *Food Chem. Toxicol.* **2019**, *125*, 1–9. [CrossRef]
57. Kim, J.; Kim, M.M. Effect of Lapachol on the Inhibition of Matrix Metalloproteinase Related to the Invasion of Human Fibrosarcoma Cells. *Curr. Mol. Pharmacol.* **2021**, *14*, 620–626. [CrossRef] [PubMed]
58. Kim, H.; Kong, C.S.; Seo, Y. Salidroside, 8(E)-Nuezhenide, and Ligustroside from Ligustrum japonicum Fructus Inhibit Expressions of MMP-2 and -9 in HT 1080 Fibrosarcoma. *Int. J. Mol. Sci.* **2022**, *23*, 2660. [CrossRef]
59. Jo, H.W.; Kim, M.M. beta-Caryophyllene oxide inhibits metastasis by downregulating MMP-2, p-p38 and p-ERK in human fibrosarcoma cells. *J. Food Biochem.* **2022**, *46*, e14468. [CrossRef]
60. Jo, A.I.; Kim, M.M. Silibinin Inhibits Cell Invasion through the Inhibition of MMPs, p-p38, and IL-1beta in Human Fibrosarcoma Cells. *Front. Biosci. (Landmark Ed.)* **2023**, *28*, 64. [CrossRef]
61. Zari, A.T.; Zari, T.A.; Hakeem, K.R. Anticancer Properties of Eugenol: A Review. *Molecules* **2021**, *26*, 7407. [CrossRef] [PubMed]
62. Ulanowska, M.; Olas, B. Biological Properties and Prospects for the Application of Eugenol-A Review. *Int. J. Mol. Sci.* **2021**, *22*, 3671. [CrossRef] [PubMed]
63. Roomi, M.W.; Kalinovsky, T.; Niedzwiecki, A.; Rath, M. Modulation of u-PA, MMPs and their inhibitors by a novel nutrient mixture in adult human sarcoma cell lines. *Int. J. Oncol.* **2013**, *43*, 39–49. [CrossRef] [PubMed]
64. Jeong, J.H.; An, J.Y.; Kwon, Y.T.; Rhee, J.G.; Lee, Y.J. Effects of low dose quercetin: Cancer cell-specific inhibition of cell cycle progression. *J. Cell Biochem.* **2009**, *106*, 73–82. [CrossRef] [PubMed]
65. Ma, H.; Yue, G.G.; Lee, J.K.; Gao, S.; Yuen, K.K.; Cheng, W.; Li, X.; Lau, C.B. Scutellarin, a flavonoid compound from Scutellaria barbata, suppresses growth of breast cancer stem cells in vitro and in tumor-bearing mice. *Phytomedicine* **2024**, *128*, 155418. [CrossRef] [PubMed]
66. Sithranga Boopathy, N.; Kathiresan, K. Anticancer drugs from marine flora: An overview. *J. Oncol.* **2010**, *2010*, 214186. [CrossRef] [PubMed]
67. Bae, M.J.; Karadeniz, F.; Ahn, B.N.; Kong, C.S. Evaluation of Effective MMP Inhibitors from Eight Different Brown Algae in Human Fibrosarcoma HT1080 Cells. *Prev. Nutr. Food Sci.* **2015**, *20*, 153–161. [CrossRef] [PubMed]
68. Barletta, E.; Ramazzotti, M.; Fratianni, F.; Pessani, D.; Degl'Innocenti, D. Hydrophilic extract from Posidonia oceanica inhibits activity and expression of gelatinases and prevents HT1080 human fibrosarcoma cell line invasion. *Cell Adh. Migr.* **2015**, *9*, 422–431. [CrossRef]
69. Bae, M.J.; Karadeniz, F.; Lee, S.G.; Seo, Y.; Kong, C.S. Inhibition of MMP-2 and MMP-9 Activities by Limonium tetragonum Extract. *Prev. Nutr. Food Sci.* **2016**, *21*, 38–43. [CrossRef]
70. Lee, S.G.; Karadeniz, F.; Oh, J.H.; Yu, G.H.; Kong, C.S. Inhibitory Effect of Hizikia fusiformis Solvent-Partitioned Fractions on Invasion and MMP Activity of HT1080 Human Fibrosarcoma Cells. *Prev. Nutr. Food Sci.* **2017**, *22*, 184–190. [CrossRef]
71. Karadeniz, F.L.S.; Oh, J.H.; Kim, J.A.; Kong, C.S. Inhibition of MMP-2 and MMP-9 activities by solvent-partitioned Sargassum horneri extracts. *Fish. Aquat. Sci.* **2018**, *21*, 1–7. [CrossRef]
72. Karadeniz, F.L.S.; Oh, J.H.; Yu, G.H.; Jang, M.S.; Seo, Y.; Kong, C.S. Solvent-partitioned fractions from Ishige okamurae extract inhibit MMP-2 and MMP-9 activities in human fibrosarcoma cells in vitro. *J. Appl. Phycol.* **2018**, *30*, 121–127. [CrossRef]
73. Al-Khayri, J.M.; Sahana, G.R.; Nagella, P.; Joseph, B.V.; Alessa, F.M.; Al-Mssallem, M.Q. Flavonoids as Potential Anti-Inflammatory Molecules: A Review. *Molecules* **2022**, *27*, 2901. [CrossRef] [PubMed]
74. Tuli, H.S.; Sak, K.; Adhikary, S.; Kaur, G.; Aggarwal, D.; Kaur, J.; Kumar, M.; Parashar, N.C.; Parashar, G.; Sharma, U.; et al. Galangin: A metabolite that suppresses anti-neoplastic activities through modulation of oncogenic targets. *Exp. Biol. Med.* **2022**, *247*, 345–359. [CrossRef]
75. Balahbib, A.; El Omari, N.; Hachlafi, N.E.; Lakhdar, F.; El Menyiy, N.; Salhi, N.; Mrabti, H.N.; Bakrim, S.; Zengin, G.; Bouyahya, A. Health beneficial and pharmacological properties of p-cymene. *Food Chem. Toxicol.* **2021**, *153*, 112259. [CrossRef] [PubMed]
76. Chen, L.; Wang, X.; Huang, B. The genus Hippocampus—A review on traditional medicinal uses, chemical constituents and pharmacological properties. *J. Ethnopharmacol.* **2015**, *162*, 104–111. [CrossRef] [PubMed]
77. Makino, T.; Izumi, K.; Hiratsuka, K.; Kano, H.; Shimada, T.; Nakano, T.; Kadomoto, S.; Naito, R.; Iwamoto, H.; Yaegashi, H.; et al. Anti-proliferative and anti-migratory properties of coffee diterpenes kahweol acetate and cafestol in human renal cancer cells. *Sci. Rep.* **2021**, *11*, 675. [CrossRef] [PubMed]
78. Mendes Miranda, S.E.; Alcantara Lemos, J.; Fernandes, R.S.; Silva, J.O.; Ottoni, F.M.; Townsend, D.M.; Rubello, D.; Alves, R.J.; Cassali, G.D.; Ferreira, L.A.M.; et al. Enhanced antitumor efficacy of lapachol-loaded nanoemulsion in breast cancer tumor model. *Biomed. Pharmacother.* **2021**, *133*, 110936. [CrossRef]
79. Fidyt, K.; Fiedorowicz, A.; Strzadala, L.; Szumny, A. beta-caryophyllene and beta-caryophyllene oxide-natural compounds of anticancer and analgesic properties. *Cancer Med.* **2016**, *5*, 3007–3017. [CrossRef]
80. Ray, P.P.; Islam, M.A.; Islam, M.S.; Han, A.; Geng, P.; Aziz, M.A.; Mamun, A.A. A comprehensive evaluation of the therapeutic potential of silibinin: A ray of hope in cancer treatment. *Front. Pharmacol.* **2024**, *15*, 1349745. [CrossRef]
81. Tavakoli, J.; Miar, S.; Majid Zadehzare, M.; Akbari, H. Evaluation of effectiveness of herbal medication in cancer care: A review study. *Iran. J. Cancer Prev.* **2012**, *5*, 144–156. [PubMed]
82. Fasinu, P.S.; Rapp, G.K. Herbal Interaction With Chemotherapeutic Drugs-A Focus on Clinically Significant Findings. *Front. Oncol.* **2019**, *9*, 1356. [CrossRef] [PubMed]

83. Hoelder, S.; Clarke, P.A.; Workman, P. Discovery of small molecule cancer drugs: Successes, challenges and opportunities. *Mol. Oncol.* **2012**, *6*, 155–176. [CrossRef] [PubMed]
84. Zhong, L.; Li, Y.; Xiong, L.; Wang, W.; Wu, M.; Yuan, T.; Yang, W.; Tian, C.; Miao, Z.; Wang, T.; et al. Small molecules in targeted cancer therapy: Advances, challenges, and future perspectives. *Signal Transduct. Target. Ther.* **2021**, *6*, 201. [CrossRef] [PubMed]
85. Anand, U.; Dey, A.; Chandel, A.K.S.; Sanyal, R.; Mishra, A.; Pandey, D.K.; De Falco, V.; Upadhyay, A.; Kandimalla, R.; Chaudhary, A.; et al. Cancer chemotherapy and beyond: Current status, drug candidates, associated risks and progress in targeted therapeutics. *Genes. Dis.* **2023**, *10*, 1367–1401. [CrossRef] [PubMed]
86. Roomi, M.W.; Kalinovsky, T.; Monterrey, J.; Rath, M.; Niedzwiecki, A. In vitro modulation of MMP-2 and MMP-9 in adult human sarcoma cell lines by cytokines, inducers and inhibitors. *Int. J. Oncol.* **2013**, *43*, 1787–1798. [CrossRef]
87. Takahashi, N.; Watanabe, Y.; Maitani, Y.; Yamauchi, T.; Higashiyama, K.; Ohba, T. p-Dodecylaminophenol derived from the synthetic retinoid, fenretinide: Antitumor efficacy in vitro and in vivo against human prostate cancer and mechanism of action. *Int. J. Cancer* **2008**, *122*, 689–698. [CrossRef] [PubMed]
88. Takahashi, N.; Takeda, K.; Imai, M. Inhibitory effects of p-dodecylaminophenol on the invasiveness of human fibrosarcoma cell line HT1080. *Bioorg Med. Chem.* **2013**, *21*, 6015–6021. [CrossRef] [PubMed]
89. Nanjan, P.; Nambiar, J.; Nair, B.G.; Banerji, A. Synthesis and discovery of (I-3,II-3)-biacacetin as a novel non-zinc binding inhibitor of MMP-2 and MMP-9. *Bioorg Med. Chem.* **2015**, *23*, 3781–3787. [CrossRef]
90. Liang, S.; Sun, Y.; Dai, X. A Review of the Preparation, Analysis and Biological Functions of Chitooligosaccharide. *Int. J. Mol. Sci.* **2018**, *19*, 2197. [CrossRef]
91. Hong, S.; Ngo, D.N.; Kim, M.M. Inhibitory effect of aminoethyl-chitooligosaccharides on invasion of human fibrosarcoma cells. *Environ. Toxicol. Pharmacol.* **2016**, *45*, 309–314. [CrossRef] [PubMed]
92. Bar, J.; Lukaschuk, N.; Zalcenstein, A.; Wilder, S.; Seger, R.; Oren, M. The PI3K inhibitor LY294002 prevents p53 induction by DNA damage and attenuates chemotherapy-induced apoptosis. *Cell Death Differ.* **2005**, *12*, 1578–1587. [CrossRef] [PubMed]
93. Wang, K.; Hou, Y.; Gu, C.; Zhao, D.; Duan, Y.; Ran, Z.; Li, Q.; Li, X. Inhibitory effect of the mitogen activated protein kinase specific inhibitor PD98059 on Mtb-Ag-activated gammadeltaTau cells. *Int. J. Clin. Exp. Pathol.* **2017**, *10*, 9644–9648.
94. Sreekanth, G.P.; Chuncharunee, A.; Sirimontaporn, A.; Panaampon, J.; Noisakran, S.; Yenchitsomanus, P.T.; Limjindaporn, T. SB203580 Modulates p38 MAPK Signaling and Dengue Virus-Induced Liver Injury by Reducing MAPKAPK2, HSP27, and ATF2 Phosphorylation. *PLoS ONE* **2016**, *11*, e0149486. [CrossRef]
95. Yu, S.M.; Kim, S.J. Salinomycin causes migration and invasion of human fibrosarcoma cells by inducing MMP-2 expression via PI3-kinase, ERK-1/2 and p38 kinase pathways. *Int. J. Oncol.* **2016**, *48*, 2686–2692. [CrossRef]
96. Bayat Mokhtari, R.; Homayouni, T.S.; Baluch, N.; Morgatskaya, E.; Kumar, S.; Das, B.; Yeger, H. Combination therapy in combating cancer. *Oncotarget* **2017**, *8*, 38022–38043. [CrossRef]
97. Saga, R.; Monzen, S.; Chiba, M.; Yoshino, H.; Nakamura, T.; Hosokawa, Y. Anti-tumor and anti-invasion effects of a combination of 4-methylumbelliferone and ionizing radiation in human fibrosarcoma cells. *Oncol. Lett.* **2017**, *13*, 410–416. [CrossRef]
98. Alford, V.M.; Kamath, A.; Ren, X.; Kumar, K.; Gan, Q.; Awwa, M.; Tong, M.; Seeliger, M.A.; Cao, J.; Ojima, I.; et al. Targeting the Hemopexin-like Domain of Latent Matrix Metalloproteinase-9 (proMMP-9) with a Small Molecule Inhibitor Prevents the Formation of Focal Adhesion Junctions. *ACS Chem. Biol.* **2017**, *12*, 2788–2803. [CrossRef]
99. Basak, D.; Arrighi, S.; Darwiche, Y.; Deb, S. Comparison of Anticancer Drug Toxicities: Paradigm Shift in Adverse Effect Profile. *Life* **2021**, *12*, 48. [CrossRef]
100. Lei, Z.N.; Tian, Q.; Teng, Q.X.; Wurpel, J.N.D.; Zeng, L.; Pan, Y.; Chen, Z.S. Understanding and targeting resistance mechanisms in cancer. *MedComm (2020)* **2023**, *4*, e265. [CrossRef]
101. Beck, H.; Harter, M.; Hass, B.; Schmeck, C.; Baerfacker, L. Small molecules and their impact in drug discovery: A perspective on the occasion of the 125th anniversary of the Bayer Chemical Research Laboratory. *Drug Discov. Today* **2022**, *27*, 1560–1574. [CrossRef]
102. Newman, D.J.; Cragg, G.M. Natural Products as Sources of New Drugs over the Nearly Four Decades from 01/1981 to 09/2019. *J. Nat. Prod.* **2020**, *83*, 770–803. [CrossRef]
103. Li, Y.; Meng, Q.; Yang, M.; Liu, D.; Hou, X.; Tang, L.; Wang, X.; Lyu, Y.; Chen, X.; Liu, K.; et al. Current trends in drug metabolism and pharmacokinetics. *Acta Pharm. Sin. B* **2019**, *9*, 1113–1144. [CrossRef] [PubMed]
104. Vadevoo, S.M.P.; Gurung, S.; Lee, H.S.; Gunassekaran, G.R.; Lee, S.M.; Yoon, J.W.; Lee, Y.K.; Lee, B. Peptides as multifunctional players in cancer therapy. *Exp. Mol. Med.* **2023**, *55*, 1099–1109. [CrossRef]
105. Ebrahimi, S.B.; Samanta, D. Engineering protein-based therapeutics through structural and chemical design. *Nat. Commun.* **2023**, *14*, 2411. [CrossRef]
106. Mirzaei, H.; Kazemi, B.; Bandehpour, M.; Shoari, A.; Asgary, V.; Ardestani, M.S.; Madadkar-Sobhani, A.; Cohan, R.A. Computational and nonglycosylated systems: A simpler approach for development of nanosized PEGylated proteins. *Drug Des. Devel Ther.* **2016**, *10*, 1193–1200. [CrossRef] [PubMed]
107. Bruno, B.J.; Miller, G.D.; Lim, C.S. Basics and recent advances in peptide and protein drug delivery. *Ther. Deliv.* **2013**, *4*, 1443–1467. [CrossRef] [PubMed]
108. Kim, S.R.; Eom, T.K.; Byun, H.G. Inhibitory effect of the carnosine-gallic acid synthetic peptide on MMP-2 and MMP-9 in human fibrosarcoma HT1080 cells. *J. Pept. Sci.* **2014**, *20*, 716–724. [CrossRef]

109. Hsiao, C.C.; Wang, W.C.; Kuo, W.L.; Chen, H.Y.; Chen, T.C.; Hamann, J.; Lin, H.H. CD97 inhibits cell migration in human fibrosarcoma cells by modulating TIMP-2/MT1- MMP/MMP-2 activity--role of GPS autoproteolysis and functional cooperation between the N- and C-terminal fragments. *FEBS J.* **2014**, *281*, 4878–4891. [CrossRef]
110. Ko, S.C.; Heo, S.Y.; Choi, S.W.; Qian, Z.J.; Heo, S.J.; Kang, D.H.; Kim, N.; Jung, W.K. A heptameric peptide isolated from the marine microalga suppresses PMA-induced secretion of matrix metalloproteinase-9 through the inactivation of the JNK, p38, and NF-κB pathways in human fibrosarcoma cells. *J. Appl. Phycol.* **2018**, *30*, 2367–2378. [CrossRef]
111. Zani, M.L.; Moreau, T. Phage display as a powerful tool to engineer protease inhibitors. *Biochimie* **2010**, *92*, 1689–1704. [CrossRef] [PubMed]
112. Shoari, A.; Rasaee, M.J.; Kanavi, M.R.; Daraei, B. Functional mimetic peptide discovery isolated by phage display interacts selectively to fibronectin domain and inhibits gelatinase. *J. Cell Biochem.* **2019**, *120*, 19699–19711. [CrossRef] [PubMed]
113. Shoari, A.; Khalili, S.; Rasaee, M.J.; Löwik, D.W.P.M. A Phage Display Derived Cyclized Peptide Inhibits Fibrosarcoma Cells Invasion via Suppression of MMP-9 Activity. *Int. J. Pept. Res. Ther.* **2022**, *28*, 136. [CrossRef]
114. Winer, A.; Adams, S.; Mignatti, P. Matrix Metalloproteinase Inhibitors in Cancer Therapy: Turning Past Failures Into Future Successes. *Mol. Cancer Ther.* **2018**, *17*, 1147–1155. [CrossRef]

MDPI AG
Grosspeteranlage 5
4052 Basel
Switzerland
Tel.: +41 61 683 77 34

Targets Editorial Office
E-mail: targets@mdpi.com
www.mdpi.com/journal/targets

www.ingramcontent.com/pod-product-compliance
Lightning Source LLC
Chambersburg PA
CBHW040757250926
51850CB00007BA/1016
9783725877577